Alkaliphiles

To my wife, Sachiko

To my wife, Sadako

Preface

In autumn 1990 I had the chance to see one of the largest exhibitions of Claude Monet's most famous series, at the Royal Academy of Arts in London. As Kenneth Clark explains in his famous book *Civilization*, Monet attempted a kind of color symbolism to express the changing effects of light. For example, he painted a series of cathedral facades in different lights—pink, blue and yellow—which seem to me too far from my own experience. The colors of these objects depend on the physical environment, such as sunlight, snow, time of day, season, etc. Under different conditions, one object may show quite different properties. Who can be sure what the absolute property is? The microbial world involves the same uncertainty.

Three hundred years ago Antony van Leewenhoek first observed microorganisms through his microscope. In the middle of the 19th century Louis Pasteur conducted one of the most important experiments in the field of microbiology, as a result of which he was able to refute the theory of spontaneous generation. Alexander Fleming made his famous serendipitous discovery of the first antibiotic, penicillin, in 1928 which has lengthened our average life span. The industrial production of penicillin has resulted in the development of basic microbiology, such as physiology and genetics, as well as industrial microbiology. And in 1977, only two decades ago, the first DNA sequence of the virus SV40 was determined by Maxam and Gilbert. The decoding of the three and a half billion-year history of life has only just begun.

Not too many years ago almost all biologists believed that life could survive only within a very narrow range of temperature, pressure, acidity, alkalinity, salinity and so on, in so-called "moderate" environments. So when microbiologists looked around for interesting bacteria and other life forms, they attempted to isolate microorganisms only from moderate environments.

Nature, however, contains many extreme environments, such as acidic or hot springs, saline lakes, deserts, alkaline lakes and the ocean bed. All of these environments would seem to be too harsh for life to survive.

However, in recent times, many organisms have been found in such extreme environments. Moreover, some of them cannot survive in so-called "moderate" environments. For example, thermophilic bacteria grow in environments with extremely high temperatures, but will not grow at $20°$ to $40°C$. Some alkali-loving bacteria cannot grow in a nutrient broth at pH 7.0, but flourish at pH 10.5. If a moderate environment for conventional organisms such as *Escherichia coli* were superimposed on that for thermophilic organisms, for example, the "moderate" environment would seem very cold for thermophiles. Thus the idea of an extreme environment is relative, not absolute. Clearly we have been too anthropocentric in our thinking. We should therefore extend our consideration to other environments in order to isolate and cultivate new microorganisms.

In 1968 I was visiting Florence, Italy. I was looking at the Renaissance buildings, which are so very different from Japanese architecture. About 500 years ago no Japanese could have imagined this Renaissance culture. Then suddenly a voice whispered in my ear, "There might be a whole new world of microorganisms in different unexplored cultures.

Could there be an entirely unknown domain of microorganisms at alkaline pH ?" The acidic environment was being studied, probably because most food is acidic. However, hardly any work had been done in the alkaline region.

Science, just as much as the arts, relies upon a sense of romance and intuition. Upon my return to Japan I prepared an alkaline medium containing 1% sodium carbonate, placed a small amount of soil in it, and incubated it overnight at 37 °C. To my surprise, various microorganisms flourished in all 30 test tubes. Here was a new alkaline world which was utterly different from the neutral world discovered by Pasteur. This was my first encounter with alkaliphiles[*1].

When I started experiments on alkaliphilic bacteria I found only 16 scientific papers on the subject[*2]. In Japan, since ancient times, indigo has been naturally reduced in the presence of sodium carbonate. Indigo from indigo leaves can be reduced by bacteria that grow under high alkaline conditions. We call this process "indigo fermentation." The most important factor in this process is control of the pH value. Indigo reduction was controlled only by the skill of the craftsman. Takahara and his colleagues isolated the indito-reducing bacterium from an indigo ball. They then improved the indigo fermentation process by adding alkaliphilic *Bacillus* sp. No. S-8. Indigo fermentation is the first industrial application of alkaliphiles anywhere in the world. Further microbiological studies, however, were not conducted until the author's systematic investigation. Many of these alkaliphiles remained little more than interesting biological curiosities. No industrial application was attempted at all before 1968.

I named these microorganisms which thrive in alkaline environments "alkaliphiles," and conducted systematic microbial physiological studies of them. The results showed that these microorganisms, which are completely different from any previously reported, were widely distributed throughout the globe and they produced new products. My colleagues and I have isolated a great number of alkaliphilic microorganisms since 1968 and purified many alkaline enzymes. The first paper concerning an alkaline protease was published in 1971. Over the past two decades our studies have focused on the enzymology, physiology, ecology, taxonomy, molecular biology and genetics of alkaliphilic microorganisms to establish a new microbiology of alkaliphilic microorganisms. Industrial applications of these microorganisms have also been investigated extensively and some enzymes, such as alkaline proteases, alkaline amylases and alkaline cellulases, have been put to use on an industrial scale. Subsequently, many microbiologists have published numerous papers on alkaliphilic microorganisms in various fields.

The first book on alkaliphilic microorganisms was published by Springer-Verlag in 1982. My coworkers and I described the microbiology of alkaliphiles, which had hardly been studied until we began our systematic research. We mainly discussed the isolation procedures for alkaliphiles, their physiology, and the properties and industrial applica-

---

[*1] In the previous books (Horikoshi, 1982; 1991), the author used "alkalophilic" or "alkalophiles." In this book, however, alkaliphilic or alkaliphiles are used for grammatical reasons.

[*2] Johnson, 1928; Downie and Cruickshank, 1928; Vedder, 1934; Jenkin, 1936; Bornside and Kallio, 1956; Chesbro and Evans, 1959; Kushner and Lisson, 1959; Takahara and Tanabe 1960; Chislett and Kushner, 1961; Shislett and Kushner, 1961*b*; Takahara *et al.*, 1961; Takahara and Tanabe, 1962; Wiley and Stokes, 1962; Wiley and Stokes, 1963; Barghoorn and Tyler, 1965; Siegel and Giumarro, 1966.

tions of the novel enzymes of alkaliphiles. The second book, *Microorganisms in Alkaline Environments* (Kodansha-VCH), described works on alkaliphiles conducted from 1981 to 1990, and focused on genetic investigation of enzymes and applications of active DNA fragments, such as promoters, and secretion vectors. Alkaliphiles are one of the most interesting sources of DNA.

This new volume deals with gene manipulation technologies, which is a new field of alkaliphilic microbiology. Genes of alkaline enzymes produced by alkaliphiles have been cloned and analyzed. Furthermore, we can analyze the alkaliphily of alkaliphiles by using genetic engineering technology even though we do not yet have answers to crucial problems. Alkaliphiles thrive in alkaline environments. The cell surface of alkaliphiles can maintain the neutral intracellular pH values in alkaline environments of pH 10–13. How the pH homeostasis is maintained is one of the most fascinating aspects of alkaliphiles. Recently, new host-vector systems have been developed using alkaliphilic *Bacillus* C-125 mutants that are alkaline-sensitive, and genes responsible for alkaliphily have been analyzed. Alkaliphiles have clearly evolved large amounts of information and developed the ability to cope with particular environments. And unexplored genes await further exploration by biotechnologists.

The author wishes to extend his thanks to Dr. Makio Kitada for his help in preparing portions of the manuscript for Chapter 4 Physiology and is also grateful to Ms. Cecilia M. Hamagami for her assistance in finalizing the English manuscripts.

Koki Horikoshi
Chiswick, London

Contents

7. Extracellular Enzymes   147

1

Introduction

1.1 Definition of Alkaliphilic Organisms

There are no precise definitions of what characterizes an alkaliphilic or an alkalitolerant organism. The following definitions are given by Kroll (1990).

Alkaliphile: An organism whose optimum rate of growth is observed above at least two pH units above neutrality. Alkalitolerant: An organism capable of growth or survival at values more than 9.0 but whose optimal rate of growth is around neutrality or less.

However, some alkaliphiles exhibit more than one pH optimum. Some organisms can appear to be either alkalitolerant or alkaliphilic depending on the growth conditions in the presence or absence of sodium ions. *Bacillus alcalophilus* has a more alkaline pH optimum for growth when grown on malate than on lactose. As a result this volume will use a much simpler and less strict definition. The term "alkaliphile" will be applied only to microorganisms that grow optimally or very well at pH values above 9.0, but cannot grow or only grow slowly at neutral pH values of 6.5.

Almost all alkaliphilic *Bacillus* strains require sodium ions for growth. In the early stages of our study, alkaliphiles were thought to require only higher pH values for growth. In conventional classification experiments the author found a strange result. *Bacillus* sp. No. Ku-1, which was thought to be an alkaliphilic strain, could grow at pH 7.0 in a nutrient broth containing 5% NaCl, but could not grow in the absence of NaCl (Kurono and Horikoshi, 1973). This was the beginning of physiological studies of NaCl on the growth of alkaliphiles.

1.2 Early Studies on Alkaliphiles

As described in the preface, in 1968 when the author started experiments on alkaliphilic bacteria only 16 scientific papers (Johnson, 1923; Downie and Cruickshank, 1928; Vedder, 1934; Jenkin, 1936; Bornside and Kallio, 1956; Chesbro and Evans, 1959; Kushner and Lisson, 1959; Takahara and Tanabe, 1960; Chislett and Kushner,1961*b*; Takahara, Takasaki and Tanabe, 1961; Chislett and Kushner, 1961*a*; Wiley and Stokes, 1962; Takahara and Tanabe, 1962; Wiley and Stokes, 1963; Barghoorn and Tyler, 1965; Siegel and Giumarro, 1966) were found.

In Japan, since ancient times, indigo has been naturally reduced in the presence of sodium carbonate. Indigo from indigo leaves can be reduced by bacteria that grow under high alkaline conditions. We call this process "indigo fermentation". The most important factor in this process is control of the pH value. Indigo reduction was controlled only by the skill of the craftsman. Takahara and his colleagues isolated the indigo-reducing bacterium from an indigo ball. They then improved the indigo fermentation process by adding alkaliphilic *Bacillus* sp. No. S-8 during the fermentation process (Takahara and Tanabe, 1960; Takahara, Takasaki and Tanabe, 1961; Takahara and Tanabe, 1962). Indigo fermentation was the first industrial application of alkaliphiles anywhere in the world. Further systematic microbiological studies, however, were not conducted until the present author's investigations.

The author and his coworkers have isolated a great number of alkaliphilic microorganisms since 1968 and purified many alkaline enzymes. The first paper concerning an alkaline protease was published in 1971 (Horikoshi, 1971*b*). Over the past two decades our studies have focused on the enzymology, physiology, ecology, taxonomy, molecular biology and genetics of alkaliphilic microorganisms to establish a new microbiology of alkaliphilic microorganisms. Industrial applications of these microorganisms have also been investigated extensively and some enzymes, such as alkaline proteases, alkaline amylases and alkaline cellulases, have been put to use on an industrial scale. Subsequently, many microbiologists have published numerous papers on alkaliphilic microorganisms in various fields. Gene manipulation technologies have developed a new field of alkaliphilic microbiology. We have obtained some information on the alkaliphily of alkaliphiles using genetic engineering technology, but we have not yet found the answers to a number of crucial problems. The cell surface of alkaliphiles is assumed to be responsible for maintaining the intracellular pH values neutral in alkaline environments of pH 10-13. Several host-vector systems have been developed using alkaliphilic *Bacillus* strains, and the gene(s) responsible for alkaliphily has been studied.

2

Isolation, Distribution and Taxonomy
of Alkaliphilic Microorganisms

2.1   Isolation and Distribution of Alkaliphiles

Alkaliphilic microorganisms have been isolated in alkaline media containing sodium carbonate, sodium bicarbonate and potassium carbonate. Sodium hydroxide is also used in large-scale fermenters using a pH control device. The recommended concentration of these compounds is about 0.5 % to 2 %, depending on the microorganisms used, and the pH of the medium held between about 8.5 and 11. Table 2.1 shows the standard media in our laboratory containing 1% sodium carbonate. It is most important that the sodium carbonate be sterilized separately; otherwise the microorganisms may show poor growth. Isolation of the microorganisms is conducted by conventional means: a small amount of sample, such as soil or feces, is suspended in 1 ml of sterile water and one drop of the suspension is spread on a Petri dish containing Horikoshi-I or Horikoshi-II medium. The Petri dishes are incubated at 37 °C for several days and the colonies that appear are isolated by the usual method. The isolated microorganisms are then kept at room temperature on slants (Fig. 2.1).

Enrichment culture is well known in the isolation of specific microorganisms, such as thermophiles, etc. Some alkaliphiles grow very well under enrichment culture in alkaline conditions and are predominantly isolated. Very few slow growers are isolated. In the author's laboratory direct isolation of alkaliphiles from soil samples has been carried out, but other strains having specific microbial properties (except those growing under extreme temperatures) have never been isolated.

Table 2.1   Basal media for alkaliphilic microorganisms

Ingredients	Horikoshi-I (g/l)	Horikoshi-II (g/l)
Glucose	10	—
Soluble starch	—	10
Polypeptone	5	5
Yeast extract	5	5
KH_2PO_4	1	1
$Mg_2SO_4 \cdot 7H_2O$	0.2	0.2
Na_2CO_3	10	10
Agar	20	20

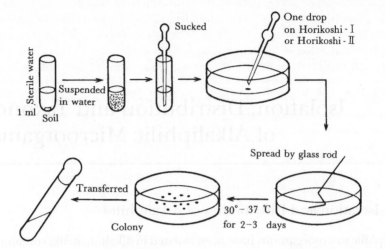

Fig. 2.1 Isolation of alkaliphilic microorganisms.
(Reproduced with permission from Horikoshi and Akiba, *Alkalophilic Microorganisms: A New Microbial World*, p.10, Springer-Verlag: Japan Scientific Societies Press (1982))

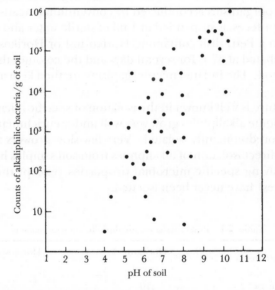

Fig. 2.2 Distribution of alkaliphilic bacteria in soil.
(Reproduced with permission from Horikoshi and Akiba, *Alkalophilic Microorganisms: A New Microbial World*, p.11, Springer-Verlag: Japan Scientific Societies Press (1982))

Alkaliphilic microorganisms are widely distributed throughout the world. They can be found in garden soil samples. In addition to bacteria, various kinds of microorganisms, including actinomycetes, fungi, yeast, and phages, have been isolated. Our studies indicate that there may be as many microorganisms in the alkaline pH region as in the neutral region. Alkaliphiles can be considered to be a newfound microbial world in nature which may be as large as the neutrophilic microbial world. The number of alkaliphilic bacteria found in soil was about 1/10 to 1/100 that of neutrophilic bacteria (Fig. 2.2). Alkaliphilic bacteria have very interesting properties, *e.g.*, the ability to change the pH of their environment to suit their growth, as described in section 4.1 on growth.

Using alkaliphilic microorganisms we established an alkaline fermentation process which is entirely different from conventional fermentation processes. Hundreds of new enzymes have been isolated by this process and their enzymatic properties investigated. Some of these enzymes have been produced in industrial scale plants as shown in Chapter 7 on extracellular enzymes.

2.2   Cultivation and Preservation of Alkaliphiles

2.2.1   Cultivation Conditions

The most popular media for alkaliphiles are Horikoshi-I and -II. Many modified media including industrial production media have been developed and reported, but essentially the ingredients of the media are the same as those for growing neutrophiles, except for the pH values. Alkaliphiles require various nutrients for growth, but a few alkaliphilic *Bacillus* strains can grow in simple minimal media containing glycerol, glutamic acid, citric acid, etc. One of these is alkaliphilic *Bacillus* C-125, and many mutants have been made by conventional mutagenesis. Some of the media are listed in Table 2.2. Cultivation temperatures are in the range of 20-55 °C for general alkaliphiles. However, some thermophilic alkaliphiles that can grow at 60-70 °C are found in garden soils. Furthermore, many haloalkaliphiles isolated from alkaline lakes can grow in alkaline environments containing 20% NaCl. Thermophilic anaerobic spore-forming alkaliphiles, alkaliphilic *Clostridium* strains, have been isolated from sewage plants (Li *et al.*, 1994). Recently, Engle *et al.* (1995) isolated and characterized a novel alkalitolerant thermophile, *Anaerobranca horikoshii* gen. nov., sp. nov., from Yellowstone National Park. A few anaerobic alkaliphiles have been isolated and characterized, and industrial applications of these microorganisms will be developed soon.

2.2.2   Preservation

Most laboratories maintain stock cultures for research, industrial or other purposes. The proper preservation of cultures is extremely important and should be given the same attention as equipment and chemical compounds. The primary aim of culture preservation is to maintain the organisms alive, uncontaminated, and without variation or mutation. Many methods have been used to preserve bacteria, but not all species respond in a similar manner to a given method.

A.   Short-term Preservation

The traditional preservation method is by periodic serial transfer to fresh medium. The interval between such transfers varies with the organism, the medium used, and the

Table 2.2 Media for alkaliphiles

Horikoshi minimal for medium *Bacillus* sp. C-125	
Glutamate	0.2%
Glycerol	0.5%
K₂HPO₄	1.4%
KH₂PO₄	0.4%
MgSO₄·7H₂O	0.02%
Na₂CO₃	1.0%
w/v	

Aono and Horikoshi synthetic medium for *Bacillus* C-125	
Citric acid	0.034%
Glucose	0.5%
Ammonium sulfate	0.1%
KNO₃	0.1%
K₂HPO₄	1.37%
KH₂PO₄	0.59%
MgSO₄·7H₂O	0.05%
Na₂CO₃	1.0%
v/w	

Modified DM-3 protoplast regeneration medium	
Agar	1.0%
Sodium succinate	0.5 M
Casamino acid (Difco)	0.5%
Yeast extract (Difco)	0.5%
K₂HPO₄	0.35%
KH₂PO₄	0.15%
Glucose	2.0%
MgCl₂	0.1 M
CaCl₂	1.25 mM
Bovine serum albumin	0.04%
Chloramphenicol	2.5 mg/ml
pH	7.3-6.8

Protease	
Soluble starch	5.0%
Soya meal	2.0%
Ground barley	5.0%
Na-caseinate	1.0%
Polyglycol	0.01%
Na₂HPO₄·10H₂O	0.9%
Na₂CO₃	1.0%

Protease	
Soluble starch	4.0%
Soya meal	3.0%
Ground barley	10.0%
Soya bean oil	0.55%
CaCO₃	0.5%
Na₂CO₃	1.0%

Protease	
Soya meal	3.0%
Ground barley	10.0%
Polyglycol	0.01%

Amylase	
Soluble starch	1.5%
Soya meal	2.0%
Na₂HPO₄·10H₂O	0.9%
CaCl₂	0.1%
NaHCO₃	1.0%

Amylase	
Soluble starch	8.0%
Soya bean extract	1.0%
Polypeptone	1.0%
Na₂CO₃	1.0%

Amylase	
Soluble starch	8.0%
Soya bean extract	1.0%
Polypeptone	2.0%
Fish extract	2.0%
NaCl	0.25%
MgSO₄·7H₂O	0.02%
Na₂CO₃	1.0%

Cellulase	
CMC	2.0%
Polypeptone	0.5%
Yeast extract	0.5%
MgSO₄·7H₂O	0.02%
Congo red	0.03%
Na₂CO₃	1.0%

β-1,3-Glucanase	
Pachyman	2.0%
Polypeptone	0.5%
Yeast extract	0.5%
KH₂PO₄	0.1%
MgSO₄·7H₂O	0.02%
Na₂CO₃	1.0%

Glowth medium for protoplast formation	
Glucose	0.2%
Polypeptone	0.5%
Yeast extract	0.5%
Citric acid	0.034%
K₂HPO₄	1.37%
KH₂PO₄	0.59%
MgSO₄·7H₂O	0.005%
Na₂CO₃	1.06%
w/v	

Protoplast regeneration medium	
Agar	1.0%
Sodium succinate	0.5 M
Casamino acid (Difco)	0.5%
Yeast extract (Difco)	0.5%
Glucose	2.0%
MgCl₂	30 mM
CaCl₂	1.25 mM
Bovine serum albumin	0.04%
Tris	30 mM
pH	7.3-6.8

external conditions. For instance, *Bacillus* strains so far isolated in our laboratory can be preserved for at least one to two years in Horikoshi-II medium in screw-cap test tubes. However, some gram-negative, non-spore forming strains isolated must be transferred once a week.

Storage in a refrigerator (5 to 8 °C) is the preferred method. However, since some microorganisms are sensitive to lower temperatures, important strains are maintained at room temperature for back-up.

B. Long-term Storage

Freeze-drying is one of the most preferable methods for long-term preservation of alkaliphiles. Successful freeze-drying is conducted by using 20% (w/v) skim milk containing 1% (w/v) monosodium glutamate as a cryoprotective agent. Freeze-dried cultures are stored at 3 to 10 °C. Recovery has been done conventionally by using the Horikoshi-I or -II medium. By this method our first isolate, *Bacillus* sp. No. 221, has been surviving since 1968.

Long-term preservation is also possible in the frozen state at the temperature of liquid nitrogen. The cells are cultivated in an appropriate medium (Horikoshi-II medium, for example) until late log phase and the pelt is suspended with sterile fresh medium containing 10% (v/v) glycerol. The cell suspensions containing at least 10^8 cells per ml in plastic presterilized screw-cap vials are placed in a mechanical freezer at -60 °C for 1 h and then kept in liquid nitrogen. Rapid thawing of frozen cultures is best for the greatest recovery of alkaliphiles. Our first cultures in liquid nitrogen exhibit very good recovery after 15 years.

2.3   Morphological, Cultural and Biochemical Properties of Alkaliphiles

2.3.1   Gram-positive Bacteria

A. *Bacillus* sp.

The first, albeit brief, description of alkaliphilic *Bacillus* isolated feces was reported in 1934 by Vedder.

> This microbe does not grow in selective cholera culture media, nor by enrichment in peptone water of "ordinary" pH. It is typical of this bacillus that it only grows in and on highly alkaline culture media, and not in cases where the medium is of weak alkalinity or still lower pH. As this micro-organism has not, as far as we know, been previously described, we proposed to name it "*Bacillus alcalophilus* n. sp." The *Bacillus* (NCTC4553) is motile and digests albumin (gelatin, haemoglobin) in a strong alkaline medium (Vedder, 1934).

The optimum pH for the growth of *B. alcalophilus* was 8.6 to 10.0. Chislett and Kushner (1961 *a*; 1961 *b*) described a strain with the ability to grow in highly alkaline medium of around pH 10, and classified it as *B. circulans.*

Takahara and Tanabe (1960; 1961; 1962) isolated alkaliphilic strain No. S-8 from a vat of fermenting indigo leaves used for reducing indigo dye. They proposed a new species name, *Bacillus alkaliphilus* nov. sp. The optimal pH for growth was 10.0 to 11.5. Another alkaliphilic bacterium was also isolated from an indigo ball (*aidama*). It grew well at pH 10.0-10.5 but did not grow at pH 7-8 (Ohta *et al.*, 1975). This bacterium is similar to the *B. alkaliphilus* nov. sp. described above.

B. alkaliphilus nov. sp. (Takahara and Tanabe, 1962) grew well on *aidama* medium which contained (per liter): *aidama* (fermented indigo leaves), 50 g; wheat flour, 5 g; NaOH, 2 g; lime, 0.05 g and agar, 9 g. This basic medium composition was employed in cultures used for taxonomical tests.

Description of *Bacillus alkaliphilus* nov. sp. strain No. S-8:

Gram-positive rods 0.9-1.0 by 2.0-3.5 μm, motile with peritrichous flagella. Spores 1.0×1.5 μm are formed, ellipsoidal to cylindrical, central to subterminal, and sporangia not definitely swollen.

Aidama agar: Growth slow, thin, opaque, and white to light brown. Colonies are thin, inconspicuous, round, opaque and whitish. *Aidama* gelatin stab:Liquefied.

Facultatively aerobic, reducing action observed under anaerobic conditions.

Ammonia, indole, and hydrogen sulfide not produced.

Nitrate reduced to nitrite.

Methylene blue reduced.

Acetylmenthylcarbinol not produced.

Starch hydrolyzed.

Catalase test positive.

Litmus milk: Decoloration and reduction of litmus but no coagulation.

Acid and gas produced from glucose, fructose, maltose, sucrose, dextrin, starch, lactose, arabinose, and galactose but not from glycerol, mannitol, xylose, or raffinose.

Temperature relationship: Good growth at 30 °C, maximum temperature for growth is 50 °C.

pH relationship: Good growth at pH above 11.0, optimum pH 10.0 to 11.5.

A heptapeptide is essential for growth.

Source: Isolated from dyeing mash in indigo fermentation vat.

Habitat: Not distributed in soil or other separated samples but only in *aidama* (fermented indigo plant at 50 °C).

The pH values for optimum growth are probably lower than that described in Takahara and Tanabe's paper, because glass electrodes in the 1960's in Japan did not show accurate pH values higher than 10. It was revealed that indigo-reducing bacteria were isolated from soil samples in various places. This was the second description of an alkaliphile isolated for industrial application, although this strain was not deposited in any culture collection.

Since 1968, many alkaliphilic bacteria have been isolated. Most of them were gram-positive, spore-forming, motile, catalase-positive, and aerobic rods, so are classified under the genus *Bacillus*.

In 1973, Boyer, Ingle and Mercer (1973) characterized the taxonomical characteristics of an alkaliphilic *Bacillus* strain, NRRL B-3881, which produced an alkaline amylase (Boyer and Ingle, 1972), and compared it with our *Bacillus* sp. No. A-59 (ATCC21591)

(Yamamoto,Tanaka and Horikoshi, 1972) and *Bacillus alcalophilus* Vedder strain NCTC4553 (ATCC27647), which was designated as the type strain.

Description of *Bacillus alcalophilus* subsp. *halodurans* subsp. nov. strain NRRL B-3881 (Gr. noun *hals, halis,* salt;L. part. adj. *durans* hardening, resisting; M. L. Part. adj. *halodurans* salt-resisting):

> Cells are gram-positive (rarely gram-negative), straight, motile rods (0.9-1 by 3-4 μm) with rounded ends and peritrichous flagella. Spores are oval(0.5-0.8 by 11.5 μm), terminal and subterminal, and refractile; sporangia are swollen and clavate.
> Nutrient broth (Difco; pH 9): Flocculent with a slimy sediment.
> Plate count agar (Difco; pH 9.8) colonies: White, smooth, shiny, and convex with entire margins.
> Facultatively anaerobic.
> Glucose broth (pH 10): The final pH was 8.98.
> NaCl broth: Good growth in 7% NaCl, slow growth in 15%.
> Acetylmenthylcarbinol, indole, urease, and crystalline dextrins are not produced.
> Starch, gelatin, and casein are hydrolyzed.
> Citrate is utilized.
> Nitrate reduced to nitrite.
> Methylene blue reduced.
> Sucrose, D-glucose, lactose, maltose, D-mannitol, D-xylose, L-arabinose, glycerol, sorbitol, and salicin are fermented without gas production.
> Temperature relationship: Optimum temperature, 48 to 50 °C; maximum temperature for growth is 50 to 55 °C; without aeration, slow growth occurs at 26 °C.

The differences in the taxonomical characteristics among these three strains were studied, and a new subspecies, *B. alcalophilus* subsp. *halodurans* subsp. nov.(ATCC27557), was proposed for the strain NRRL B-3881. Strain NRRL B-3881 and ATCC21591 are different from *B. alcalophilus* (ATCC27647) in several points, as shown in Table 2.3; G + C content for ATCC27647, 37.0, ATCC21591, 43.0, ATCC27557, 42.6. Later, Gordon and Hyde (1982) reported that ATCC27647 and 21591 belong to *B. firmus–B. lentus* intermediates and that ATCC27557 is *B. lentus* group 3.

Gordon and Hyde started taxonomic studies of alkaliphilic strains of *Bacillus* species that had been isolated mainly from soil, dung and water and grew optimally in the alkaline region. Since some of the physiological tests for the identification of *Bacillus* species could not be done at this higher pH value, the alkaliphilic strains could not be compared with the descriptions of recognized species (Gordon and Hyde, 1982). By successive transfer on media with decreasing pH values, variants of the alkaliphilic strains that would grow on media at pH 7.0, and could, therefore, be compared with known strains, were obtained. They reported that pH 7.0 variants of all 174 alkaliphilic strains isolated by Horikoshi *et al.* and Aunstrup *et al.* resembled strains of the *B. firmus–B. lentus* complex.

Although Gordon and Hyde's experiments are somewhat classic, the adaptation experiment is suggestive from an evolutionary point of view. Development of pH 7.0 variants was carried out as follows. At intervals of 5 to 10 days, the alkaliphilic cultures were grown successively on soil extract agar (Gordon, 1973) at pH 9.0, pH 8.5, pH 8.0, pH 7.5

Table 2.3 Differentiation of NRRL B-3881, ATCC21591, and NCTC4553[†1]

Characteristic	NRRL B-3881	ATCC21591	NCTC4553
Maximum temperature for growth in broth with aeration	$> 53 < 54$	$> 54 < 57$	$> 44 < 46$
Optimal growth temperature in broth with aeration	48	50.5	$33 - 35$
Growth in NaCl[†2] 15%	2 +	1 +, Very little growth	−
Hydrolysis of gelatin[†3]	+, Good growth	+, Good growth	+, Little growth
Hydrolysis of casein[†3]	+, Good growth	Sl +, Good growth	Sl +, Scant growth
Anaerobic growth	1 +	2 +	Scant growth
Reduction of methylene blue	+	+ +	Sl +
Nitrate reduction	+	+ +	−
Growth in soybean broth	1 +, Uniformly turbid	2 +, Pellicle	1 +, Uniformly turbid
Hydrolysis of starch[†3]	+	+	+
Production of crystalline dextrins	−	−	−
Voges-Proskauer	−	−	−
Production of indole	−	−	−
Production of urease[†3]	−	−	−
Utilization of citrate[†3]	+	+	−

[†1] Symbols : ND, not done ; Sl +, slight reaction ; +, positive ; + +, more positive ; −, negative.
[†2] Grown without shaking for 48 h at 37 °C, pH 9.8.
[†3] 1 % $Na_4P_2O_7 \cdot 10H_2O$ buffer (pH 8.8) instead of 1 % $NaHCO_3$ buffer (pH 9.8) was used.

Table 2.4 Assignment of pH 7.0 variants of alkaliphilic strains to groups of the *B. firmus–B. lentus* complex

Group	No. of strains	Strain numbers
B. firmus	47	BB 20, 22, 24, 40 ; BC 1, 2, 8, ; C 324, 326, 338, 340, 342, 343, 346 to 350, 352, 353, 355, 379a, 387, 395, 399 to 402, 404 ; PB 20, 33, 35, 38(1), 39 ; SB 14a, 14b, 16, 37; VB 1 to 3, 5, 6, 11 ; ATCC21592, 21593, 21596
B. firmus–B. lentus intermediates	17	BB 16, ; BC 3, 4, 7, 10 ; C 325, 373, 403, 413 ; PB 9, 19, 38(2); ATCC21591, 21594, NCIB9218 ; NRS1548, 1554
B. lentus Group 1	18	BB 45 : C 301, 302, 304, 311, 323, 334 to 337, 339, 368, 369, 392; PB 25 ; SB 5 ; ATCC21595
Group 2		BB 3, 7, 23, 30 ; C 300, 303, 312 to 314, 360, 365, 372, 374 ; SB 32; TB 13, VB 7 ; ATCC21522
Group 3		BB 1, 10, 16(1), 17 to 19, 21, 25, 31, 34 to 37, 41, 43, 49 ; BC 6, 9, 11 to 17 ; C 351, 354, 356 to 358, 364, 366, 367, 370, 371, 375 to 378, 380a, 382b, 383 to 386, 388 to 390, 393, 394, 396 to 398, 410 to 412; PB 16, 17, 40, 41; TB 2; 4, 6, 9 to 12, 14 to 19; ATCC27557; NRRL B-3881

(Reproduced with permission from Gordon and Hyde, *J. Gen. Microbiol.*, **128**, 1112 (1982))

Table 2.5 Some physiological properties of strains of the *B. firmus-B. lentus* complex

| | NRS and ATCC reference strains | | | pH 7.0 variants of alkaliphilic strains | | *B. lentus* groups | | |
	B. firmus	Intermediates	*B. lentus*	*B. firmus*	Intermediates	1	2	3
No. of strains	20	16	10	47	17	18	17	75
Property[†]								
Hydrolysis of hippurate	80	31	80	0	59	28	100	100
Growth in 7% NaCl	95	81	80	100	100	100	100	100
Utilization of :								
Citrate	5	31	10	4	24	44	100	100
Propionate	0	13	0	0	0	28	83	69
Growth at 50°C	0	0	0	0	6	0	100	100
Acid from :								
Arabinose	0	0	100	0	12	100	100	100
Glucose	100	100	100	100	94	100	100	100
Mannitol	90	100	100	89	88	94	100	100
Mannose	0	69	100	0	65	100	100	100
Melibiose	0	50	100	0	35	100	100	100
Raffinose	0	56	100	0	35	100	100	100
Sorbitol	0	6	30	0	12	56	82	0
Sucrose	100	100	100	100	94	100	100	100
Xylose	0	31	100	0	41	100	100	100
Sensitivity to								
bacteriophage	100	31	0	100	88	0	0	100
Decomposition of :								
Casein	100	94	30	91	76	83	94	91
Gelatin	90	94	40	79	65	94	100	100
Tyrosine	15	13	0	0	0	0	0	0
Production of urease	0	13	60	0	6	0	0	0
Reduction of nitrate								
to nitrite	65	13	40	0	18	0	94	3
Deamination of								
phenylalanine	85	19	20	100	35	6	0	1

[†] Positive properties of all strains: formation of catalase, hydrolysis of starch and growth at 35 °C.
Negative properties: production of acetoin and dihydroxyacetone, anaerobic growth, egg-yolk reaction, growth in Sabouraud dextrose broth and agar and resistance to lysozyme.
(Reproduced with permission from Gordon and Hyde, *J. Gen. Microbiol.*, **128**, 1112 (1982))

and pH 7.0 (media at the higher pH values were prepared by the addition of a sterile solution of 1 M sodium sesquicarbonate to the autoclaved media). The resulting cultures were transferred monthly onto soil extract agar (pH 7.0) for six months or more before they were examined for microscopic appearance and physiological reaction. In a few instances, growth and viability were enhanced by the addition of 3 % (w/v) NaCl to the soil extract agar. All strains whose maximum temperature for growth was 45 °C or higher were incubated at 37 °C; others were incubated at 28 °C. Examination of the variants of these 162 isolates and of the 12 alkaliphilic reference strains, all of which grew and were maintained for five to eight years on media at pH 7.0, resulted in their assignment to the *B. firmus–B. lentus* complex (Tables 2.4 and 2.5). Among the pH 7.0 variants of 174 alkaliphilic strains,

Table 2.6 G + C contents of the alkaliphilic *Bacillus* strains and phenotypic groupings

Taxon	Strain	G + C content (mol%)	DNA group	Phenotypic group — This study	Phenotypic group — Study of Gordon and Hyde[†]
Bacillus sp.	DSM 2519	34.0	A	5	*B. lentus* I
	NCIB 10318	35.0	A	6	Intermediate
	PB38 (2)	35.1	A	5	Intermediate
	DSM 2518	35.2	A	5	Intermediate
	NCIB 10327	35.4	A	5	Intermediate
	NCIB 10284	35.4	A	5	Intermediate
	M5	35.5	A	6	
	DSM 2528	35.8	A	6	Intermediate
	DSM 2522	36.1	A	6	
	DSM 2526	36.2	A	5	Intermediate
	M8	36.3	A	6	
B. alcalophilus	DSM 485^T	36.5	A	5	Intermediate
Bacillus sp.	PB9/1	36.9	A	6	Intermediate
	WMI1	36.9	A	5	
	WMI3	37.1	A	5	
	DSM 2521	37.2	A	6	
	PB9/2	37.5	A	6	Intermediate
	DSM 1972	38.2	B	5	*B. lentus* I
	NCIB 10291	38.4	B	2	*B. lentus* I
	NCIB 10289	38.6	B	2	*B. lentus* I
	NCIB 10282	38.8	B	2	*B. lentus* I
	O3	38.8	B	6	
	NCIB 10299	39.0	B	1	*B. firmus*
	NCIB 10286	39.1	B	1	*B. lentus* I
	NCIB 10287	39.2	B	2	*B. lentus* I
	NCIB 10288	39.2	B	2	*B. lentus* I
	NCIB 10290	39.3	B	1	*B. firmus*
	NCIB 10314	39.3	B	2	*B. lentus* I
	ATCC21592	39.4	B	1	*B. firmus*
	NCIB 10294	39.4	B	1	*B. firmus*
	NCIB 10300	39.4	B	1	*B. firmus*
	NCIB 10296	39.5	B	1	*B. firmus*
	NCIB 10292	39.5	B	1	*B. firmus*
	NCIB 10293	39.5	B	2	*B. lentus* I
	DSM 2517	39.5	B	1	*B. firmus*
	NCIB 10302	39.5	B	1	*B. firmus*
	DSM 2523	39.6	B	2	*B. firmus*
	O2	39.6	B	1	
	NCIB 10283	39.6	B	1	*B. firmus*
	NCIB 10285	39.6	B	1	*B. firmus*
	NCIB 10303	39.6	B	1	*B. firmus*
Bacillus sp.	NCIB 10298	39.7	B	1	*B. firmus*
	NCIB 10297	39.7	B	1	*B. firmus*
	DSM 2524	39.7	B	1	
	RAB	39.7	B	1	
	NCIB 10305	40.2	B	1	*B. firmus*
	BC3	40.3	B	1	Intermediate
	NCIB 10295	40.4	B	1	*B. firmus*
	DSM 2520	40.4	B	1	*B. firmus*
	PB19	40.8	B	2	Intermediate
	NCIB 10307	42.1	C	4	*B. lentus* III
	NCIB 10308	42.2	C	4	*B. lentus* III
	NCIB 10316	42.3	C	4	*B. lentus* III
	NCIB 10313	42.5	C	4	*B. lentus* III
	NCIB 10312	42.5	C	4	*B. lentus* III
	BB16	42.5	C	4	Intermediate
"*B. alcalophilus* subsp *halodurans*"	NCIB 10304	42.5	C	4	*B. lentus* III
	DSM 497	42.5	C	4	*B. lentus* III
Bacillus sp.	NCIB 10306	42.6	C	4	*B. lentus* III
	NCIB 10301	42.6	C	4	*B. lentus* III
	NCIB 10322	42.6	C	4	*B. lentus* III
	NCIB 10325	42.6	C	4	Intermediate
	BC4	42.7	C	4	*B. lentus* III
	NCIB 10321	42.7	C	4	*B. lentus* III
	NCIB 10309	42.8	C	3	*B. lentus* II
	NCIB 10310	42.8	C	4	*B. lentus* III
	NCIB 10311	42.8	C	3	*B. lentus* II
	NCIB 10324	42.8	C	4	*B. lentus* III
	NCIB 10323	42.8	C	4	*B. lentus* III
"*B. alcalophilus* subsp *halodurans*"	DSM 2513	42.8	C	4	Intermediate
Bacillus sp.	NCIB 10320	43.0	C	4	*B. lentus* III
	DSM 2514	43.2	C	3	*B. lentus* II
	NCIB 10319	43.3	C	3	*B. lentus* II
	NCIB 10317	43.3	C	3	*B. lentus* II
	DSM 2525	43.5	C	3	
	DSM 2515	43.6	C	3	
	NCIB 10326	43.9	C	3	*B. lentus* III
	DSM 2512	43.9	C	3	*B. lentus* II

† See reference Gordon and Hyde (1982). (Reproduced with permission from Fritze, Flossdorf and Claus, *Int. J. Syst. Bacteriol.*, **40**, 94 (1990))

B. firmus was represented by 47, including ATCC21592, 21593 and 21596. Of the pH 7.0 variants, 17 were assigned as intermediates between *B. firmus* and *B. lentus* on the basis of their formation of acid from seven carbohydrates and by their sensitivity to bacteriophage for the type strain of *B. firmus*. The pH 7.0 variants representing *B. lentus* were divided into three groups by growth at 50 °C, sensitivity to bacteriophage and reduction of nitrate to nitrite. Those of Group I were most like the reference strain of *B. lentus*. Group 2 could be distinguished from Group 1 by growth at 50 °C and reduction of nitrate to nitrite. Group 3 strains grew at 50 °C and were sensitive to bacteriophage.

Fritze, Flossdorf and Claus (1990) have systematically worked on taxonomical studies. The DNA base composition of 78 alkaliphilic *Bacillus* strains for taxonomical studies was analyzed. These strains were grouped as follows: DNA group A, guanine + cytosine (G + C) content of 34.0–37.5 mol% (17 strains); DNA group B, 38.2–40.8 mol% (33 strains); and DNA group C, 42.1–43.9 mol% (28 strains), as shown in Table 2.6. DNA group A includes the type strains *Bacillus alcalophilus* Vedder 1934. DNA-DNA hybridization studies with DNA group A strains revealed that only one strain, strain DSM 2526, exhibited a high level of DNA homology with *B. alcalophilus* DSM 485^T (T = type strain). Neither strain DSM485^T nor any other DNA group A strain is homologous to any of the *Bacillus* type strains with comparable base compositions. Six strains formed a distinct group containing three highly homologous strains and three strains exhibiting >50% DNA homology.

Spanka and Fritze (1993) isolated and characterized 20 alkaliphilic *Bacillus* strains. They proposed *Bacillus cohnii* for the newly isolated strains. A group of 20 alkaliphilic *Bacillus* strains in which all strains revealed the same unique combination of properties— obligate alkaliphily, oval spores distending the sporangium, and ornithine and aspartic acid instead of diaminopimeric acid in the cell wall—was examined (Fig. 2.3). Most of the strains had been isolated by a five-step enrichment and isolation procedure. The G + C content was determined to span a range from 33.5 to 35.0 mol%. Unsaturated fatty acids amounted to 17 to 28% of the total cellular fatty acids. Through DNA-DNA hybridization experiments 11 strains could be grouped in one species. The type strain of the new species is strain RSH (= DSM6307).

Nielsen *et al.* (1994) analyzed 16S rDNA of 14 alkaliphiles. They conducted comparative sequence analysis on about 1520 nucleotides, corresponding to 98% of the whole 16S

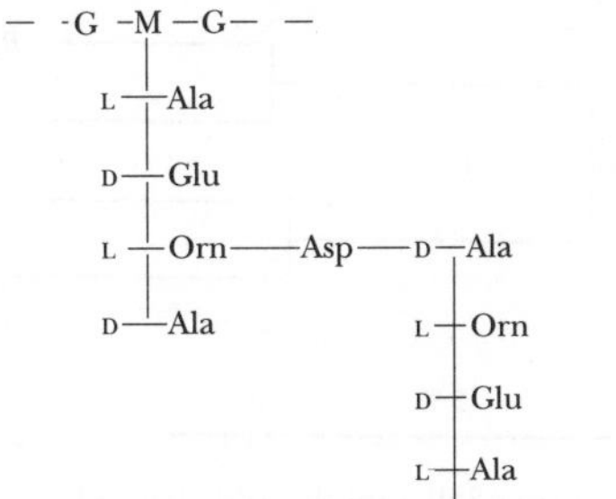

Fig. 2.3 Structure of the cell wall. G, *N*-acetylglucosamine; M, *N*-acetylmuramic acid; L-Ala, L-alanine; Asp, aspartic acid; D-Glu, D-glutamic acid; L-Orn, L-ornithine.

rDNA of 14 alkaliphilic or alkalitolerant, gram-positive, aerobic, endospore-forming bacterial strains. *Bacillus alcalophilus* DSM 485^T and *Bacillus cohnii* DSM 6307^T were included to represent the two validly described alkaliphiles assigned to the genus *Bacillus*. The majority of isolates (8 strains) clustered with *B. alcalophilus* DSM 485^T form a distinct phy-

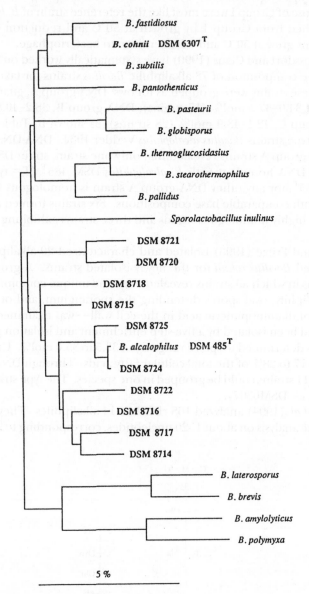

5 %

Fig. 2.4   Phylogenetic position of alkaliphilic and alkalitolerant *Bacillus* strains within
the genus *Bacillus* and related taxa. The scale bar indicates five substitutions
per 100 nucleotides.
(Reproduced with permission from Nielsen *et al.*, *FEMS Microbiol. Lett.*, **117**,
63 (1994))

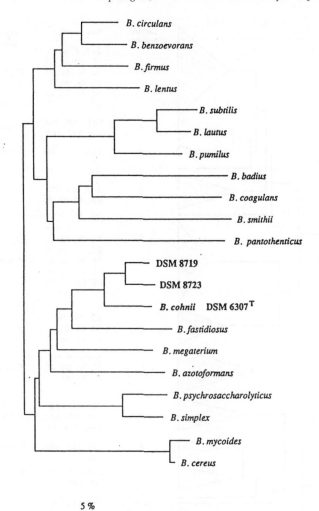

Fig. 2.5   Phylogenetic position of the alkaliphilic and alkalitolerant strains clustering in the *Bacillus* group 1. The scale bar indicates five substitutions per 100 nucleotides.
(Reproduced with permission from Nielsen *et al.*, *FEMS Microbiol. Lett.*, **117**, 63 (1994))

logenetic group (rRNA group 6) within the radiation of the genus *Bacillus* and related taxa. *Bacillus cohnii* DSM 6307[T] and two of the isolates, DSM 8719 and DSM 8723, were grouped with *B. fastidiosus* and *B. megaterium* and allocated to rRNA group 1. The remaining two strains, DSM 8720 and DSM 8721, show an equidistant relationship to both groups, as shown in Figs. 2.4 and 2.5.

Recently, Nielsen, Fritze and Priest reported phenetic diversity of alkaliphilic *Bacillus* strains isolated from various sources and proposed nine new species (Nielsen, Fritze and Priest, 1995). One hundred and nineteen strains of alkaliphilic and alkalitolerant, aero-

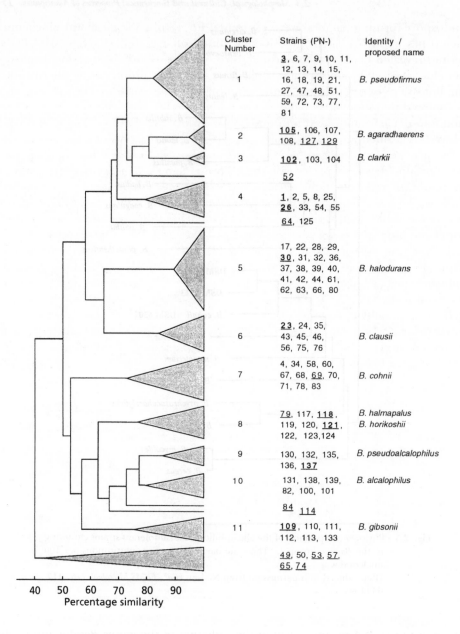

Cluster Number	Strains (PN-)	Identity / proposed name
1	**3**, 6, 7, 9, 10, 11, 12, 13, 14, 15, 16, 18, 19, 21, 27, 47, 48, 51, 59, 72, 73, 77, 81	*B. pseudofirmus*
2	**105**, 106, 107, 108, <u>127</u>, <u>129</u>	*B. agaradhaerens*
3	**102**, 103, 104	*B. clarkii*
	<u>52</u>	
4	**1**, 2, 5, 8, 25, **26**, 33, 54, 55 <u>64</u>, 125	
5	17, 22, 28, 29, **30**, 31, 32, 36, 37, 38, 39, 40, 41, 42, 44, 61, 62, 63, 66, 80	*B. halodurans*
6	**23**, 24, 35, 43, 45, 46, 56, 75, 76	*B. clausii*
7	4, 34, 58, 60, 67, 68, <u>69</u>, 70, 71, 78, 83	*B. cohnii*
8	<u>79</u>, 117, **118**, 119, 120, **121**, 122, 123,124	*B. halmapalus* *B. horikoshii*
9	130, 132, 135, 136, **137**	*B. pseudoalcalophilus*
10	131, 138, 139, 82, 100, 101	*B. alcalophilus*
	<u>84</u> <u>114</u>	
11	**109**, 110, 111, 112, 113, 133	*B. gibsonii*
	<u>49</u>, 50, <u>53</u>, <u>57</u>, <u>65</u>, <u>74</u>	

40 50 60 70 80 90
Percentage similarity

Fig. 2.6 Abbreviated dendrogram showing the allocation of strains to clusters based on the S_1/UPGMA analysis of the biochemical and physiological characteristics. Bold underlined strain numbers indicate those strains for which the full 16S rRNA sequence has been determined (Nielsen *et al.*, 1994); underlined strain numbers indicate those for which a partial 16S rRNA sequence has been determined (unpublished).

(Reproduced with permission from Nielsen *et al.*, *Microbiology-UK*, **141**, 1750 (1995))

bic endospore-forming bacteria were examined for 47 physiological and biochemical characteristics, and DNA base composition. Numerical analysis (S-J and S-SM/UPGMA clustering) revealed 11 clusters comprising three or more strains. Most of the phena were further characterized by analysis of carbohydrate utilization profiles using the API 50CH system, but strains of two taxa could not be cultured by this method. DNA reassociation studies showed that nine of the phena were homogeneous, but strains of phenon 4 and phenon 8 were each subdivided into two DNA hybridization groups. The strains could therefore be classified into 13 taxa plus a number of unassigned single-membered clusters. Two taxa were equated with *Bacillus cohnii* and *B. alcalophilus* and nine of the remainder were proposed as new species with the following names: *B. agaradhaerens* sp. nov., *B. clarkii* sp. nov., *B. clausii* sp. nov., *B. gibsonii* sp. nov., *B. halmapalus* sp. nov., *B. halodurans* comb. nov., *B. horikoshii* sp. nov., *B. pseudoalcalophilus* sp. nov. and *B. pseudofirmus* sp. nov. Two taxa were insufficiently distinct to allow confident identification and these have therefore not been proposed as new species (Fig. 2.6). They described *Bacillus horikoshii* as follows: *Bacillus horikoshii* (ho.ri.ko'shi.i) sp. nov. ML gen. n. horikoshii of Horikoshi. This description is taken from their study based on strains of DNA hybridization group 8b. Colonies are small, circular with an entire margin, shiny surface and a cream-white color. Cells are rod-shaped ($0.6-0.7 \times 2.0-4.0$ µm) with ellipsoidal spores ($0.5-0.7 \times 0.7-1.2$ µm) located subterminally in a sporangium, which may be slightly swollen. Strains of this species hydrolyze casein, hippurate, gelatin, pullulan and starch. Three of the four strains hydrolyze Tween 40 and 60. Growth is observed at pH 7.0, with an optimum at about pH 8.0. Strains grow between 10 and 40 °C. Salt tolerance is moderate, with a maximum at

Table 2.7   Genus *Bacillus* reference species used

B. acidocaldarius (acidophile, thermophile)	
B. alcalophilus (alkaliphile)	
B. stearothermophilus (thermophile)	
B. brevis	*B. lictheniformus*
B. cereus	*B. macerans*
B. circulans	*B. megaterium*
B. coagulans	*B. polymyxa*
B. globisporus	*B. pumilus*
B. laterosporus	*B. sphaericus*
B. lentus	
B. subtilis (Type species)	

Genus *Bacillus* alkaliphilic strains used

(Alkaline amylase)				
Bacillus sp. 13	17-1	27-1	124-1	135
169	A-40-2	A-59	38-2 (CGTase)	
(Alkaline protease)				
Bacillus sp. 221	8-1	D-1		
(Alkaline cellulase)				
Bacillus sp. N-1	N-4			
(Other enzymes)				
Bacillus sp. 9	202-1	C-3	C-125	N-6
17	2b-2	C-11	IC	O-4
170	AH 101	C-59-2	K-12-5	Y-76

8–9% NaCl. Strains do not hydrolyze MUG or Tween 20, do not deaminate phenylalanine and do not reduce nitrate to nitrite. No growth is observed on ribose, D-xylose, L-arabinose, galactose, rhamnose, sorbitol, lactose, melibiose, melizitose, D-raffinose, sorbitol, lactose melibiose, melizitose, D-raffinose or D-tagatose. The G + C content of chromosomal DNA is between 41.1 and 42 mol%. Source: soil samples. Type strain is PN-121 (= DSM 8719).

Furthermore, three novel alkaliphilic bacteria were isolated and characterized (Agnew *et al.*, 1996). Novel alkaliphilic bacteria were isolated and characterized following

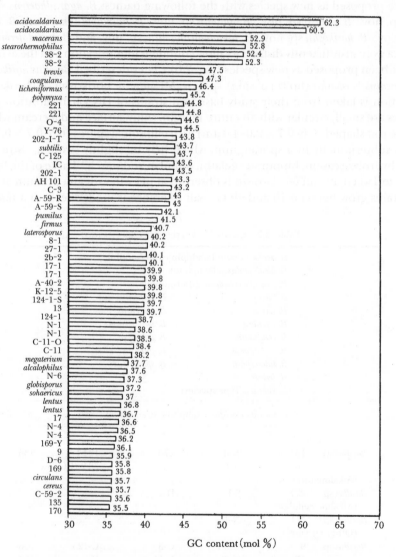

Fig. 2.7   Analysis of GC content in alkaliphilic *Bacillus* strains.

Table 2.8 Analysis of menaquinone (MK) in *Bacillus* strains

	GC	MK-6	MK-7	MK-8	MK-9
C-59-2	35.7	3	96	1	
cereus	35.7	1	99		
169	35.8	1	99		
N-4	36.5	1	99		
17	36.7		100		
sphaericus	37.2	2	98		
globisporus	37.3	4	96		
alcalophilus	37.7		100		
megaterium	38.2	5	95		
N-1	38.7	11	86	3	
13	39.7	1	99		
K-12-5	39.8	1	99		
17-1	39.9	3	97		
laterosporus	40.7	2	98		
pumilus	42.1	1	99		
A-59	43.0	13	86	1	
C-3	43.2	2	98		
subtilis	43.7	1	98	1	
Y-76	44.5		100		
coagulans	47.3		96	4	
brevis	47.5	3	97		
38-2	52.3	3	96	1	
stearothermophilus	52.8	1	98	1	
acidocaldarius	62.3	11	89		
124-1	39.7	35	65		
A-40-2	39.8	57	43		
2b-2	40.1	66	34		
27-1	40.2	44	56		
8-1	40.2	59	41		
221	44.8	41	59		
lentus	36.8		48	43	9
170	35.5				
135	35.6				
circulans	35.8				
D-6	35.9				
9	36.1				
N-6	37.6				
C-11	38.4				
AH101	43.3				
202-1	43.5				
IC	43.6				
G-125	43.7				
O-4	44.6				
polymyxa	45.2				
licheniformus	46.4				
macerans	52.9				

enrichment at pH 10 using mud from a bauxite-processing red mud tailing pond as source material. The isolates, designated strains JaA, JaD and JaH, were aerobic, gram-positive, spore-forming, mesophilic rods which were motile by means of peritrichously arranged flagella. Strains JaD and JaH could also grow anaerobically. All three isolates had pH optima for growth of approximately 10, with strains JaD and JaH obligate alkaliphiles. Strains JaA and JaD grew in the presence of 10% (w/v) NaCl while strain JaH could grow in the presence of up to 7.5% (w/v) NaCl. The mole% G+C content of purified DNA was 41.3% for strain JaA, 41.8% for strain JaD and 38.3% for strain JaH. The new isolates appear to be members of the genus *Bacillus* but 16S rDNA sequence analysis and sodium dodecyl sulfate-polyacrylamide gel electrophoresis comparisons of whole cell protein profiles, as well as nutritional, biochemical and morphological traits indicate that the new isolates are not members of any validly published *Bacillus* species. The epithet *Bacillus vedderi* is proposed for strain JaH.

Our research team has just started study of the alkaliphilic *Bacillus* strains isolated in our laboratory. Tamaoka (personal communication) has analyzed G+C contents, isoprenoide-quinone, and fatty acids of alkaliphilic *Bacillus* strains. As shown in Table 2.7, 17 strains of genus *Bacillus* as reference strains and 29 alkaliphilic *Bacillus* strains were investigated. As shown in Fig. 2.7, 30 strains of alkaliphilic *Bacillus* strains isolated in our laboratory and 20 strains of known *Bacillus* strains as reference strains were analyzed by the method described above. G+C contents of alkaliphilic *Bacillus* strains were distributed in the range of 52.4% to 35.5%. It is of interest that alkaliphilic *Bacillus* sp. No. 38-2, which is a CGTase producer, has almost the same G+C content as that of *B. macerans*, which is also a CGTase producer. And both CGTases exhibited very high homology not only in amino acid sequence, but also in nucleotide sequence. Such a similarity was also observed for *B. cereus* and alkaliphilic *Bacillus* sp. No. 170. They produce essentially the same penicillinases in the culture broth. As Gordon and Hyde (1982) reported, *Bacillus* sp. No. A-40-2 (ATCC21592), No. 124-1 (ATCC21593) and No. 27-1 (ATCC 21596) have intermediate G+C contents of *B. firmus* (G+C content, 41.5%) and *B. lentus* (G+C content, 37.0%). However, analysis of menaquinone (Table 2.8) showed these strains to be definitely different from either *B. firmus* or *B. lentus*, because alkaliphilic *Bacillus* sp. No. 124-1, No. A-40-2, No. 27-1, No. 221, No. 8-1 and No. 2b-2 have menaquinone-6 (having six isoprenoid units) and menaquinone-7(having 7 isoprenoid units) in their cells. No

Table 2.9 DNA GC content, menaquinone composition, and amylase type of alkaliphilic strains

Strain No.	GC mol%	Menaquinone	Amylase
27-1	40.2	MK-6, 7	Type I
124-1	39.7	MK-6, 7	Type I
A-40-2	39.8	MK-6, 7	Type I
135	35.6	MK-7	Type II
169	35.8	MK-7	Type II
38-2	52.3	MK-7	Type III
13	39.7	MK-7	Type IV
17-1	39.9	MK-7	Type IV

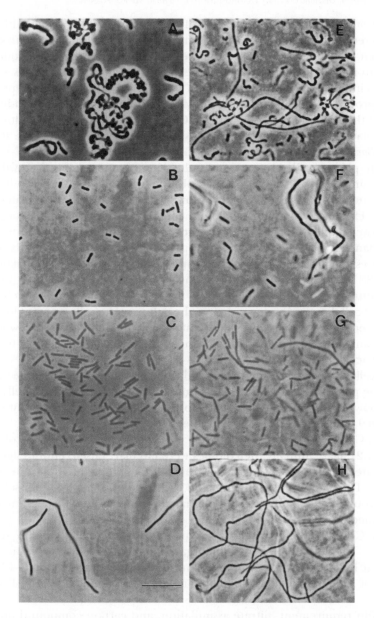

Fig. 2.8 Cellular morphology of strain C-125. Alkaliphilic *Bacillus* sp. strain C-125 was
grown aerobically in complex medium at 37 °C, and cellular morphology was
observed at the late exponential phase (A to D) or stationary phase (30 h) (E
to H) of growth. The culture pH was maintained at pH 6.8 (A and E), pH
7.5 (B and F), pH 9.0 (C and G), or pH 10.8 (D and H). The same magnifi-
cation was used for all micrographs. Bar = 10 µm.
(Reproduced with permission from Aono, *Int. J. Syst. Bacteriol.*, **45**, 583
(1995))

menaquinone-6 was detected in the cells of either *B. firmus* or *B. lentus*. Although further analysis is required, it is reasonable to assume that these alkaliphilic *Bacillus* strains are different from *B. lentus* and *B. firmus*. We have classified alkaline amylase-producing *Bacillus* strains into four groups. Surprisingly, there are very good coincidences among several properties including amylases, as shown in Table 2.9. Further study is in progress.

Our group has been extensively studying microbial properties including molecular biology of the alkaliphilic *Bacillus* sp. No. C-125 as one of the representative alkaliphiles that we isolated. After characterization of strain C-125, Aono assigned the *Bacillus* sp. No. C-125 to *Bacillus lentus* group 3 (Aono, 1995). The morphology of strain C-125 varied with the culture pH controlled by a pH-stat, cultivation temperature, and growth phase. The cells grew singly and in pairs at 37 °C and pH 7.5 to 10 (Fig. 2.8). A few cells formed long chains during the exponential phase of growth. The organism grew in filaments at pH 6.8 and 10.8. These pH values were close to the lower and upper pH limits for growth. Most filaments were coiled at pH 6.8 and straight at pH 10.8; this coiling was more pronounced at 30 °C than at 37 °C. Most cells grown at pH 7.0 to 7.5 formed coiled filaments at 30 °C. Old cultures contained many long chains and filaments; extremely long filaments formed at pH 10.8. The rate of cross-linking among peptide moieties of peptidoglycans depends on the culture pH; low cross-linking rates may be one explanation for coil formation at pH 7. The cell dimensions depended on the culture pH. Cells grown at lower pH values were shorter and wider than cells grown at higher pH values. Cells grown at pH values greater than 8 were straight rods with round ends, and cells grown at pH 6.8 appeared to be coccobacillary (Table 2.10).

The cells had numerous peritrichous flagella and were actively motile when grown at an alkaline pH. Flagellin synthesis depends on the culture pH. Most cells grown at pH values below 7 had no flagella or a few short flagella. Strain C-125 sprouted abundantly at pH 7 to 9, but less effectively at pH values above 10.

Strain C-125 formed circular, convex, entire, smooth, cream-white colonies on neutral (pH 7.2) agar. Initially the colonies were circular, raised, entire edge, rough, and cream-white on alkaline (pH 10) agar; these colonies gradually became irregular and undulate. The colonies were butyrous on each type of agar. No water-soluble pigment formed around the colonies. Thus, colonial appearance varied depending on the pH of the agar medium.

Strain C-125 grew at relatively high temperatures (up to 53 °C at pH 7 and up to 57 °C at pH 10). The doubling times of the organism grown in the complex medium at 37 °C were about 1, 0.5, and 2 h at pH 6.8, 9.0 and 10.8, respectively. Strain C-125 grew in the presence of 10 % (wt/vol) NaCl in neutral medium containing 0.1 % KH_2PO_4 (initial pH, 8).

Vitamin requirement, nitrate assimilation, and carbon compound utilization tests were performed using a modified alkaline synthetic basal medium. Strain C-125 grew by assimilating nitrate instead of ammonium ions, but did not exhibit anaerobic denitrification. This organism was strictly aerobic and positive for oxidase and catalase activities. The carbon sources assimilated by the organism included several pentoses, hexoses hexitol, and oligosaccharides (Table 2.11).

The G + C content of the chromosomal DNA was 42 mol%. No ubiquinone-like pigment was found in the lipid fraction. The length of the isoprene side chains of the quinone ranged from five to seven isoprene units. The predominant quinone was

Table 2.10 Characteristics of alkaliphilic *Bacillus* sp. strain C-125[†1]

Characteristic	Strain C-125
Morphological characteristics	
Cell shape	Rods, short rods, coccobacilli
Cell size at pH 7.5 to 10 (μm)	0.4–0.7 + 1.9–3.4
Gram strain	+
Motility at pH 7	−
Motility at pH 10	+
Fragella at pH 7	None or few
Fragella at pH 10	+
Sporulation	+
Spore shape	Ellipsoidal
Spore size (μm)	0.5–0.7 + 1.0–1.2
Sporangium shape	Swollen
Position of sporangia	Central to subterminal
Cultural characteristics	
Reaction to oxygen	Strictly aerobic
Oxidation-fermentation test	Oxidative
Growth temp. range at pH 7 (°C)	30–53
Growth temp. range at pH 10 (°C)	20–57
Growth pH range at 37°C	6.8–10.8
NaCl tolerance (%)	≦ 10
Assimilation of nitrate	+
Reduction of nitrate to nitrite	−
Indole production	−
Levan formation	−
Splitting of MUG	+
Hydrolysis of gelatin	+
Hydrolysis of casein	+
Hydrolysis of starch	+
Hydrolysis of pullulan	+
Hydrolysis of Tween 60	+
Hydrolysis of Tween 80	−
Hydrolysis of hippurate	+
Vitamin requirement	−
Biochemial characteristics	
Oxdase activity	+
Catalase activity	+
Arginine dihydrolase activity	−
Phenylalanine deaminase acitvity	−
Lecithinase activity	−
Urease activity	−
G + C content of DNA (mol%)	42.2
Menaquinones[†2]	MK-7, MK-6, MK-5
mDAP in peptidoglycan	+ (A1γ type)

[†1] +, positive; −, negative ; MUG, 4-methyl-umbelliferyl-β-D-glucuronide; MK, menaquinone ; mDAP, *meso*-diaminopimelic acid.

[†2] The percentages of the menaquinones are as follows: menaquinone 7, 68%; menaquinone 6, 25%; menaquinone 5, 7%
(Reproduced with permission from Aono *et al.*, *Int. J. Syst. Bacteriol.*, **45**, 584 (1995))

Table 2.11 Comparison of properties of strain C-125 and *B. firmus*-*B. lentus* intermediates

Taxon	Carbohydrate utilization[†1]														Decomposition of:		Production of urease	Reduction of nitrate to nitrite	Deamination of phenylalanine	Hydrolysis of hippurate	Growth at 50°C	Sensitivity to *B. firmus* ATCC 5475 bacteriophage[†2]
	Ara	Rib	Xyl	Glc	Fru	Gal	Man	Mal	Mel	Lac	Suc	Tre	Raf	Sor	Casein	Gelatin						
B. firmus	−[†3]	NR	NR	NR	NR	NR	NR	NR	NR	NR	+	+	−	−	+	±	−	±	+	−	−	+
Intermediates	±	±	±	+	NR	NR	NR	NR	NR	NR	±	±	±	−	±	±	±	±	±	±	±	±
B. lentus group 1	+	+	+	+	NR	NR	NR	NR	NR	NR	+	+	+	±	±	±	±	±	−	±	−	−
B. lentus group 2	+	+	+	+	NR	NR	NR	NR	NR	NR	+	+	+	±	+	+	−	+	−	+	+	−
B. lentus group 3	+	+	+	+	NR	NR	NR	NR	NR	NR	+	+	+	−	+	+	−	−	−	+	+	+
Strain C-125	+	+	+	+	+	+	+	+	+	+	+	+	+	−	+	+	−	−	−	−	+	NT

†1 Abbreviations: Ara, L-arabinose; Rib, D-ribose; Glc, D-glucose; Fru, Fructose; Gal, Galactose; Man, D-mannose; Mal, maltose ; Mel, melibiose; Lac, lactose; Suc, sucrose; Tre, trehalose; Raf, raffinose; Sor, sorbitol; The utilization of carbohydrates by strain C-125 was determined in this study; the utilization of carbohydrates by other strains was determined by examining acid production from carbohydrates in another study.

†2 See Gordon and Hyde (1982)

†3 +, strains are positive; −, strains are negative; ±, some strains are positive and some strains are negative; NR, not reported; NT, not tested.

(Reproduced with permission from Aono, *Int. J. Syst. Bacteriol.*, **45**, 584(1995))

Table 2.12   Comparison of the properties of strain C-125 and the six groups classified by Fritze et al.[†]

Taxon	G + C content (mol%)	Colony color	Growth at 50 °C	Deamination of phenyl-alanine	Reduction of nitrate
Group 1	38.2–40.8	Bright yellow	−	+	−
Group 2	38.2–40.8	Dull yellow	−	−	−
Group 3	42.1–43.9	Cream-white	+	−	+
Group 4	42.1–43.9	Cream-white	+	−	−
Group 5	34.0–37.5	Cream-white	−	−	−
Group 6	34.0–37.5	Cream-white	−	−	+
Strain C-125	42.2	Cream-white	+	−	−

[†] The properties of strain C-125 were determined in this study, and the data for the other taxa were obtained from Fritz et al. (1990)
(Reproduced with permission from Aono et al., Int. J. Syst. Bactriol., **45**, 585 (1955))

menaquinone-7. The quinone composition depended to some extent on the culture pH.

The peptidoglycan type of strain C-125 was type A1γ. The γ-carboxyl group of L-glutamic acid ω-carboxyl group of meso-diaminopimelic acid was not amidated. The glycan chain was composed of N-acetylglucosamine and N-acetylmuramic acid. This subtype of type A1γ peptidoglycan is found in the cell walls of *Bacillus firmus*, *B. lentus* and *B. megaterium*, but not in the cell walls of *Bacillus licheniformis* or *B. subtilis*.

A wide range of DNA base composition has been found in the alkaliphiles. The alkaliphilic *Bacillus* strains have been arbitrarily divided into three groups on the basis of their DNA G + C contents as follows: group A, 34.0 to 37.5 mol%; group B, 38.2 to 40.8 mol%; and group C, 42.1 to 43.9 mol%. Each group has been further subdivided into two phenotypic groups, as shown in Table 2.12. Aono's observations indicated that the strain C-125 is related to *B. lentus* group 3 of Gordon and Hyde (Aono, 1995). Furthermore, he mentions that the strain C-125 belongs to the *Bacillus halodurans* cluster according to the classification of Nielsen, Fritze and Priest (1995), and that the strain is different from *Bacillus firmus*.

B.   Anaerobic Sporeformers and Non-sporeformers

Many anaerobic sporeforming alkaliphiles have been isolated by conventional procedures, although few industrial applications have been developed (Podkovyrov and Zeikus, 1992). Wiegel and his colleagues (Li et al., 1994; Engle et al., 1995) isolated thermophilic anaerobic alkaliphiles and characterized them. An anaerobic and thermophilic alkaliphile, strain JW/YL23-2[T] (T = type strain), was isolated from sewage sludge obtained from a sewage plant in Atlanta, GA. At pH 10.1 and 50 °C, the doubling time of this strain was 19 min. Strain JW/YL23-2[T], a motile rod-shaped bacterium with 2 to 12 peritrichous flagella, exhibited a negative gram stain reaction but was gram-type positive as judged by the polymyxin B test. No heat-stable (85 °C, 15 min) endospores were detected. At 50 °C, growth occurred at pH values ranging from 7.0 to 11.0; the optimum pH was 9.6 to 10.1. The temperature range for growth ranged from 27 to 57.5 °C; the optimum temperature was 48 to 51 °C (pH 10.1). Dissimilatory sulfate reduction was not detected. The organism utilized glucose, fructose, sucrose, maltose, cellobiose, and casamino acids. The DNA

G + C content was 32 mol%. A 16S rRNA sequence analysis revealed a 2% inferred evolutionary distance to *Clostridium paradoxum*. However, the cell wall type of strain JW/YL23-2^T was A4 β (L-Orn-D-Asp), while that of *C. paradoxum* was A1τ (*m*-diaminopimelic acid direct). On the basis of the alkaline pH values and high temperatures for optimal growth, the inability to form spores, and other characteristics different from *C. paradoxum* characteristics, strain JW/Yl-23-2 was placed in a new species, *Clostridium thermoalcaliphilum*; JW/YL23-2 (= DSM 7309) is the type strain of this new species.

Furthermore, nine moderately alkalitolerant thermophilic bacteria with similar properties were isolated from water and soil samples obtained from Yellowstone National Park. These gram-positive, rod-shaped bacteria produce cells with primary branches. The cells are peritrichous and exhibit only slight tumbling motility. At 60 °C the pH range for growth is 6.9 to 10.3, and the optimum pH is 8.5. At pH 8.5 the temperature range for growth is 34 to 66 °C, with an optimum temperature of 57 °C. The strains are mainly proteolytic. The fermentation products from yeast extract are acetate, CO_2, and H_2. Fumarate added to minimal medium containing yeast extract is stoichiometrically converted to succinate, indicating that it is used as an alternative electron acceptor. The DNA G + C content is 33 to 34 mol%, as shown in Table 2.13. On the basis of its unique properties, such as branch formation, growth at alkaline pH values at elevated temperatures, and the relative distance of its 16S rRNA sequence from those of other known bacteria, Engle *et al.* (1995) propose that strain JW/YL-138^T and eight similar strains represent a new genus and species, *Anaerobranca horikoshii*. Strain JW/YL-138 is designated the type

Table 2.13 Characteristics that differentiate *Anaerobranca horikoshii* from other phylogenetically related organisms

Characteristic	*Anaerobranca horikoshii*	*Moorella thermoacetica*	*Clostridium thermocellum*	*Thermoanaerobacterium thermosulfurigenes*	*Clostridium purinolyticum*
G + C content (mol%)	33–34	54	38–39	33	29
pH range	6.5–10.3	5.4–7.8	6.0–8.0	4.0–7.6	6.5–9.0
Optimum pH	8.5–8.8	6.6–6.8	6.8–7.2	5.5–6.5	7.3–7.8
Optimum temp. (°C)	57	55–60	60–64	60	36
Maximum temp. (°C)	66	65	68	75	42
Spore formation	–	+	+	+	+
Branch formation	+	–	–	–	–
$S_2O_3 \rightarrow S^0$	–	–	–	+	–
Fumarate → succinate	+	–	–	–	–
Proteolytic	+	–	–	–	–
Utilization of:					
Glucose	–	+	$(+)^\dagger$	+	–
Cellulose	–	–	+	–	–
Xylose	–	+	–	–	–
Pyruvate	–	+	+	–	+
Pectin	–	–	–	+	–
Purine	–	–	–	–	+
Glycine	–	–	–	–	+
H_2-CO_2	–	+	–	–	–

† Parentheses indicate that the organism must be adapted for growth.
(Reproduced with permission from Engle *et al.*, *Int. J. Syst. Bactriol.*, **45**, 458 (1995))

strain of the type species, *A. horikoshii*. Description of *Anaerobranca horikoshii* is as follows: The rod-shaped cells, which are 8 to 22 μm long and 0.5-0.65 μm in diameter, form one to three branches at a frequency of 1 to 10%, and these branches can form daughter cells by restriction. The cells are peritrichous and, during growth in low-nutrient medium, exhibit slight tumbling motility. Spores have not been detected. The cells are chemoorganotrophic and mainly proteolytic. Yeast extract is required for growth and can be used as the sole carbon source. During growth on yeast extract, if fumarate is added, it is reduced stoichiometrically to succinate. All isolates are moderately alkaliphilic thermophiles. The temperature range for growth is 30 to 66 °C, with an optimum temperature of 57 °C, and the pH range for growth is 6.5 to 10.3, with an optimum pH of 8.55. The DNA base composition is 33 to 34 mol% G + C. 16S rRNA analysis data place this organism in the gram-positive *Clostridium–Bacillus* subphylum, but there are no closely related species.

Recently, Engel *et al.* (1996) reported a thermophilic alkaliphilic obligate anaerobe, *Thermobrachium celere* gen. nov., sp. nov. More than 40 isolates of a novel, ubiquitous, proteolytic, moderately alkaliphilic, thermophilic obligate anaerobe were obtained from geothermally and anthropogenically heated environments and mesobiotic environments located on three continents. Whole-cell protein sodium dodecyl sulfate gel electrophoresis revealed that most of these organisms are very similar. Eight of the isolates were characterized in detail; this analysis included 16S ribosomal DNA sequence analysis. The cells of those organisms are 0.5 to 0.87 μm in diameter, 1.5 to 13 μm long, exhibit tumbling motility, and are gram-positive. The temperature range for growth is 43 to 75 °C (optimum temperature, 66 °C), and the pH range for growth is 5.4 to 9.5 (optimum pH, 8.2); the shortest doubling time is around 10 min. Yeast extract is required for growth, and glucose, sucrose, fructose, galactose, and ribose are utilized. The fermentation products from glucose in the presence of yeast extract are CO_2, H_2, acetate, formate, and ethanol. The G + C content is 30 to 31 mol%. On the basis of these properties, Engle *et al.* propose that the isolates will be placed in a single species of the new genus *Thermobrachium*; strain JW/YL-NZ35 is the type strain of the type species, *Thermobrachium celere*.

Three strains of an anaerobic thermophilic organoheterotrophic lipolytic alkalitolerant bacterium were isolated (Svetlitshnyi *et al.*, 1996). *Thermosyntropha lipolytica* gen nov., sp. nov. (type strain JW/VS-265[T]; DSM 11003) was isolated from alkaline hot springs of Lake Bogoria (Kenya). The cells were nonmotile, non-spore forming, straight or slightly curved rods. At 60 °C the pH range for growth was 7.15 to 9.5, with an optimum between 8.1 and 8.9. At a pH of 8.5 the temperature range for growth was from 52 to 70 °C, with an optimum between 60 and 66 °C. The shortest doubling time was around 1 h. In pure culture the bacterium grew in a mineral base medium supplemented with yeast extract, tryptone, casamino acids, betaine, and crotonate as carbon sources, producing acetate as a major product and constitutively a lipase. During growth in the presence of olive oil free long-chain fatty acids were accumulated in the medium but the pure culture could not utilize olive oil, triacylglycerols, short- and long-chain fatty acids, and glycerol for growth. In syntrophic coculture (methanobacterium strain JW/VS-M29) the lipolytic bacteria grew on triacylglycerols and linear saturated and unsaturated fatty acids with 4 to 18 carbon atoms, but glycerol was not utilized. Fatty acids with even numbers of carbon atoms were degraded to acetate and methane, while from odd-numbered fatty acids 1 mol of propionate per mol of fatty acid was additionally formed. 16S rDNA sequence analysis identified *Syntrophospora* and *Syntrophomonas* spp., as closest phylogenetic neighbors.

C. *Micrococcus* sp.

Alkaliphilic strain No. 31-2 is gram-positive, catalase-positive, and an aerobic bacterium. The cells are spheres, occurring singly, in pairs, and in tetrads; the colonies are yellow in color. According to this morphology, strain No. 31-2 was classified as a *Micrococcus* species (Akiba and Horikoshi, 1976*a*). The strain can actively grow and produce intracellular α-galactosidase at alkaline pH between 9.0 and 11.0.

Kimura and Horikoshi (1988) isolated an alkalipsychrotrophic bacteria strain, No. 207, which showed the highest growth rate at 0 °C. This bacterium is a gram-positive, catalase-positive, non-sporeforming, aerobic coccus. It is motile with a few flagella (Fig. 2.9). Citrate was not utilized. Starch was hydrolyzed but not casein. Neither oxidase nor urease was produced. Indol and hydrogen sulfide were not produced. Its colonies changed color from dark yellow to orange with elapse of culture time. It required proline, methionine, thiamine, and biotinic acid for growth in defined medium. Although growth was observed with NaCl up to 8%, it was inhibited considerably with 1% NaCl. From the results, this strain was classified as a member of the genus *Micrococcus*.

Kim, Lee and Kim (1993) isolated *Micrococcus* sp. Y-1 grown on starch under alkaline conditions and it effectively secreted extracellular pullulanases. The isolate was extremely alkaliphilic since bacterial growth and enzyme production occurred at pH values ranging from pH 6.0 to 12.0. Both strains secrete enzymes that possess amylolytic and pullulanolytic activities (Kim, Choi and Lee, 1993*a*).

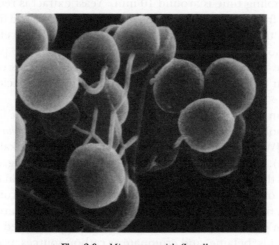

Fig. 2.9 *Micrococcus* with flagella.

D. *Corynebacterium* sp.

Kobayashi and Horikoshi isolated two strains of alkaliphilic *Corynebacterium* species which were maltose dehydrogenase producers (Kobayashi and Horikoshi, 1980*a, b, c, d*). Strain No. 93-1 and strain No. 150-1 have similar taxonomic characteristics in many respects. Both strains are aerobic, gram-positive, non-sporeforming, not acid-fast, catalase-positive, and rod-shaped bacteria which undergo snapping-type cell division. In addition to the similarities in morphology, both strains grew in nutrient broth and nutrient glucose broth at pH 10, grew in 10 % NaCl and Christensen citrate medium, liquefied gelatin, peptonized litmus milk, reduced nitrate to nitrite, hydrolyzed starch and casein; and had a wide pH range for growth (from 7 to 11). The optimum pH for both strains was 9.6. The main amino acid contained in the cell wall of the two strains is DL-diaminopimelic acid. Metachromatic granules are not found in the cells of the two strains. On the other hand, neither strain hydrolyzed cellulose, produced indole, or utilized ammonium salts. However, the two strains differed in such characteristics as those summarized in Table 2.14. The morphological characteristics indicate that both strains are coryneform bacteria. Ikeda, Nakajima and Yumoto (1994) reported that a novel facultatively alkaliphilic bacterium that grows on a chemically defined medium containing *n*-alkanes as the sole carbon source was isolated from soil. The isolate was obligately aerobic, non-motile, gram-positive, and formed metachromatic granules. It was not acid-fast and did not form endospores. The cell wall contained *meso*-diaminopimelic acid, arabinose, and galactose; the glycan moiety of the cell wall contained acetyl residues. The bacterium was catalase-positive, oxidase-negative, and the G + C content of DNA was 70.8 mol%. According to these tests, the isolate was assigned to the genus *Corynebacterium.* The bacterium grew well between pH 6.2 to 10.2 and the doubling time in this pH range was 4–6 h. Na$^+$ added to the culture medium stimulated growth, but was not indispensable at pH 7.2 or pH 10.2. In addition to hydrocarbons, the isolate was able to grow on a chemically defined medium containing acetate, glucose, or fructose as the sole carbon source. Analysis of reduced minus oxidized difference spectra of whole cells showed that the bacterium possessed less than one tenth the amount of total cytochromes as compared with *Bacillus alcalophilus.* The above results suggest that the bacterium has characteristics different from those of the alkaliphilic *Bacillus* previously described.

Table 2.14   Differences in characteristics of alkaliphilic *Corynebacterium* sp. No. 93-1 and No. 150-1

Characteristic	Strain No. 93-1	Strain No. 150-1
Cell size, μm	0.8–1.0 × 2.0–3.0	0.7–0.8 × 1.0–3.0
Motility	Motile by peritrichous flagella	Nonmotile
Color of colonies	Yellow	
H$_2$S production	−	+
Denitrification	+	−
Urease	+	−
GC content, mol%	65.8	52.0
Temperature for growth, °C	20–40	15–40
Acid production from carbohydrate	Glucose; fructose; xylose; sorbose; lactose; maltose; cellobiose; dextrin; and starch	Arabinose; xylose; glucose; fructose; mannose; galactose; lactose; maltose; cellobiose; trehalose; dextrin; starch; glycerol; and mannitol

2.3.2 Gram-negative Bacteria

A. *Pseudomonas* sp.

Two strains, 26.1B and 22.39B, which produced alkaline lipases, have been isolated from soil (Watanabe *et al.*, 1977). Strain 26.1B grew over a pH range of 5.5 to 9.0 with optimum pH between 8.2 and 8.7. The other strain, 22.39B, grew over a range of 4.5 to 9.0 with optimum pH between 6.6 and 7.0. For the most part, strain 26.1B resembles *P. nitroreducens*, which grows at temperatures below 37 °C. On the other hand, strain 26.1B can grow at 42 °C. In addition to this, strain 26.1B is different in terms of acid production from glucose and xylose, assimilation of benzoate and protocatechuate, tolerance to NaCl, and lipase production. These differences led to the proposal of a new species, *P. nitroreducens* nov. var. *thermotolerans*, for strain 26.1B. The other strain, 22.39B, was identified as *P. fragi* on the basis of its taxonomical characteristics.

Dees *et al.* characterized a gram-negative cellulase-producing bacterium NCIMB 10462 previously named *Pseudomonas fluorescens* subsp. or var. *cellulosa* (Dees *et al.*, 1995) because of renewed interest in cellulose-degrading bacteria for use in the bioconversion of cellulose to chemical feed stocks and fuels. Metabolic and physical characterization of NCIMB 10462 revealed that this is an alkaliphilic, nonfermentative, gram-negative, oxidase-positive, motile, cellulose- degrading bacterium. The aerobic substrate utilization profile of this bacterium has few characteristics consistent with the classification of *P. fluorescens* and a very low probability match with the genus *Sphingomonas*. However, total lipid analysis did not reveal any sphingolipid bases produced by this bacterium. The strain NCIMB 10462 grows best aerobically, but also grows well in complex media under reducing conditions. NCIMB 10462 grows slowly under anaerobic conditions on complex media, but growth on cellulosic media occurred only under aerobic conditions. Total fatty acid analysis of NCIMB 10462 failed to group this bacterium with a known *Pseudomonas* species. However, fatty acid analysis of the bacteria when grown at temperatures below 37 °C suggest that the organism is a pseudomonad. Since a predominant characteristic of this bacterium is its ability to degrade cellulose, the authors suggest that it be called *Pseudomonas cellulosa*.

Vazquez *et al.* (1995) isolated 34 psychrotrophic bacteria during the austral summers of 1989/90 and 1991/92 from different sources of the Antarctic ecosystem. Thirty-two of the selected strains were gram-negative bacteria and *Pseudomonas* was the predominant genus. Three *Pseudomonas maltophilia* strains showed the highest levels of proteolytic activity at 20 °C.

B. *Flavobacterium* sp.

Souza *et al.* (1974) reported that an aerobic and an anaerobic alkaliphilic bacteria were isolated from highly alkaline springs (pH 10.8–12.1) in California. The aerobic bacterium is gram-negative, strictly aerobic, non-sporeforming, motile rods and forms orange-colored colonies on tryptic soy agar adjusted to pH 10.5 with 0.1 N NaOH. The pH range for growth is 8.0 to 11.4 with an optimum pH of 9.0 to 10.0: no growth occurs at pH 7.6 or 11.6, although it survives for long periods of time even at pH 11.5 to 12.0. The bacterium has been tentatively classified as a *Flavobacterium* species.

C. Actinomycetes

Taber (1959) first reported that a *Streptomyces*, which was identified as *Streptomyces caeruleus*, was an "alkaline requirement species." The species grew in media adjusted to pH above 7.2 but not below 6.9; good growth was observed at pH 8.2. The following year Taber (1960) isolated from alkaline and acid Canadian soils many strains of "acid-sensitive" actinomycetes, which showed the lower pH limit for growth at pH 6.1 to 6.8 and abundant growth at pH 8.5. The taxonomic characterization of the isolates was, however, not reported. The first taxonomic description of "alkaliphilic *Actinomycetes*" was as follows.

1. *Streptomyces* sp.

Nakanishi (Nakanishi and Yamamoto, 1974; Nakanishi *et al.*, 1974; Nakanishi and Yamamoto, 1975) isolated an alkaliphilic strain of a *Streptomyces* species which grew well and produced alkaline protease at pH over 9 while no growth occurred at neutral pH. The aerial mycelium of the strain is branched and bent; the strain forms chains of spores at pH over 10. Spores are smooth on the surface and about 1 μm in size. The strain hydrolyzes starch and gelatin; does not utilize L-arabinose, D-xylose, D-glucose, sucrose, inositol, rhamnose, raffinose, or mannitol; only growth on D-fructose is observed. From its morphological, cultural, and physiological characteristics, the strain resembles *S. griseus*. However, on the basis of its alkaliphilic properties the strain was deposited as *S. alkaliphilis* (FERM No. 2086).

Mikami *et al.* (Mikami, Miyashita and Arai, 1982; Miyashita, Mikami and Arai, 1984) widely investigated alkaliphilic *Streptomyces* in soils and in a culture collection, and isolated twenty strains from soils. These strains showed good growth with alkaline media between pH 9.0 and 9.5 and little or no growth below pH 7.0, suggesting that the most naturally occurring source for isolating alkaliphilic bacteria is soil. Among 420 strains from the International *Streptomyces* Project, six strains, *Streptomyces caeruleus* ISP 5103, *S. alborubidus* ISP 5465, *S. canescens* ISP 5001, *S. autotrophicus* ISP 5001, *S. càvourensis* ISP 5300, and *S. hydrogenans* ISP 5586, were found to grow at pH 11.5.

2. *Nocardiopsis* sp.

Alkaliphilic actinomycete strains isolated from Japanese soil have been shown to belong to the genus *Nocardiopsis* (Miyashita, Mikami and Arai, 1984). Four isolates, strains 153, 161, 164, and 208, were examined in detail for their taxonomic characteristics and identified as a new subspecies of *Nocardiopsis dassonvillei, N. dassonvillei* subsp. *prasina* subsp. nov. (JCM 3336). These organisms were capable of growing between pH 7.0 and 11.0, but not at pH 6.65. The best growth was observed between pH 9.0 and 10.0 These four strains, 153, 161, 164, and 208, formed abundant aerial mycelia with fragmented straight spore chains having about 50 spores on various agar media at pH 9.5. The aerial mycelium had a zig-zag shape at the beginning of sporulation. It subsequently subdivided into spores of varying lengths, and elongated spores were also formed. This mode of spore formation is somewhat different from that of typical streptomycetes but coincident with the characteristic feature of *Nocardiopsis*. The color of the aerial mycelium was white to pale pink, occasionally showing a greenish shade. The vegetative mycelium was colorless. They formed soluble pigment of faint brown on some media but no melanoid pigment. The main components of fatty acids of the alkaliphilic strains were $C_{18:1}$ straight-chain fatty acids, $C_{15:0}$ and $C_{17:0}$ anteiso fatty acids, and $C_{18:0}$ iso fatty acids. The distribution of the fatty acids also coincided with that of *Nocardiopsis* type strains.

All strains contained menaquinones having ten isoprene units as major components, a characteristic common to *N. dassonvillei*. The phospholipid contained in the cells was phosphatidylcholine.

On the basis of the taxonomic characteristics strains 153, 161, 164, and 208 were considered to be a new subspecies of *N. dassonvillei*, for which *N. dassonvillei* subsp. *prasina* was proposed (Pra. si' na, Gradj. *prasina* leek green, referring to the mature aerial mycelium color). This new subspecies is easily differentiated from other strains of *N. dassonvillei* by its higher optimum pH for growth and by its incapability of utilizing maltose and rhamnose.

2.3.3 Yeasts and Fungi

Two strains of previously undescribed pleomorphic species assignable to the black, yeast-like hyphomycetes and which grew even at pH 10.4 were isolated from soil in Wako-shi, Saitama Prefecture, Japan on 23 March 1978, by the author. The cultures of strains No. 11 and No. 21 have been deposited in the culture collection of the Institute of Applied Microbiology, University of Tokyo, as IAM 12519 and 12520, respectively. Goto *et al.* (1981) examined their microbial properties and proposed as new species, *Exophiala alcalophila* Goto et Sugiyama sp. nov., with an accompanying new yeast morph *Phaeococcomyces alcalophilus* Goto et Sugiyama sp. nov. (Fig. 2.10). These strains grow between pH 5.4 and 10.4 on YM medium (5 g peptone, 3 g malt extract, 3 g yeast extract, 15 g glucose, and 15 g agar in 1,000 ml distilled water).

Aono (1990*a*) studied taxonomic distribution of alkali-tolerant yeasts in his laborato-

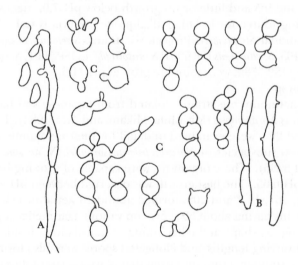

Fig. 2.10 *Exophiala alcalophila* strains No. 11 and No. 21. A colony cultured on SABA for 21 days at room temperature (17–20 °C). Strain No. 11 after one month on PDA at room temperature. The sympodial type of fertile hyphae (A), and conidial apparatus (B). *Phaeococcomyces alcalophilus* strain No. 11. Chains of globose cells of variable length; several cells are connected by narrow isthmi after one month on PDA at room temperature.
(Reproduced with permission from Goto *et al.*, *Trans. Mycol. Soc. Japan*, **22**, 429 (1981))

ry. Yeasts and yeast-like fungi (433 strains, 296 species, 53 genera) were examined to determine the upper limit of pH for growth. Among them, 135 strains of 86 species were found
to be capable of growth at a pH above 10. These alkali-tolerant strains belonged to 27 genera of yeasts, of which 10 genera contained only alkali-tolerant species. Furthermore,
Aono (1992*a*) measured upper pH limits for growth of yeasts belonging to the genus
Hansenula and some related strains. As shown in Fig. 2.11, alkali-tolerant species occupied
particular positions of a phylogenetic tree proposed for the genus. Many filamentous
fungi which could grow on Horikoshi-II medium have been isolated, and some of them
(*Fusarium* sp., etc.) exhibited good growth in the range of pH 5 to 10.

It is noteworthy that microorganisms isolated on solid alkaline media can not always
grow in liquid media having the same pH values. They usually need a lower pH value
(about 1 pH unit) in liquid media than in solid media. Therefore, details of culture conditions (temperature, culture media, liquid or solid, shaking, etc.) should always be cited.

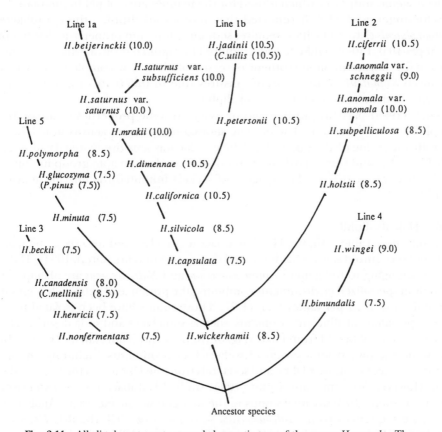

Fig. 2.11 Alkali-tolerant yeasts on a phylogenetic tree of the genus *Hansenula*. The
number in parentheses shows the upper limit pH for growth of a given strain.
Species related to *Hansenula* are shown in parentheses.
(Reproduced with permission from Aono, *Syst. Appl. Microbiol.*, **15**, 589
(1992))

2.3.4   Alkaliphiles Showing Other Properties

A.   Cyanobacteria

Recently, several alkaliphilic cyanobacteria have been isolated and studied for pH homeostasis, but no industrial application has been reported (Bakels *et al.*, 1993; Miller, Tukrpin and Canvin, 1984; Prasad and Kashyap, 1991; Dwivedi *et al.*, 1994; Xu *et al.*, 1994). Buck and Smith (1995) reported a Na^+/H^+ electrogenic antiporter in alkaliphilic *Synechocystis* sp. An alkaliphilic cyanobacterium characterized as a *Synechocystis* species was isolated from a soil sample taken from a village in Java, Indonesia, by its preferential growth at elevated pH; it grew optimally at pH 9.5. Phosphorus nuclear magnetic resonance studies showed that the organism can maintain an intracellular pH value of 8 at an external pH of 10. It was observed that the viability of the organism in the dark was dependent on sodium ions. Evidence from experiments in which the extrusion of Na^+ was measured from cells subjected to an alkali shock suggests that the organism possesses a Na^+/H^+ electrogenic antiporter which is used for the maintenance of pH homeostasis.

Schlesinger *et al.* (1996) reported that another alkaliphile, the cyanobacterium *Spirulina platenis* (Norst.) Geitler, requires sodium to function properly at elevated pH values. At pH 10.0, 150–250 mM Na^+ was required for optimal growth, whereas 2.5 mM was sufficient for short-term photosynthetic oxygen evolution. The complete absence of sodium caused *S. platenis* to deteriorate, possibly due to inactivity of a sodium/proton antiporter, as demonstrated for other alkaliphiles.

Singh (1995) partially purified and investigated some properties of urease from an alkaliphilic cyanobacterium. The enzyme showed its optimum activity at pH 7.5 and at 40 °C with a K_m value of 120 µm mol/l. The enzyme was sensitive to metal cations, particularly Hg^{2+}, Ag^+ and Cu^{2+}. Hydroxymercuribenzoate (a mercapto-group inhibitor) and acetohydroxamic acid (a chelating agent of nickel) inhibited the enzyme activity completely.

B.   Haloalkaliphiles

As shown in Table 2.15, the Halobacteriaceae are classified into six genera, and the number of recognized species has increased during recent years (Oren, 1994). Halophilic archaea belonging to the genera *Natronobacterium* and *Natronococcus* are alkaliphiles that grow only in specialized environments combining the presence of extremely high salt concentrations with high pH values (8.5–11.0). Those strains have been isolated from samples of hypersaline-alkaline environments, such as soda lakes and soda deserts. The original isolates were obtained from Lake Magadi (Kenya) and Wadi Natrun (Egypt), but isolates are now known from a variety of soda-rich environments, including Owens Lake (California), Lake Zhabuye (Tibet) and soda soils from the Ukraine (Horikoshi and Grant, 1991). However, McGenity and Grant (1995) studied Halobacteriaceae extensively and transferred several *Halobacterium* strains to the new genus, *Haloruburum*. Analyses of 16S rRNA gene sequences from *Halobacterium saccharovorum* NCIMB2081, *Hb. sodomense* ATCC33755, *Hb. lacusprofundi* ACAM34, and *Hb. trapanicum* NRC34021 indicate that these four species are phylogenetically distinct from *Hb. salinarium*, the only species in the genus *Halobacterium sensu stricto*. In addition, these four species comprise a phylogenetic group distinct from all other halobacteria. Accordingly, they propose that *Halobacterium saccharovorum*, *Hb. sodomense*, *Hb. trapanicum* NRC34021 and *Hb. lacusprofundi* should be classi-

Table 2.15 The genera and species of halophilic archaea

Genus and species	Source
Genus *Halobacterium*	
Halobacterium	Salted hides, salted fish
Species *incertae sedis* :	
Halobacterium halobium	Salted fish
Halobacterium cutirubrum	Salted buffalo hide
Halobacterium trapanicum	Solar salt
Validly described species currently placed within the genus *Halobacterium* but deserving placement in another taxon	
Halobacterium saccharovorum	Saltern, California
Halobacterium sodomense	Dead Sea
Halobacterium lacusprofundi	Deep Lake, Antarctica
Halobacterium distributum	Saline soil
Genus *Haloarcula*	
Haloarcula vallismortis	Salt pools, Death Valley
Haloarcula marismortui	Dead Sea
Haloarcula japonica	Saltern, Japan
Haloarcula hispanica	Saltern, Spain
Species *incertae sedis* :	
Haloarcula californiae	Saltern, California
Haloarcula sinaiiensis	Salt pool, Sinai
Genus *Haloferax*	
Haloferax volcanii	Dead Sea
Haloferax denitrificans	Saltern, California
Haloferax gibbonsii	Saltern, Spain
Haloferax mediterranei	Saltern, Spain
Genus *Halococcus*	
Halococcus morrhuae	Salted fish
Halococcus saccharolyticus	Saltern, Spain
Halococcus turkmenicus	Saline soil
Genus *Natronobacterium*	
Natronobacterium pharaonis	Wadi Natrun, Egypt
Natronobacterium magadii	Lake Magadi, Kenya
Natronobacterium gregoryi	Saltern-Lake Magadi
Natronobacterium vacuolatum	Lake Magadi, Kenya
Genus *Natronococcus*	
Natronococcus occultus	Lake Magadi, Kenya

(Reproduced with permission from Oren, *FEMS Microbiol. Rev.,* **13**, 4134 (1994))

fied in the new genus, *Halorubrum.*

Grant *et al.* (Tindall, Mills and Grant, 1980; De Rosa *et al.*, 1982; Tindall, Ross and Grant, 1984) have worked extensively on these haloalkaliphiles isolated from hypersaline-alkaline environments. They reported on classification of their strains by analyzing 16S rRNA (Lodwick *et al.*, 1991; McGenity and Grant, 1993; Mwatha and Grant, 1993). The ribosomal RNA operon of *Natronobacter magadii* was cloned into plasmid vector. Southern analysis of the cloned genes revealed an organization similar to that of other halobacteria. The 16S rRNA gene is flanked by three sets of inverted repeats which may be involved in 16S rRNA maturation. Then the 16S rRNA gene from the alkaliphilic halobacterium *Natronococcus occultus* was sequenced directly following PCR amplification. Comparisons with 16S rDNA sequences from four additional halobacteria and other four archaea indicate that *Natronococcus occultus* represents a distinct lineage within the halobacteria. It is most closely related to *Natronobacterium magadii* (92.1 % similarity). The secondary structure of the 16S rRNA from *Natronococcus occultus* closely resembles those of other halobacteria except for the variable region between positions 119-153.

A novel haloalkaliphilic archaeon was isolated from Lake Magadi (Mwatha and Grant, 1993). Cells of the organism contain large gas vacuoles in the stationary phase of growth, and colonies produced by these archaea are bright pink in appearance (Table 2.16). The major polar lipids of these organisms are $C_{20}C_{20}$ and $C_{20}C_{25}$ derivatives of phosphatidyl-

Table 2.16 Characteristics of natronobacteria isolated from Lake Magadi

Characteristic	Result for :			
	M24	*N. pharaonis*	*N. gregoryi*	*N. magadii*
Gram strain	V [†]	−	−	−
pH range	8.5–10.5	8.0–11.0	8.5–11.0	8.5–11.5
Salt concn. range (%)	15–30	12–30	12–30	12–30
Quinone(s)	MK8	MK8, MK8(H2)	MK8, MK8(H2)	MK8, MK8(2)
Mol% G + C	62.7	64.3	65.0	63.0
Motility	−	+	−	−
Nitrate reduction	+	+	−	−
Gelatin liquefaction	−	+	−	−
Presence of gas vacuoles	+	−	−	−
Sensitivity to erythromycin	−	−	+	+
Simulation of growth by:				
Glucose	+	−	+	−
Fructose	−	−	+	−
Fumarate	+	−	+	−
Succinate	+	−	+	+
Citrate	+	+	+	−
Acetate	+	+	+	−
Sucrose	+	−	−	−
Galactose	+	+	+	−
Mannitol	−	−	+	−
Pyruvate	−	+	+	+
Proline	+	−	−	−
Lysine	+	−	−	+

[†] V, gram variable.

(Reproduced with permission from Mwatha and Grant, *Int. J. Syst. Bactriol.*, **43**, 403 (1993))

glycerol phosphate and phosphatidylglycerol, and the organisms contain an unidentified phospholipid as a minor component. The G + C content of the DNA is 62.7 mol%. The name *Natronobacterium vacuolata* sp. nov. has been proposed. The type strain is designated NCIMB 13189. Lodwick, McGenity and Grant (1994) determined the nucleotide sequence of the 23S and 5S rRNA genes from the haloalkaliphilic archaeon *Natronobacterium magadii*. A phylogenetic tree for the halobacteria was produced by alignment of the 23S rRNA sequence with those of other archaea (Fig. 2.12). This analysis confirms that the genus *Natronobacterium* forms a distinct lineage within the halobacteria, and that the halobacterial genera diverged over a relatively short time interval. Extensive works on ecology, physiology and taxonomy of haloalkaliphiles have been reviewed by Jones *et al.* (1994). Works on ecology, physiology and taxonomy of haloalkaliphiles have been reported by Jones *et al.* (1994) and Duckworth *et al.* (1996). 16S rRNA genes from a range of aerobic chemoorganotrophic, alkaliphilic soda lake Bacteria and Archaea have been sequenced and subjected to phylogenetic analysis. Gram-negative alkaliphiles were found to be confined to the gamma 3 subdivision of the Proteobacteria, with many isolates related to the *Halomonas/Deleya* group. Gram-positive alkaliphiles were found in both high % G + C and low % G + C divisions of the gram-positive lineage, with many isolates being related to the *Bacillus* group, others to *Arthrobacter* spp. Alkaliphilic Archaea were relatively closely related to members of the genera *Natronococcus* and *Natronobacterium*. An

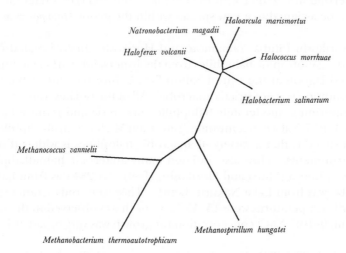

Fig. 2.12 Phylogenetic tree (using 23S rRNA) of halobacteria and representatives of three orders of methanogens. The bar represents a phylogenetic distance of 0.1 average nucleotide substitutions per site. Data used in preparing this figure were derived from the Ribosomal Database Project (RDP) on the anonymous FTB server at the University of Illinois in Urbana, Illinois, updated on October 24, 1993.
(Reproduced with permission from Lodwick, McGenity and Grant, *Syst. Appl. Microbiol.*, **17**, 403 (1994))

anaerobic, thermophilic isolate has been assigned to a new genus within the *Thermotogales*.

Romano *et al.* (1996) isolated and characterized a haloalkaliphilic strictly aerobic bacterium from hard sand of the lake of Venere on Pantelleria Island (Italy). The isolate, growing optimally at pH 9.0 in the presence of 10% NaCl at 33–35 °C, was a gram-negative pleomorphic rod, strictly aerobic, capable of accumulating polyhydroxybutyrate. The isolate was able to grow on different carbon sources in presence of thiamine or biotin. The G + C content of DNA was 65.02 %. The isolate clustered phylogenetically within the gamma subclass of the Proteobacteria and is closely related to the other species on the *Halomonas-Daleya* branch. The haloalkaliphilic strain should be placed in a new species of the genus *Halomonas* for which the name *Halomonas pantelleriense* is proposed.

Furthermore, a new alkaliphilic halotolerant and denitrifying bacterium was isolated from the Göttingen sewage plant (Berendes *et al.* (1996). Enrichment cultures from samples of the Göttingen municipal sewage treatment plant resulted in the isolation of an alkaliphilic organism capable of dissimilatory nitrate reduction. The organism expressed all enzymes necessary for the complete reduction of nitrate to N_2. Optimal denitrification, which was induced by oxygen limitation, as well as optimal growth appeared between pH 9 and 10. Under these conditions N_2 was the main denitrification product. A combination of chemical (polar lipids, respiratory lipoquinones, and fatty acids) and 16S rDNA analysis showed that the isolate was a member of the *Halomonas/Deleya* complex within the γ-subclass of the Protobacteria. Based on the data obtained from a biochemical and physiological characterization together with the chemical and 16S rDNA data they proposed that the organism be assigned to a new species within the genus *Halomonas* as *Halomonas desiderata* sp. nov.

Shiba and Horikoshi (Shiba, Yamamoto and Horikoshi, 1989) isolated four strictly anaerobic, chemoorganotrophic halophiles from the hypersaline surface sediments of the evaporating closed lagoon at the rim of Salton Sea, California, and of Big Soda Lake, Nevada, whose condition was not strictly anaerobic. All of the isolates were gram-negative, motile, non-sporeforming, moderately halophilic eubacteria and required a minimum concentration of 3–10% NaCl concentration, and strain M-20 was an alkaliphile. Isolation of these bacteria suggests that a variety of anaerobic halophiles is widely distributed in hypersaline environments. They also isolated many anaerobic haloalkaliphiles from hypersaline lakes of Kenya (Shiba and Horikoshi, 1989). KY-284 was from Lake Magadi, Kenya, and KY-364 was from Lake Nakuru, Kenya. They were rods, gram-negative, and anaerobic. Growth temperatures were 15–45 °C. Growth was observed in the range of pH 8.0–11.0 (optimum 9–10). NaCl concentration for growth was very broad: 0.1–25% (optimum 3–5).

Kobayashi *et al.* (1992; 1994) also isolated *Natronococcus* Ah-36 from Lake Magadi. The bacterium produces maltotriose-forming α-amylase. Strain Ah-36 was an aerobic, non-motile, coccoid-shaped bacterium (1 to 2 μm in diameter) occurring in irregular clusters, pairs, and single cells. Cells did not lyze in distilled water. The temperature range for growth extended from 20 to 55 °C, with an optimal temperature of 40–45 °C. Growth of the strain occurred within a pH range of 8.0 to 10.0 (an optimal pH of 9.0). The molar ratio of G + C was 63.5 mol%.

Kanai *et al.* (1995) carried out a taxonomical study on this halophile. Most cells of strain Ah-36 occurred in irregular clusters, and the colonies were orange-red. The polar lipids of this organism were composed of C-20, C-20, and C-20, C-25 derivatives of phosphatidylglycerol and phosphatidylglycerophosphate. Phosphatidylglycero-(cyclo-) phosphate, which is characteristic of *Natronococcus occultus*, was not detected. The complete nucleotide sequence of the 16S rRNA gene revealed that the closest relative of strain Ah-36 is *N. occultus* ATCC43101^T (level of similarity 96.4%), an extremely halophilic archaeon. However, strain Ah-36 did not exhibit a significant level of DNA homology to *N. occultus* ATCC43101^T, which represents the only previously described species in the genus *Natronococcus*. Therefore, they propose the name *Natronococcus amylolyticus*.

C. Methanogens

Nakatsugawa isolated several alkaliphilic methanogens from various samples. Strain No. NY-218 (FERM P-2080) was the first motile haloalkaliphilic methanogen discovered. The strain was isolated from the hypersaline Salton Sea in California, U.S.A. The isolate was archaebacterium and a gram-negative, non-sporeforming, anaerobic coccus (0.5-1.5 μm in diameter) with flagella. The optimum NaCl concentration, pH and temperature for growth and methanogenesis were 2.5-3.0 M, 8.0-8.5 and 30-40 °C, respectively. Methanogenesis occurred even in the presence of 3.5-4.0 M NaCl. Methanol, methylamine, dimethylamine and trimethylamine were very good substrates from methane production. The G + C contents was 42%. Strain NY-728 (FERM P-9920), another new isolate, is an alkaliphilic methanogen which was isolated from Japanese lake sediment. The bacterium is gram-positive, psudosarcina or coccus. The optimum pH and temperature for growth and methanogenesis were at 8.0-8.7 and 34-42 °C. Hydrogen and carbon dioxide, acetate, methanol and methylamine were used for growth and methanogenesis. Methanol was the best substrate for growth. The G + C content was 52%. Since NY-728 cross-reacts weakly with antisera raised against *Methanosarcia* sp., this isolate is probably *Methanosarcina*. The strain NY-218, however, has no similarity to standard stains so far tested.

Boone *et al.* (1993) isolated six strains of coccoid, halophilic methanogens from various salinaria and natural hypersaline environments. These strains grew fastest at temperatures near 40°C and, in medium containing 0.5 to 2.5 M NaCl, at pH values near 7. A strain, SD-1, was slightly alkaliphilic (fastest growth occurred at pH 7.8). Zhilina and Zavarzin (1994) studied the bacterial community of an alkaline lake, and the diversity of bacteria found there indicates that both conditions of autonomy and phylogenetic variety are fulfilled for anaerobic bacteria developing at pH 10. Major functional groups in the trophic network were present. Representatives of proteolytic, bacteriolytic, cellulolytic, saccharolytic, dissipotrophic, acetogenic, sulfate-reducing, methanogenic bacteria were isolated.

Cell Structure

3.1 Flagella

Several alkaliphilic bacteria demonstrate vigorous motility at alkaline pH. Flagella exposed to alkaline environments and devices for movement may operate at alkaline pH.

3.1.1 Flagella Formation and Flagellin

Aono, Ogino and Horikoshi investigated pH-dependent flagella formation by facultative alkaliphilic *Bacillus* sp. C-125 (1992). This strain grown at alkaline pH had many sinuous peritrichous flagella and was highly motile. However, most of the cells grown initially at pH 7 were non-motile and possessed few straight flagella. The amount of flagellin was low when the organism was grown at pH 7, suggesting that non-motility is due to poor synthesis of flagellin. Sakamoto *et al.* cloned the flagellin protein gene (*hag* gene) of *Bacillus* sp. C-125 and expressed it in the *Escherichia coli* system (Sakamoto *et al.*, 1992). Flagellin protein from this strain was purified to homogeneity and the N-terminal sequence determined. Using the *hag* gene of *Bacillus subtilis* as a probe, the *hag* gene of *Bacillus* sp. C-125 was identified and cloned into *E. coli*. As shown in Fig. 3.1, sequencing this gene revealed that it encodes a protein of 272 amino acids (Mr 29995). The alkaliphilic *Bacillus* sp. C-125 flagellin shares homology with other known flagellins in both the N- and C-terminal regions. The middle portion, however, shows considerable differences from that of flagellin of *Bacillus subtilis*.

3.1.2 Flagellar Motor

Flagellar movement of neutrophilic bacteria is not caused by ATP but by H^+-driven motors. Hirota, Kitada and Imae (1981), however, reported on flagellar motors of alkaliphilic *Bacillus* powered by an electrochemical potential gradient of sodium ions, because the protein motive force is too low to drive flagellar motors in an alkaline environment. Cells of alkaliphilic *Bacillus* sp. No. YN-1 (Koyama and Nosoh, 1976; Koyama, Kiyomiya and Nosoh, 1976) in growth medium consisting of rich broth and NaCl showed vigorous motility between pH 8.5–11.5. *B. subtilis* showed motility between pH 6–8. The YN-1 cells were washed and resuspended in TG medium consisting of 25 mM Tris-HCl buffer (pH 9.0), 0.1 mM EDTA and 5 mM glucose; no translational swimming cells were observed. The addition of NaCl to the medium, however, caused quick recovery of translational swimming. The swimming speed increased with an increase of NaCl concentration up to 50 mM. Other cations such as Li^+ and K^+ had no effect (Table 3.1). Similar results were

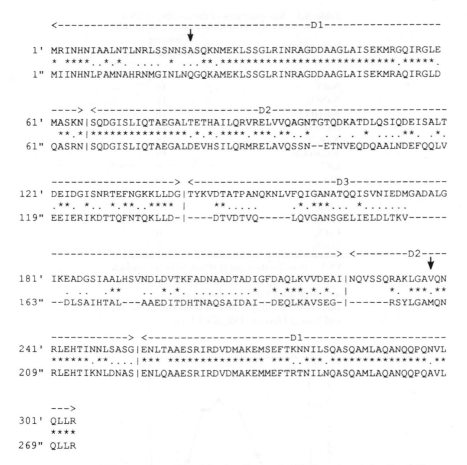

```
        <------------------------------------------D1-----------------
                           ▼
    1'  MRINHNIAALNTLNRLSSNNSASQKNMEKLSSGLRINRAGDDAAGLAISEKMRGQIRGLE
        * ****..*.*. .... *  ...**.*********************************.*****.
    1"  MIINHNLPAMNAHRNMGINLNQGQKAMEKLSSGLRINRAGDDAAGLAISEKMRAQIRGLD


        ----> <-----------------------------D2--------------------------
   61'  MASKN|SQDGISLIQTAEGALTETHAILQRVRELVVQAGNTGTQDKATDLQSIQDEISALT
        **.*|**************.*.*.****.***.**..*  ...  *  ....**. .*.
   61"  QASRN|SQDGISLIQTAEGALDEVHSILQRMRELAVQSSN--ETNVEQDQAALNDEFQQLV


        -------------------> <----------------------D3---------------
  121'  DEIDGISNRTEFNGKKLLDG|TYKVDTATPANQKNLVFQIGANATQQISVNIEDMGADALG
        .**. *.. *.**..**** |   **.....     .*.***... *.......
  119"  EEIERIKDTTQFNTQKLLD-|----DTVDTVQ-----LQVGANSGELIELDLTKV------


        -------------------------------------------> <---------D2----
                                                              ▼
  181'  IKEADGSIAALHSVNDLDVTKFADNAADTADIGFDAQLKVVDEAI|NQVSSQRAKLGAVQN
        . ..   .**   .. *.*. ..........*  *.***.*.*. |    *. ***.**
  163"  --DLSAIHTAL---AAEDITDHTNAQSAIDAI--DEQLKAVSEG-|------RSYLGAMQN


        ------------> <----------------------D1---------------------
  241'  RLEHTINNLSASG|ENLTAAESRIRDVDMAKEMSEFTKNNILSQASQAMLAQANQQPQNVL
        ******.**....|*** **************.***..***.**************.**
  209"  RLEHTIKNLDNAS|ENLQAAESRIRDVDMAKEMMEFTRTNILNQASQAMLAQANQQPQAVL


        --->
  301'  QLLR
        ****
  269"  QLLR
```

Fig. 3.1   Alignment of the alkaliphilic *Bacillus* sp. C-125 (lower) and the *B. subtilis* (upper) flagellin amino acid sequences. Identical amino acids and homologous amino acids are highlighted by asterisks and dots, respectively. Amino acids conserved in neutrophilic flagellins but altered in the alkaliphilic flagellin are indicated by the vertical arrows.
(Reproduced with permission from Sakamoto *et al.*, *J. Gen. Microbiol.*, **138**, 2164 (1992))

obtained for alkaliphilic *Bacillus* sp. No. 8-1 and C-125, as shown in Fig. 3.2 (Sakamoto *et al.*, 1992). Such a Na$^+$ requirement is a unique property of alkaliphilic *Bacillus* strains. In order to clarify the role of Na$^+$ in motility, the effects of various ionophores, valinomycin, CCCP (carbonyl cyanide *m*-chlorophenylhydrazone), nigericin, and monensin were studied (Table 3.2). The results suggest that the membrane potential is a component of the energy source for the motility of alkaliphilic *Bacillus*. Monensin, which catalyzes Na$^+$/H$^+$ exchange, caused strong inhibition of motility, indicating that the flux of Na$^+$ by the chemical potential gradient of Na$^+$ is coupled with the motility of alkaliphilic bacteria.

Hirota and Imae (1983) quantitatively measured the chemical potential of Na$^+$ and motility of *Bacillus* strain YN-1 and found that the swimming speed changed as a function

Table 3.1 Ion specificity of the motility of YN-1

Salts added	Swimming speed(μm/sec)
None	0
NaCl	16
NaNO$_3$	17
Na$_2$HPO$_4$	15
Na$_2$SO$_4$	18
NaSCN	18
Na-acetate	16
LiCl	0
KCl	0
NH$_4$Cl	0
RbCl	0
CsCl	0
CaCl$_2$	0
MgCl$_2$	0

YN-1 cells in TG medium (pH 9.0) were mixed with 15 mM of various salts, and the swimming speed was measured after 5 min incubation at 30 °C.
(Reproduced with permission from Hirota, Kitada and Imae, *FEBS Lett.*, **132**, 279(1981))

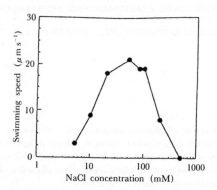

Fig. 3.2 Effect of Na$^+$ concentration on swimming speed of alkaliphilic *Bacillus* sp. C-125 cells. Data shown are mean values of the swimming speed of more than 28 cells for each concentration.
(Reproduced with permission from Sakamoto *et al.*, *J. Gen. Microbiol.*, **138**, 2161 (1992))

of the chemical potential difference of sodium ions across the cell membrane with a threshold value of −100 mV (Fig. 3.3). They concluded that the sodium motive force should be the energy source for the flagellar motors. Imae and Atsumi (1989) reported very interesting properties of flagellar motors. Bacterial flagellar motors are the reversible rotary engine which propels the cell by rotating a helical flagellar filament as a screw propeller. The motors are embedded in the cytoplasmic membrane, and the energy for rotation is supplied by the electrochemical potential of specific ions across the membrane.

Table 3.2 Effect of ionophores on the motility of YN-1

Additions	Swimming speed (μm/sec)
None	22
Valinomycin (10 μM)	22
Valinomycin (10 μM) + KCl (60 mM)	0
KCl (60 mM)	22
CCCP (20 μM)	10
Nigericin (2 μM)	0
Monensin (6 μM)	6

Cells in TG medium (pH 9.0) containing 50 mM NaCl were mixed with ionophores as indicated, and the swimming speed was measured at 30°C within 1 min.
(Reproduced with permission from Hirota, Kitada and Imae, *FEBS Lett.*, **132**, 279 (1981))

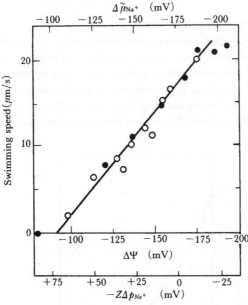

Fig. 3.3 Equality of the chemical potential of Na^+ with the membrane potential of the cell as the energy source for motility of YN-1 cells. The chemical potential of Na^+ in the cell was calculated from the concentration difference of Na^+ between the inside and outside of the cell based on an internal Na^+ concentration of 30 mM. The value of $\Delta\tilde{\mu}_{Na^+}$ was calculated as the sum of the chemical potential of Na^+ ($-Z\Delta p_{Na^+}$) and the membrane potential ($\Delta\Psi$) and is given at the top of the figure. ●, swimming speed at various chemical potentials of Na^+, $-Z\Delta p_{Na^+}$. The membrane potential of the cells under these conditions -176 mV. ○, swimming speed at various membrane potentials, $\Delta\Psi$. The chemical potential of Na^+ under these conditions was -12 mV.
(Reproduced with permission from Hirota and Imae, *J. Biol. Chem.*, **258**, 10580 (1983))

Thus the analysis of motor rotation at the molecular level is aligned to an understanding of how the living system converts chemical energy into two types: one is the H^+-driven type found in neutrophiles such as *Bacillus subtilis* and *Escherichia coli* and the other is the Na^+-driven type found in alkaliphilic *Bacillus* and marine *Vibrio*. Furthermore, Iwazawa, Imae and Kobayashi (1993) studied the torque of the bacterial flagellar motor using a rotating electric field. In this study the torque generated by the flagellar motor was measured in tethered cells of a smooth-swimming *E. coli* strain by using rotating electric fields to determine the relationship between the torque and speed over a wide range. By measuring the electric current applied to the sample cell and combining the data obtained at different viscosities, the torque of the flagellar motor was estimated up to 55 Hz, and also at negative rotation rates. By this method it was found that the torque of the flagellar motor linearly decreases with rotation rate from negative through positive rate of rotation. In addition, the dependence of torque upon temperature was stronger than at the low speeds encountered in tethered cells. From these results, the activation energy of the proton transfer ratio in the torque-generating unit was calculated to be about 7.0×10^{-20} J. Further work on alkaliphilic bacteria in this field is expected to yield more results of strong interest.

3.1.3 Inhibitors for Flagellar Motors

These Na^+-driven flagellar motors were inhibited by amiloride (Sugiyama, Cragoe and Imae 1988). The inhibition was rather specific and other biological functions such as pH homeostasis, ATP synthesis and membrane potential formation were not inhibited. From kinetic analysis of the data, it is evident that amiloride inhibits the rotation of the Na^+-driven flagellar motor by competing with Na^+ at the force-generating site of the motor. The Na^+ interacting site of the motors is somewhat similar to that of the Na^+ chan-

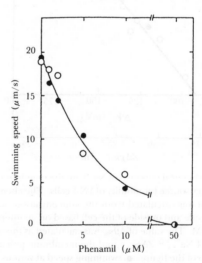

Fig. 3.4 Inhibition by phenamil of motility of RA-1 under conditions of growth. The swimming speeds of the cells were measured after incubation for 5 min (○) or 1 h (●) at 35°C.
(Reproduced with permission from Atsumi *et al.*, *J. Bacteriol.*, **172**, 1636 (1990))

Table 3.3 Effect of phenamil on the motility of various bacterial species

Strain	Swimming speed (μm/s)[1]		
	$-$Na$^+$	$+$Na$^+$	$+$Na$^+$ + phenamil
Motility tightly Na$^+$–dependent			
Alkaliphilic *Bacillus* strains :			
RAB (RA-1)[2]	0	22	0
202-1	0	41	0
8-1	0	19	0
YN-1	0	28	6
YN-2000	0	21	6
Marine *Vibrio* sp., *V. alginolyticus*	0	69	1
Motility weakly or not Na$^+$–dependent			
Alkaliphilic *Bacillus* strains :			
M-29	8	11	11
ATCC 27647	7	11	7
C-59-2	3	4	2
N-6	7	7	4
Neutrophiles			
E. coli	28	29	33
B. subtilis	22	21	22
B. sphaericus	35	37	36

[1] Motility was measured in TG medium. The pH of the medium was 9.5 for alkaliphiles and 7.5 for neutrophiles and the *Vibrio* sp. For alkaliphiles and neutrophiles, the medium was supplemented with 50 mM KCl ($-$Na$^+$) of 50 mM NaCl ($+$Na$^+$). For the *Vibrio* sp., the medium was supplemented with 400 mM KCl ($-$Na$^+$) or 400 mM NaCl ($+$Na$^+$). Phenamil was added to a final concentration of 100 μM.
[2] Motility at pH 7.5 was also tightly Na$^+$ dependent and was completely inhibited by 100 μM phenamil.
(Reproduced with permission from Atsumi *et al.*, *J. Bacteriol.*, **172**, 1638 (1990))

nels. Phenamil, an amiloride analog, inhibited motor rotation without affecting cell growth (Atsumi *et al.*, 1990). A concentration of 50 μM phenamil completely inhibited the motility of *Bacillus* RA-1 (Fig. 3.4), but showed no effect on the membrane potential, the intracellular pH, or Na$^+$-coupled amino acid transport, which was consistent with the fact that there was no effect on cell growth. Kinetic analysis of the inhibition of motility by phenamil indicated that the inhibition was noncompetitive with Na$^+$ in the medium. A motility mutant was isolated as a swarmer on a swarmer agar plate containing 50 μM phenamil. The motility of the mutant showed increased resistance to phenamil but normal sensitivity to amiloride. As shown in Table 3.3, phenamil was found to be a specific and potent inhibitor for the Na$^+$-driven flagellar motors not only in various strains of alkaliphilic *Bacillus* sp., but also in marine *Vibrio* sp.

3.1.4 Comparative Studies

Bogachev *et al.* compared the role of protonic and sodium potentials in motility of *E. coli* and alkaliphilic *Bacillus* sp. FTU (Bogachev *et al.*, 1993). It was found that *Bacillus* sp.

FTU motility (a) requires Na$^+$, (b) is resistant to the protonophorous uncoupler pentachlorophenol if cells grow at high pH, and is sensitive to the uncouplers at neutral pH, (c) is sensitized to the uncouplers with the addition of monensin, (d) sensitive to amiloride and (e) can be supported by an artificially imposed Na$^+$ gradient in the presence of uncouplers cyanide and arsenate. On the other hand, *E. coli* motility (a) does not require Na$^+$, (b) is always uncoupler-sensitive, (c) is amiloride-resistant, and (d) can be supported by an artificially-imposed gradient of H$^+$, not Na$^+$. It was concluded that the motilities of *Bacillus* sp. FTU and *E. coli* are due to the operation of the Na$^+$ and the H$^+$ motors, respectively. In *Bacillus* sp. FTU growing at alkaline pH, the Na$^+$ motors are assumed to be energized by $\Delta\mu_{Na^+}$ produced by the Na$^+$-motive respiratory chain, and therefore $\Delta\mu_{Na^+}$ is not involved in the motility process. As for *Bacillus* sp. FTU growing in a neutral medium, $\Delta\mu_{Na^+}$ is produced secondarily, via the Na$^+$/H$^+$-antiporter at the expense of $\Delta\mu_{Na^+}$ formed by the H$^+$-motive respiratory chain.

3.1.5   Flagella of Other Alkaliphiles

Packer *et al.* (1994) reported the effect of pH on the growth and motility of *Rhodobacter sphaeroides* WS8 and the nature of the driving force of the flagellar motor. *Rhodobacter sphaeroides* WS8 grew, and swam vigorously, over the pH range 6 to 9. Sustained motility was, however, observed in populations of cells resuspended at pH values between 4.9 and 10.4, although the mean run speed was reduced at the extremes of pH. The ability of *R. sphaeroides* to swim in strong alkaline conditions prompted the question of whether motility at alkaline pH was powered by a sodium motive force, as has been found in the facultative alkaliphilic *Bacillus* and *Vibrio* species, particularly as motility was found to be sensitive to the sodium channel inhibitor amiloride. It was found that *R. sphaeroides* was motile over the same pH range in the absence and presence of sodium ions. The protonophore CCCP was found to inhibit motility under all conditions, whereas monensin, an inhibitor of sodium pumps, had no effect upon motility in the presence or absence of sodium. Amiloride, a specific inhibitor of the sodium-driven flagellar motor in alkaliphilic *Bacillus* and *Vibrio*, was shown to act non-specifically on the proton-driven motor of *R. sphaeroides*, reducing the swimming speed of this organism in media with and without sodium to the same extent and over the complete pH range. $\Delta\Psi$ and P^{31} NMR to measure ΔpH showed that the maximum Δp was about -215 mV. At pH 10 the cells swam more slowly and the Δp was about -90 mV. These data suggest that the flagellar motor of *R. sphaeroides* is proton-driven under all conditions with a threshold for motor rotation below -90 mV and saturation at above -90 mV and below -215 mV. It was therefore concluded that the Δp is the driving force for the flagellar motor in *R. sphaeroides* at all values of pH.

Flagella of the haloalkaliphilic archaebacterium *Natronobacterium magadii* were purified and characterized by Fedorov *et al.* (1994). The diameter of the flagella was 10nm. It was shown that the flagella consist of four major proteins with molecular weights of 105,000, 60,000, 59,000, and 45,000. With decreasing NaCl concentration, the flagella dissociated into protofilaments. The structure of dissociated flagella and individual flagellins was studied by limited proteolysis. It was found that proteolytic cleavage of flagellins in dissociated flagella into high molecular weight fragments (about 40,000) did not lead to protofilament degradation. The most stable fragment was formed from the 60,000 molecular weight flagellin. Cleavage of this fragment led to complete disappearance of protofilaments.

Pyatibratov *et al.* studied the structure of flagella from the haloalkaliphilic archaeon *Natronobacterium pharaonis.* Flagellins irreversibly unfold with decreasing NaCl concentration without dissociation of polymers, though the flagellar surface is disordered. Hydrophobic interactions play a principal role in stabilization of the flagellar structure. *N. pharaonis* flagellins consist of two distinctly different parts. One (N-terminal) is responsible for formation of longitudinal intersubunit contacts, the other is exposed on the surface of flagella and can be removed by proteolysis of the unfolded flagellins without destruction of the polymeric core (Pyatibratov *et al.*, 1996).

3.2   Cell Walls

3.2.1   Cell Walls of Neutrophilic Bacteria

The cell wall of bacteria not only provides shape, rigidity, and gram-stainability but is also responsible for some biochemical reactions such as serological properties and phage absorption. Thickness of cell wall of bacteria appears to be between 15 and 50 nm. The cell wall of gram-positive bacteria is usually relatively thick compared with that of gram-negative bacteria. Growth conditions, age of culture and sporulation cause variations in thickness. Cell wall composition and structure differ greatly between gram-positive and gram-negative bacteria. These differences are useful in the taxonomic classification of bacteria. The cell wall of gram-positive bacteria (*Bacillus* sp.) is composed of three main components: (1) peptidoglycan; (2) polysaccharides, teichoic acid or teichuronic acid; and (3) proteins.

Peptidoglycan is a heteropolymer composed of glycan strands bound through two

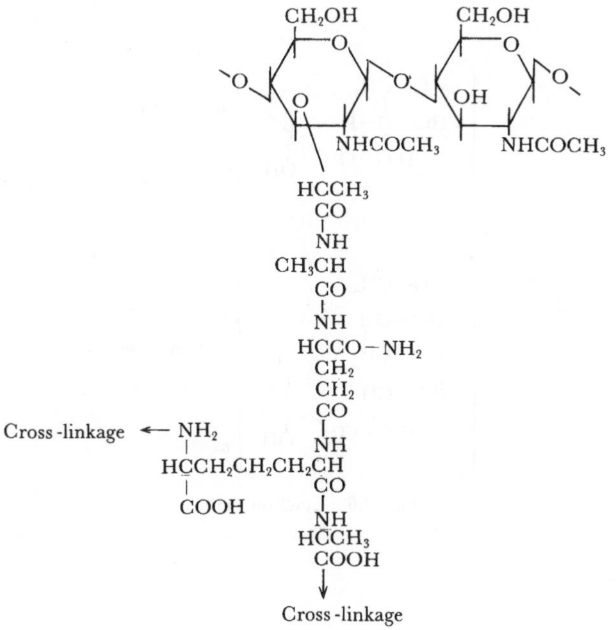

Fig. 3.5   Peptidoglycan of *Bacillus* species.

cross-linked peptide chains, as shown in Fig. 3.5. The components of the glycan strands are N-acetylglucosamine and N-acetylmuramic acid, which are alternately linked by a β-1, 4 bond. The peptide chains show great variation in amino acid composition among different genera and species of bacteria. Based on the anchoring point of the cross-linkage between the two peptide chains, the peptidoglycans have been classified into two main groups, A and B, each of which is further divided into several subgroups depending on the variation in the kind of interpeptide bridges and of the amino acids in position 3 of the peptide chain.

The general structure of peptidoglycan of the genus *Bacillus* is shown in Fig. 3.5. A cross-linkage is formed between the ε-amino group of diaminopimelic acid in position 3 of a peptide chain and the carboxyl group of D-alanine in position 4 of another adjacent peptide chain. This type of peptidoglycan is classified as $A1\gamma$-type, which is the most common structure of the genus *Bacillus*. L-Alanine is linked by its N-terminus to the carboxy group of muramic acid. The amino acid in position 2 is D-glutamic acid, which is linked by its γ-carboxyl group to *meso*-diaminopimelic acid in position 3, and position 4 is occupied by D-alanine. It is noted that amino acids with the D-configuration are present in the peptidoglycan, and these D- and L-amino acids occur in an alternating arrangement.

Teichoic acid is a polymer consisting of phosphoric acid and ribitol or glycerol, to which D-glucose and D-alanine are attached (Fig. 3.6). The phosphoric acid and ribitol or glycerol are bound by ester linkage. Polysaccharide (teichuronic acid) contains glucose, galactose, mannose, or rhamnose and their uronic acids derivatives. Teichoic acids and polysaccharides can be extracted by cold trichloroacetic acid (TCA) from wall fraction. Both components have serological activity and phage adsorption activity. Phosphate groups in the teichoic or in the teichuronic acid cause electron negative charge of the cell surface.

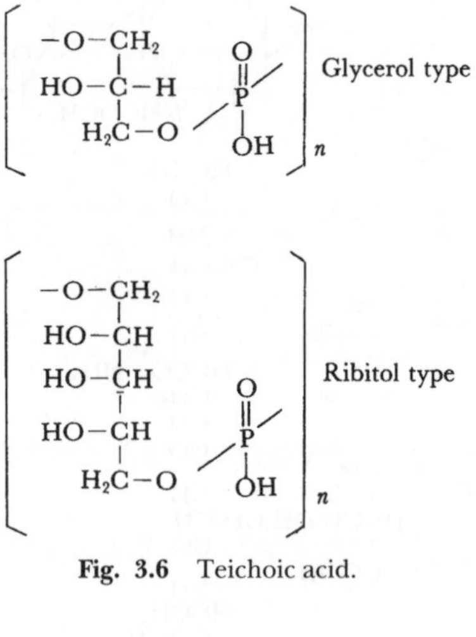

Fig. 3.6   Teichoic acid.

3.2.2   Cell Walls of Alkaliphilic *Bacillus* Strains

The cell wall is directly exposed to high alkaline environments. As shown on page 76, the intracellular pH of alkaliphilic bacteria is neutral (pH 7-8.5) so the pH difference must be due to cell surface components. It has been observed that protoplasts of alkaliphilic *Bacillus* strains lose their stability against alkaline environments (Aono, Ito and Horikoshi, 1992). This suggests that the cell wall may play some role in protecting the cell from alkaline environments.

In a comparison between the cells of alkaliphilic *Bacillus* grown at pH 8.2 and 10.0, the pH 10-grown cells were charged more negatively than the pH 8-grown cells in an alkaline environment (pH 8-10) (Koyama and Nosoh, 1976). There was considerable difference in the protein composition, suggesting that alkaliphilic bacteria may acquire alkaliphily by changing the protein composition of the membrane. More detailed studies, however, are required for further discussion. After their report, the components of cell walls of several alkaliphilic *Bacillus* bacteria have been investigated by Aono and Horikoshi (1983), Ikura and Horikoshi (1983), and Aono, Horikoshi and Goto (1984) in comparison with those of neutrophilic *Bacillus subtilis*. Table 3.4 summarizes the alkaliphilic *Bacillus* strains used for the investigation of the cell wall.

A.   Culture Conditions and Cell Walls

Cell walls were prepared as follows. The culture medium used contained, per liter of deionized water: K_2HPO_4, 13.7g; KH_2PO_4, 5.9 g; citric acid, 0.34 g; $MgSO_4 \cdot 7H_2O$, 0.05 g; glucose, 5 g; peptone, 5 g; yeast extract, 0.5 g; Na_2CO_3, 11.7 g. The pH of this medium was 10.0. The same medium containing 10.6g NaCl instead of sodium carbonate was used to cultivate the organisms which could grow at neutral pH. This medium was adjusted to pH 7.0. Each strain was grown aerobically at 30 °C. At the late exponential phase of growth, cells were harvested by centrifugation, washed once with phosphate buffer (pH 7.0) and resuspended in 2% (w/v) sodium dodecylsulfate containing 0.1 M NaCl. The suspension was incubated at 60 °C for 2 h with gentle shaking. The cells were collected by

Table 3.4   Alkaliphilic *Bacillus* strains used for investigation of cell wall

Bacillus stains (ATCC No.)	Growth at pH 7	pH 10	Na/K Requirement	Remark
A-40-2 (21592)	−	+	Na	Amylase producer
A-59 (21591)	+	+	Na	Amylase producer
2b-2	−	+	Na	High sporulation strain
C-3	+	+	Na	High sporulation strain
C-11[1]	−	+	Na	Hemicellulase producer
C-125[2]	+	+	Na	β-Galactosidase producer
Y-25	+	+	Na	Host of alkaliphilic phage
C-59-2[3]	−	+	K	Xylanase producer
M-29 (31084)	+	+	K	DNase producer
57-1	−	+	K	
B. subtilis GSY 1026	+	−		Neutrophilic *Bacillus*

[1] FERM, No. 3071.   [2] FERM, No. 7344.   [3] FERM, No. 1698.

Table 3.5 Composition of cell walls of alkaliphilic *Bacillus* strains

Group	Strain	Culture pH	Asp	Glu	L-Glu	Gly	Ala	DAP	Mur	GlcN	GalN	P	Glyc	Neutral sugars	Glc	Gal	Uronic acid
1	A-40	10		0.43			0.67	0.42	0.36	0.67	0.23	0.04		0.08			0.90
	2b-2	10		0.36			0.58	0.35	0.34	0.65	0.27	0.17		0.18	0.05		0.81
2	C-11	10		2.32	1.05		0.52	0.33	0.22	0.26	0.04	0.07		0.05			0.60
	C-125	7		1.75	1.08		0.88	0.52	0.48	0.46	0.04	0.03		0.04			0.46
		10		2.13	1.80		0.64	0.38	0.28	0.30	0.03	0.05		0.06			0.90
	Y-25	7		1.55	0.59		0.90	0.47	0.38	0.44	0.04	0.03		0.04			0.36
		10		1.97	0.86		0.54	0.34	0.29	0.32	0.03	0.05		0.06			0.79
	A-59	7	0.55	1.71	0.52		0.93	0.56	0.39	0.42	0.05	0.05		0.04			0.44
		10	0.71	1.82	0.70		0.61	0.38	0.27	0.30	0.04	0.06		0.07			0.62
	C-3	7	0.40	1.20	0.33	0.32	0.94	0.53	0.48	0.51	0.04	0.03		0.10	0.04		0.27
		10	0.72	1.50	0.54	0.52	0.80	0.39	0.36	0.37	0.06	0.06		0.22	0.08		0.49
3	C-59-2	10	0.21	0.26			0.35	0.24	0.18	0.45	0.05	1.65	0.56	0.96	0.45	0.49	0.11
	M-29	7		0.31			0.57	0.30	0.21	0.80	0.05	1.76	0.12	1.85	1.20	0.14	0.13
		10		0.33			0.59	0.31	0.30	1.01	0.05	1.70	0.13	1.73	1.21	0.11	0.12
	57-1	7	0.01	0.29			0.44	0.29	0.19	0.22	0.03	1.53	0.99	0.78		1.05	0.08
		10	0.01	0.30			0.45	0.30	0.21	0.25	0.05	1.82	1.23	0.67		1.08	0.08
	B. subtilis GSY 1026	7		0.54			0.80	0.49	0.48	0.50	0.19	1.77	0.83	1.34	1.02		0.11

The assay method for each compound is detailed in the text. Determination is uncorrected for destruction during acid hydrolysis. Each blank space represents not detected. The following abbreviations are used: Asp, aspartic acid ; Glu, glutamic acid ; Gly, glycine ; Ala, alanine ; DAP, diaminopimelic acid ; Mur, muramic acid ; GlcN, glucosamine ; GalN, galactosamine ; P, phosphorus ; Glyc, glycerol ; Glc, glucose ; Gal, galactose. (Reproduced with permission from Aono and Horikoshi, *J. Gen. Microbiol*, **129**, 1085 (1983))

centrifugation and stored at $-20\,°C$. From the frozen cells, cell walls of 10 alkaliphilic *Bacillus* strains were prepared by inactivation of autolytic enzymes with SDS, disruption with a sonic oscillator, trypsin digestion and washing with SDS. The walls were composed of peptidoglycan and acidic compounds. The peptidoglycan was similar in composition to that of *B. subtilis*. After hydrolysis, the acidic compounds detected were galacturonic acid, glucuronic acid, glutamic acid, aspartic acid and phosphoric acid (Table 3.5). The strains tested were divided into three groups as follows: Group 1, high glucuronic acid and hexosamine contents, no growth observed at pH 7.0, Na^+ essential for growth; Group 2, large amounts of glutamic acid, aspartic acid, galacturonic acid and glucuronic acid detected, growth at higher pH values increased the content of acidic compounds, growth observed at pH 7.0, Na^+ essential for growth ; Group 3, phosphoric acid found, no marked difference detected in chemical components in comparison with *B. subtilis*, growth observed in the presence of Na^+ or K^+ at pH 7.0 and 10.2.

It is well known that cells of the facultative alkaliphile *Bacillus* sp. C-125 grown at neutral pH autolyze rapidly in alkaline buffers of pH 9-10. Recently, Aono *et al.* (1996) found that a peptidoglycan lytic enzyme was extracted from the cell wall fraction suspended in 4 M LiCl. The enzyme was identified as N-acetylmuramyl-L-alanine amidase, with a molecular mass of 58 kDa. At low salinity, the enzyme formed an aggregate of high molecular mass. The peptidoglycan was lyzed at pH 9.0-10.5 at $37\,°C$ by the amidase associated with the cell walls. The enzyme was most active at $60\,°C$ when assayed at pH 9.0.

B. Composition of the Peptidoglycans

As shown in Table 3.6, the peptidoglycans appeared to be similar in composition to those of *Bacillus subtilis*. Cell walls were prepared and suspended in 5% (w/v) TCA. The mixture was incubated at $60\,°C$ overnight and centrifuged at $7,000 \times g$ for 30 min. This extraction was repeated until neutral sugars or uronic acids disappeared in the supernatant. Peptidoglycans prepared as TCA-insoluble fractions were obtained in a 20% to 50% yield of cell walls (dry weight). As shown in Table 3.6, major constituents detected commonly in hydrolysates of the peptidoglycans were glucosamine, muramic acid, D- and L-alanine, D-glutamic acid, *meso*-diaminopimelic acid, and acetic acid. Essentially, the composition of peptidoglycan was not changed whether the strain was cultured at pH 7 or 10. The peptides should be cross-linked directly between *meso*-diaminopimelic acid and D-alanine because of lack of amino acids in the hydrolysates, which are known to be involved in an interpeptide bridge. It was therefore concluded that all of the peptidoglycans of the alkaliphilic *Bacillus* strains so far examined are of the A1γ-type of peptidoglycan, which is found in the majority of strains of genus *Bacillus*. Variation was found in the amide content and was similar to the variation which is known in neutrophilic bacteria.

Aono and Sanada reported that *Bacillus* C-125 grown at neutral pH was rapidly autolyzed in alkaline buffer (Aono and Sanada, 1994). The cells of *Bacillus* sp. C-125 grown at neutral pH autolyzed rapidly in alkaline buffers of pH 9-10. By contrast, the cells grown at alkaline pH were apparently stable under the same conditions. Alkaline N-acetyl-muramyl-L-alanine amidase associated with the cell walls caused this alkali-instability, Meanwhile, cross-linkage between peptide moieties of the peptidoglycan of the cells was dependent on the culture pH. The cross-linking rate was low (31%) in the peptidoglycan of the cells grown at neutral pH, and high (52%) in the cells grown at alkaline pH. This low linking rate is one of the reasons why the cells grown at neutral pH were unstable at

Table 3.6   Composition of the peptidoglycan of alkaliphilic *Bacillus* strains[†]

Strain	Culture pH	meso-DAP content (μmol/mg)	Molar ratio to meso-DAP						Molar ratio of GlcN + Mur to acetic acid
			GlcN	Mur	D-Ala	D-Glu	L-Ala	NH₃	
A-40-2	10	0.94	0.91	0.98	0.86	0.94	0.64	0.09	0.83
2b-2	10	0.82	0.86	0.99	0.82	1.0	0.37	0.04	0.91
C-11	10	1.1	0.84	0.93	0.81	0.93	0.52	0.43	0.94
C-125	7	0.90	0.83	0.99	0.88	1.0	0.70	0.02	0.72
	10	0.93	0.87	1.0	1.0	1.1	0.54	0.01	0.80
Y-25	7	0.84	0.99	1.1	1.0	1.0	0.60	0.06	0.72
	10	0.93	0.81	0.97	0.95	1.0	0.54	0.02	1.1
A-59	7	0.83	0.91	1.1	1.0	1.1	0.62	0.03	0.84
	10	0.87	0.89	1.1	1.0	1.1	0.64	0.01	0.87
C-3	7	1.1	0.87	0.97	0.86	1.0	0.74	0.05	0.80
	10	0.99	0.96	0.84	0.80	1.1	0.60	0.04	0.74
C-59-2	10	0.94	0.91	0.90	0.75	1.0	0.75	0.08	1.0
M-29	7	0.84	0.91	1.1	1.0	1.0	0.81	0.16	0.88
	10	0.94	0.83	1.0	0.97	1.0	0.86	0.33	0.98
57-1	7	0.81	0.93	1.1	1.0	0.98	0.45	0.31	0.84
	10	0.87	1.1	1.1	0.93	1.1	0.62	0.49	1.2
B. subtilis GSY 1026	7	0.97	0.80	0.86	0.83	0.87	0.59	0.43	0.82

[†] Abbreviations: Ala, alanine ; GlcN, lucosamine ; Glu, glutamic acid ; meso-DAP, meso-diaminopimelic acid ; Mur, muramic acid.

alkaline pH. The high linkages in the peptidoglycan may be a cellular adaptation of the organism for growth in alkaline environments.

C.   Acidic Polymers in the Cell Walls of Alkaliphilic *Bacillus* Strain C-125

Most strains of group 2 can grow at neutral pH and require the presence of sodium ions. The same acidic amino and uronic acids are found in much smaller quantities in the walls prepared from bacteria grown at neutral pH. This indicates that the acidic components in the cell walls of the group 2 bacteria play a role in supporting growth at alkaline pH. One of the alkaliphilic *Bacillus* strains isolated in our laboratory, *Bacillus* sp. No. C-125, which is a producer of xylanase, grows well at neutral pH. The chemical composition of its non-peptidoglycan components was relatively simple compared with other group 2 strains (Aono, 1985).

1. Teichuronic acid in the non-peptidoglycan

The non-peptidoglycan components (TCA-soluble fraction) of alkaliphilic *Bacillus* sp. No. C-125 grown at alkaline or neutral pH contained two acidic structural polymer fractions (Fig. 3.7). The fraction Al polymer contained glucuronic acid, galacturonic acid and an amino sugar that was identified as D-fucosamine by 400 MHz NMR spectrometric analysis, measurement of optical rotation and elemental analysis. The Al polymer of alkaliphilic *Bacillus* No. C-125 was a teichuronic acid composed of glucuronic acid, galacturonic acid and *N*-acetyl-D-fucosamine in a molar ratio of 1 : 1 : 1 (Aono and Uramoto, 1986). It is noteworthy that the amount of teichuronic acid is enhanced in the cell walls of *Bacillus* sp. No. C-125 grown at alkaline pH. This teichuronic acid amounted to 390 μg per mg pep-

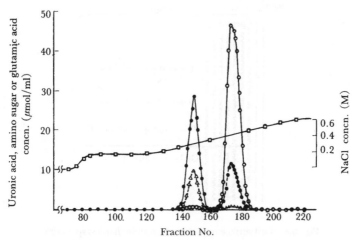

Fig. 3.7   DEAE-cellulose chromatography of the non-dialyzable fraction of TCA extracts. The non-dialyzable fraction of TCA extracts from the alkaline walls were dissolved in 50 mM-acetic acid/NaOH buffer (pH 5.0) and loaded on a column (2.5 × 85cm) of DEAE-cellulose equilibrated with the same buffer. The column was eluted as described in the text. Fractions (12 ml) were collected and assayed for uronic acids (●), amino sugars (△), L-glutamic acid (○) and NaCl (□). Fractions 142-157 (fraction A1) and 168-186 (fraction A2) were pooled.
(Reproduced with permission from Aono, *J. Gen Microbiol.*, **131**, 108 (1985))

tidoglycan in the walls of the bacterium grown at an alkaline pH 80 µg per mg at a neutral pH. The molecular weight of the teichuronic acid from the cells grown at alkaline pH was approximately 70,000 compared to a molecular weight of 48,000 at a neutral pH as estimated by gel chromatography.

2. Teichuronopeptide

The other fraction (A2) contained glucuronic acid and L-glutamic acid in a molar ratio of 1 : 5. This fraction was a copolymer of both acids. This polymer amounted to 560 µg per mg peptidoglycan in the walls of the bacterium grown at alkaline pH or 240 µg per mg at neutral pH. The molecular weight of the polymer from both wall preparations was about 21,000.

Poly (γ-L-glutamate) was prepared from the A2 polymer described above by removal of almost all of the glucuronic residues with trifluoromethane sulfonic acid treatment and purified chromatographically (Aono,1987). The molecular weight of the polyglutamate preparation was about 14,000 by gel chromatography. The molecular weight of the polyglucuronic acid prepared from the fraction by hydrazinolysis, which was free from almost all glutamic acid residues, was about 4,800. Periodate oxidation and Smith degradation of the moiety and enzymatic analysis after reduction of glucuronic acid to glucose revealed that glucuronic acid bound together with alternating α- and β-1, 4-linkages. Therefore, the acidic polymer found in the cell wall of the organism was concluded to be a complex composed of two kinds of polymers (polyglutamate and polyglucuronate). The polymer is designated as "teichuronopeptide" (Aono, 1989a; Aono, 1990b). γ-Polyglutamate has been found widely in other bacteria of the genus *Bacillus*, *e.g.*,

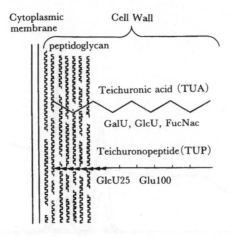

Cytoplasmic membrane   Cell Wall

peptidoglycan

Teichuronic acid (TUA)

GalU, GlcU, FucNac

Teichuronopeptide (TUP)

GlcU25   Glu100

Fig. 3.8   Cell surface model of alkaliphilic *Bacillus* sp. C-125.

B. anthracis, B. mesentericus and *B. subtilis.* These are capsular structures or extracellular mucilaginous material, and do not bind to the peptidoglycan layer of the cell wall. Moreover, they do not contain glucuronic acid as a structural copolymer.

These results show that the cells of *Bacillus* sp. No. C-125 are shaped by the A1γ-type of peptidoglycan, and the peptidoglycan is enclosed by at least two acidic polymers with highly negative charges (Fig. 3.8).

D.   Non-peptidoglycan of Other Bacteria

To obtain further information on group 2 alkaliphilic *Bacillus* sp., cell walls were prepared from four strains (A-59, C-3, C-11 and Y-25). Non-peptidoglycan components were extracted with trichloroacetic acid from the cell wall preparations and isolated by DEAE-cellulose column chromatography as described above. All the components were acidic and composed of amino acids and sugars. Although several components with different compositions were detected, the cell walls commonly contained teichuronopeptide composed of polyglucuronic acid and a polypeptide of acidic amino acids. The molar ratios of glucuronic acid and amino acids were 1 : 4.6–5.4(Fig. 3.9). Molecular weight of these substances was estimated to be 20,000 to 21,000 by gel chromatography. These results suggest that the substances were similar to one another in chemical structure (Fig. 3.10) (Aono *et al.*, 1993 *a*).

Ito *et al.* isolated D-quinovosamine (Fig. 3.11) from the hydrolysate of the acidic polymer of the cell walls of *Bacillus* Y-25 (Ito, Aono and Horikoshi, 1993). As described above, cell walls of alkaliphilic *Bacillus* strain Y-25 are composed of γ-peptidoglycan and two acidic polymers. An amino sugar, which was a main component of one acidic polymer, was identified as D-quinovosamine (2-amino-2,6-dideoxy-D-glucose) by 500 MHz NMR spectroscopic analysis and polarimetry. As far as the author knows, quinovosamine rarely appears as a structural component of cell walls. D- and L-forms are known to occur in the capsular polysaccharide of *Bacteroides fragilis.* The L- form is present in the O-specific polysaccharide of *Proteus vulgaris.* The D- form is known in the polysaccharides of *Vibrio cholerae, Salmonella* sp. This is the first report of D-quinovosamine in the cell walls of *Bacillus* strains.

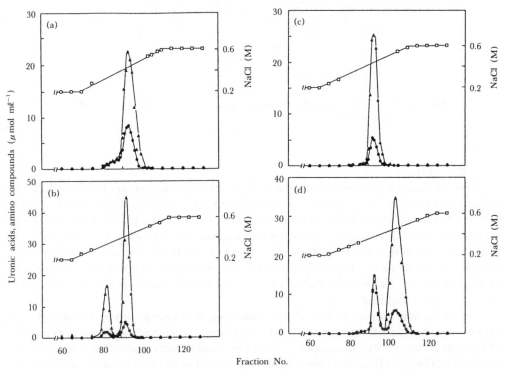

Fig. 3.9 Elution profile of DEAE-cellulose chromatography of the non-dialyzable materials of TCA extracts from strains (a) A-59, (b) C-3, (c) C-11 and (d) Y-25. Fractions (15 ml) were assayed for uronic acids (●), amino compounds (▲) and NaCl (□). The concentration of NaCl was determined from the refractive index of the fractions.

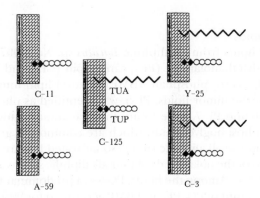

Fig. 3.10 Cell surface model of alkaliphilic *Bacillus* strain.

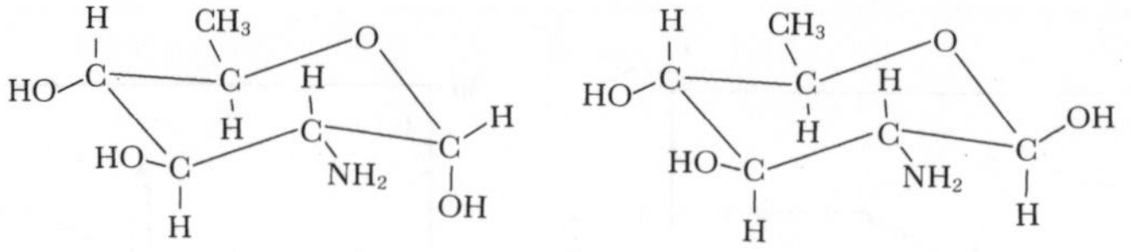

α Anomer

β Anomer

Fig. 3.11 Proposed structure for the amino sugar isolated from the acidic polymer of the cell walls.

3.2.3 Lipids

The cell wall composition of alkaliphilic bacteria differs from that of neutrophilic bacteria, as described above. Lipids are a major component of the cell membrane, the barrier that keeps the intracellular pH lower than the extracellular pH. It has been reported that thermophilic bacteria have some mechanism which produces alterations in the lipid composition of cell membranes. This suggests that there may also exist differences in lipid compositions of alkaliphilic bacteria. The first study on lipids in cell membrane of alkaliphilic bacteria was conducted by Koga, Nishihara and Mori (1982), followed by Nishihara, Mori and Koga (1982).

An alkaliphilic *Bacillus* sp. No. A-007 (Ando *et al.*, 1981) was grown aerobically at 42 °C. Lipids were extracted from cells at the late log phase. Total lipids were fractionated by a silicic acid column chromatography via stepwise elution with chloroform, acetone, and methanol. The lipid fractions eluted from the column were collected and purified by silica gel thin-layer chromatography (TLC) using the following solvent systems: Solvent A for nonpolar lipids: petroleum ether-diethylether-acetic acid (80 : 20 : 3), Solvent B for polar lipids: chloroform-methanol-28% ammonia aqueous solution (200 : 120 : 15), Solvent C for polar lipids: chloroform-methanol acetic acid (200 : 75 : 20). Purified lipid was obtained by scraping silica gel from TLC and eluting with chloroform and methanol (2 : 1).

A. Polar Lipids

Major polar lipids from alkaliphilic *Bacillus* sp. No.A-007 are phosphatidylglycerol (PG), phosphatldylethanolamine (PE), cardiolipin (CL), and PL-4, which was identified as bis-(monoacylglycero)-phosphate (BMP), a novel lipid found only in the alkaliphilic bacteria. One of the minor lipids, PL-5, was identified as phosphatidic acid (PA). The other lipids, PL-6 and PL-7, were found in trace amounts too slight for identification. Glycolipids and phosphoglycolipids, which are common to gram-positive bacteria, have not been found in the alkaliphilic strains so far tested but are present in *Bacillus subtilis*. The compositions of the polar lipids of four alkaliphilic strains and *B. subtilis* W23 are summarized in Table 3.7. Among the lipids, PG was a predominant component of five organisms. The total amount of PG, PE, and BMP accounted for 99% of the polar lipids in alkaliphilic strains. BMP was detected in all alkaliphilic strains except strain A-40-2. No BMP was detected in neutrophilic bacteria such as *B. subtilis* W23, *B. cereus* IAM 1656, *B. megaterium* M3, and *Escherichia coli* W3110 (Koga, Nishihara and Mori, 1982). The presence of

Table 3.7 Polar lipid composition (mol%) of four alkaliphilic *Bacillus* strains and *B. subtilis* W23

Polar lipid	A-007	A-40-2	A-59	No. 202-1	W23
Phospholipid					
Aminoacyl PG	ND	ND	ND	ND	3
PE	21	8	14	9	15
PG	67	87	78	79	71
CL	7	5	7	3	1
BMP	4	ND	T	9	ND
PA	T	T	1	T	1
Others	1	ND	ND	T	1
Phosphoglycolipid	ND	ND	ND	ND	4
Glycolipid					
MGDG	ND	ND	ND	ND	T
DGDG	ND	ND	ND	ND	4

Abbreviations : PE, phosphatidylethanolamine ; PG, phosphatidylglycerol ; CL, cardiolipin ; BMP, bis-(monoacylglycero)-phosphate ; PA, phosphatidic acid ; MGDG, monoglycosyldiacylglycerol ; DGDG, diglycosyldiacylglycerol ; T, trace amount ; ND, not detected

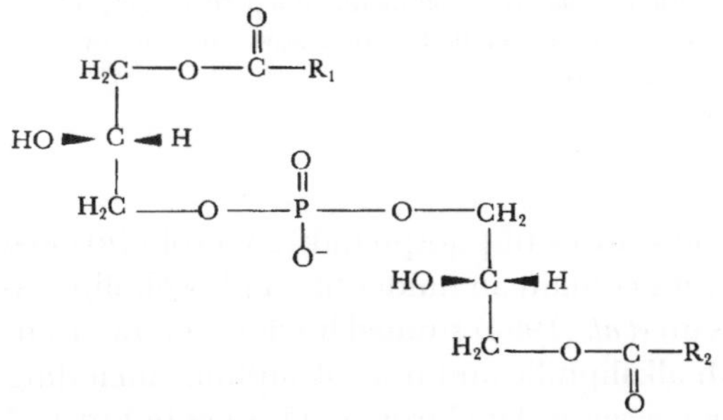

Fig. 3.12 Structure of BMP of *Bacillus* sp. No. A-007. R_1 and R_2: Alkyl groups.

BMP in microorganisms was noted for the first time in these alkaliphilic *Bacillus* strains. The role of BMP in bacteria is not known.

The structure of BMP from *Bacillus* sp. No. A-007 was determined to be 1-acyl-sn-glycero-3-phosphoryl-1 (3'-acyl)-sn-glycerol (Fig. 3.12). This is the same as the stereochemical configuration of the backbone of PG, suggesting that PG is a precursor of BMP. The distribution of fatty acids at the C1 and C3' positions in the glycerol moieties of BMP is summarized in Table 3.8 (Nishihara, Mori and Koga, 1982).

There was a characteristic pattern in the fatty acids esterified at the C1 and C3 positions of BMP. At the C1 position, C_{16} and C_{17} acids were predominant, while at the C3' position C_{15} acids were predominant. This pattern was similar to those in C1 and C2 of PE and PG of strain A-007. *In vivo* experiments on phospholipid metabolism with [14]C- and ^{3}H-labeled glycerols revealed that BMP and CL were synthesized from PG.

Mwatha and Grant (1993) isolated a novel haloalkaliphilic archaeon from Lake Magadi, an alkaline soda lake in Kenya, and proposed *Natronobacterium vacuolata* sp. nov. for the name. The type strain is designated NCIMB 13189. Cells of the organism contain large gas vacuoles in the stationary phase of growth, and colonies produced by these archaea are bright pink in appearance. The major polar lipids of these organisms are

Table 3.8   Fatty acid distribution in the Cl and C3' Positions of BMP

Product	Fatty acid in BMP(%)	Distribution of fatty acid in BMP(%) Cl	C3'
	A	B C	D E
iso $C_{14:0}$	1	1(T)	1(2)
$C_{14:0}$	2	2(T)	2(2)
iso $C_{15:0}$	30	20(23)	40(37)
anteiso $C_{15:0}$	19	12(15)	26(23)
iso $C_{16:0}$	8	10(10)	6(6)
$C_{16:0}$	17	25(24)	9(10)
$C_{16:1}$	10	16(14)	4(6)
iso $C_{17:0}$	9	12(11)	6(7)
Others	4	2(1)	6(7)

The numbers in parentheses indicate the percentage of each fatty acid, with the exception of myristic acid ($C_{14:0}$) which is present in amounts of less than 2%. The fatty acid composition at the C3' position was calculated by subtracting the percentage of each fatty acid at the C1 position from that in total fatty acids as follows:
$$D = 2A - B$$
$$E = 2A - C$$

$C_{20}C_{20}$ and $C_{20}C_{25}$ derivatives of phosphatidylglycerol phosphate and phosphatidylglycerol, and the organisms contain an unidentified phospholipid as a minor component.

Soon after Upasani *et al.* (1994) studied lipids of several strains of extremely halophilic archaebacteria, both alkaliphilic and non-alkaliphilic, including *Halobacterium*, *Haloferax* and *Natronobacterium* species isolated from salt locales in India. The major phospholipids in these strains were the $C_{20}C_{20}$-glycerol diether analogues of phosphatidylglycerol-methylphosphate (PCP-Me), phosphatidylglycerol (PG) and phosphatidic acid (PA). In addition, the *Halobacterium* strains possessed the characteristic glycolipids, sulfated triglycosyl and tetraglycosyl diethers (S-TGD-1 and S-TeGD, respectively) and the unsulfated triglycosyl diether (TGD-1); and the *Haloferax* strains had the characteristic sulfated and unsulfated diglycosyl glycerol diethers (S-DCD-1 and DGD-1, respectively). The PGP-Me, and PC components of the haloalkaliphiles each occurred as two molecular species with $C_{20}C_{20}$ and $C_{20}C_{25}$ (isopranoid) glycerol diether lipid cores as reported by Mwatha and Grant (1993). In contrast to previous reports of the absence of glycolipids in Natronobacteria, the *Natronobacterium* strains from India were found to contain small amounts of a novel glycolipid identified as glucopyranosyl-1 → 6-glucopyranosyl-1 → 1-glycerol diether (DCD-4). The lipid cores of DCD-4 also contained mainly hydroxylated or unhydroxylated $C_{20}C_{25}$ and $C_{25}C_{25}$ molecular species with unsaturated (isoprenoid) chains. Hydroxylated lipid cores have previously been identified only in some methanogenic archaebacteria. Recently, Kates (1996) reported phospholipids and glycolipids in extremely halophilic archaebacteria. The use of chemical analysis, chromatography and spectral analysis (including FTIR, NMR, FAB-MS and CI-MS) in determining the chemical structures of some phospholipids and glycolipids in extremely halophilic archaebacteria is reviewed. Examples include the recently revised structure of the major phospholipid (phosphatidylglycerolmethylphosphate) in Halobacteria and other extreme halophiles, a novel sulfated glycolipid in an extreme halophile from Japan and an unusual diglucosyl glycolipid in haloalkaliphilic archaebacteria from India.

These results strongly suggest that lipids of haloalkaliphilic archaeon will be a fascinating subject for the understanding of both halophily and alkaliphily, although host-vector systems should be established to study genetic information.

B. Nonpolar Lipids

Three major components were detected from the nonpolar lipid fraction of *Bacillus* sp. No. A-007 by TLC developed with solvent A. The third component, which had the highest value, was further separated into two components (squalene and dehydrosqualene). The results of the identification of the components and their content are summarized in Table 3.9. Three other alkaliphilic strains, No. A-40-2, No. A-59, and No. 202-1, also contained 1,2- and 1,3-diacylglycerol, and two squalenes as nonpolar lipids. On the other hand, *B. subtilis* W23 contained only 1,2-diacylglycerol as a major nonpolar lipid (Nishihara, Mori and Koga, 1982).

Table 3.9 Nonpolar lipid composition of *Bacillus* sp. No. A-007

Lipid	Content(%)
1,2-Diacylglycerol	33
1,3-Diacylglycerol	23
Squalene	6
Dehydrosqualene	18
Other (unidentified)	20

C. Fatty Acids

The fatty acid composition of total lipids extracted from four alkaliphilic strains are summarized in Table 3.10, together with that from *B. subtilis* W23. The major components were *iso* $C_{15:0}$ *anteiso* $C_{15:0}$, $nC_{16:0}$, and, *iso* $C_{17:0}$ in the alkaliphilic strains. In strain W23

Table 3.10   Fatty acid composition (%) of total lipid from four alkaliphilic *Bacillus* species and one neutrophilic, *B. subtilis* W23

Fatty acid	A-007	A-40-2	A-59	No. 202-1	W23
iso $C_{14:0}$	3	2	1	3	1
$n\,C_{14:0}$	2	1	1	6	1
iso $C_{15:0}$	27	8	21	22	6
anteiso $C_{15:0}$	23	36	22	22	44
iso $C_{16:0}$	7	8	5	5	6
$n\,C_{16:0}$	18	18	16	25	6
$n\,C_{16:1}$	7	2	12	4	7
iso $C_{17:0}$	10	13	18	6	27
Others	3	12	4	7	2

anteiso $C_{15:0}$ and *iso* $C_{17:0}$ were predominant; this is somewhat similar to strain A-40-2 (Koga, Nishihara and Mori, 1982).

The analytical study on lipids of alkaliphilic and neutrophilic bacteria described above revealed the following differences:

(1) Glycolipids and phosphoglycolipids, which commonly occur in gram-positive bacteria, are not present in alkaliphilic strains.

(2) Instead, BMP, which does not occur in neutrophilic strains of *Bacillus*, is found in alkaliphilic strains.

(3) A characteristic distribution of fatty acids is observed in alkaliphilic strains. These different features in lipid components may contribute to the alteration of the membrane properties of alkaliphilic bacteria.

Dunkley *et al.* (1991) reported that facultative alkaliphiles lacked fatty acid desaturase activity and lost the ability to grow at near-neutral pH when supplemented with an unsaturated fatty acid. Two obligate alkaliphiles (*Bacillus alcalophilus* and *B. firmus* RAB) were found to have high levels of fatty acid desaturase, whereas two facultative alkaliphiles (*B. firmus* OF1 and OF4) had no detectable activity. The obligate strain outgrows the facultative strain in a chemostat at very high pH, whereas the converse is true at a pH of 7.5, and the two strains grow equally well at pH 9.0. Thus the obligate strain is compromised at a near-neutral pH but is better adapted than a related facultative alkaliphile to an extremely alkaline pH. Obligates: *B. alcalophilus* and *B. firmus* RAB, 4.8 and 5.0 (fatty acid desaturase); facultatives *B. firmus* OF1 and OF4, less than 0.7. This was the first study on an enzyme of fatty acids in alkaliphiles.

4
Physiology

4.1 Growth Conditions

Alkaliphiles require alkaline environments and sodium ions not only for growth but also for sporulation and germination. Sodium ion-dependent uptake of nutrients has been reported in some alkaliphiles. Many alkaliphiles require various nutrients, such as polypeptone and yeast extracts, for growth; a few alkaliphilic *Bacillus* strains (*Bacillus* No. C-125 or No.A-59) can grow in simple minimal media containing glycerol, glutamic acid, citric acid, etc. One of the best strains for genetic analysis is alkaliphilic *Bacillus* C-125 and many mutants have been made by conventional mutation methods. Some of them have been developed as hosts for genetic recombination experiments. Thermophilic alkaliphiles that can grow at 60 °C are found in soils. Many haloalkaliphiles isolated from alkaline lakes can grow in alkaline environments (pH 10) containing 20% NaCl. Recently, thermophilic anaerobic spore-forming alkaliphiles, alkaliphilic *Clostridium* strains, have been isolated from sewage plants.

4.1.1 Hydrogen Ion

Alkalinity in nature may be the result of the geology and climate of the area, of industrial processes, or promoted by biological activities. The most stable alkaline environments on earth are the soda lakes and soda deserts distributed throughout the world in tropical and sub-tropical areas. These environments are about pH 10-11.5. Many microorganisms have been isolated from these alkaline environments. More transient, localized alkaline environments due to animal excreta and such may exist in soils that are not alkaline overall. Therefore, many alkaliphilic bacteria can be isolated more commonly from soil (see p. 4). Some alkaliphilic bacteria can change external pH value to a pH suitable for growth and create their own world.

A. Bacteria

Alkaline-resistant bacteria have been found among the nitrate reducers, sulfate reducers, and active ammonifiers. Meek and Lipman (1922) reported *Nitrosomonas* and *Nitrobacter* sp. which could survive pH values of 13. Some *Rhizobium* sp. closely related to *Agrobacterium radobacter* grow actively at pH 10 to 12 with an optimum pH between 6 to 9. Many of the enteric bacteria can tolerate pH values near 9 to 11. Downie and Cruickshank (1928) obtained a pure culture of *Streptococcus faecalis* which was resistant to very acid and very alkaline media. This strain was easily obtained in pure culture by direct inoculation of feces into alkaline broth (pH 11) adjusted with NaOH.

Chesbro and Evans (1959) reported that the ability to grow in media of about pH 10 was one of the characteristics of the enterococcus group. In a favorable medium buffered with carbonate rather than glycine all enterococci tested initiated growth up to pH 10.5. Such a medium adjusted to pH 10 has been shown to be superior enrichment broth for the detection of enterococci in feces samples.

Bacillus pasteurii required alkaline media containing ammonia (Gibson, 1934). Optimum conditions for growth in bouillon broth were produced by the addition of about 1% NH_4Cl at pH 9. The most suitable substrate for cultivating the organisms was urea from which the necessary ammonia was formed during sterilization, and the pH of the media was changed to a high pH value. This strain grew well at about pH 11 and growth was poor at pH 9 or less in a medium containing 1% NH_4Cl (Wiley and Stokes, 1962). However, *B.sphaericus*, *B.pantotheniticus*, and *B.rotans* did not require ammonia but were capable of growth at pH 11. Bornside and Kallio (1956) investigated 25 freshly isolated strains of mesophilic urea-hydrolyzing members of the genus *Bacillus*. Battley and Bartlett (1966) reported a convenient pH-gradient method for the determination of the maximum and minimum pH of microbial growth. Data were given for several yeasts and for *Serratia marcescens* which could grow at pH 9.0 (Table 4.1). They concluded that pH limits were unsuitable criteria in microbial classification.

The first isolation of an alkaliphilic *Bacillus* was reported by Vedder (1934), who briefly described organisms which he classified as a new species, *B.alcalophilus* (Vedder). A strain was deposited at the National Collection of Type Cultures in London (NCTC 4553, ATCC 27647). However, his work was not continued. This organism grows actively at pH 10 but not at pH 7 and does not hydrolyze urea or require ammonia (Table 4.2). In 1959 other bacilli extremely resistant to alkaline pH were isolated. A strain of *B.cereus* was obtained by Kushner and Lisson (1959) by multiple transfers in media at increasing pH values. Organisms with the highest alkaline resistance could grow at pH 10.3, and stocks of intermediate resistance were also obtained. Alkaline resistance became a stable characteristic and was not lost in further transfers. One strain obtained was quite distinct from *B. cereus* in that the rods were thinner than those of *B.cereus* and formed small translucent colonies on agar. Therefore, Chislett and Kushner (1961b) stated that the resistant

Table 4.1   Maximum and minimum pH for the growth of several microorganisms[†]

Organism	Lower pH limit of growth	Upper pH limit of growth
Candida pseudotropicalis	2.30	8.80
Hansenula canadensis	2.10	8.60
Saccharomyces cerevisiae	2.30	8.60
S. fragilis	2.40	9.05
S. microellipsoides	2.20	8.70
S. pastori	2.05	8.75
Schizosaccharomyces octosphorus	5.40	7.05
Serratia marcescens	4.00	9.00

[†] Battley and Bartlett (1966)

(Reproduced with permission from Horikoshi and Akiba, *Alkalophilic Microrganisms*, p. 33, Springer-Verlag: Japan Scientific Societies Press (1982))

Table 4.2 Differential characteristics of *Bacillus alcalophilus* strains[†]

Characteristic	B. alcalophilus NCTC 4553	B. alcalophilus subsp. halodurans	
		NRRL B-3881	ATCC 21591
Hydrolysis of			
starch	+	+	+
casein	+	+	+
Production of			
crystalline dextrins	−	−	−
indole	−	−	−
Reduction of			
nitrate	−	+	+
methylene blue	+	+	+
Voges-Proskauer	−	−	−
Urease	−	−	−
Utilization of citrate	−	+	+
Utilization of carbohydrate	Acid but no gas from sucrose, glucose, lactose, maltose, mannitol, xylose, arabinose, glycerol, sorbitol, salicin		
Growth in 15% NaCl	−	+	+
Temperature for growth			
optimum, °C	33–35	48	50.5
maximum, °C	46	54	57

[†] Boyer *et al.*, (1973)
(Reproduced with permission from Horikoshi and Akiba, *Alkalophilic Microorganisms*, p. 14, Springer-Verlag: Japan Scientific Societies Press (1982))

strain was thought to be a contaminant which entered during the training procedure. This organism was characterized as *B.circulans* Jordan. Other strains of *B.circulans* so far studied did not grow at pH 10.7. The alkali-resistant *B.circulans* showed little loss of resistance after many transfers on neutral medium (Fig. 4.1). Takahara and Tanabe (1962) described a highly alkaliphilic *Bacillus* strain isolated from indigo vats. This microorganism showed optimum growth at pH 11.0 and strongly reduced indigo blue (see p. 286).

Since 1968, Horikoshi and his coworkers have systematically studied alkaliphilic microorganisms (Horikoshi, 1971*a*). Aunstrup *et al.* also isolated a number of strains of alkaliphilic *Bacillus* species producing alkaline proteases (Aunstrup *et al.*, 1972). These bacteria have an optimum pH value for growth of about 9 to 10.5 in Horikoshi-I medium.

Gee *et al.* (1980) reported a very interesting study. Five strains of a new alkaliphilic bacterium were isolated from potato-processing effluent. These strains were gram-positive, non-sporeforming and motile rods, which formed an orange cell-bound pigment and were capable of growth in aerobic or anaerobic conditions at pH up to 11.5 and 43 °C. In experiments using shaken flasks of TSYG medium containing tryptone (Difco) 5 g; Soytone (Difco) 5 g; yeast extract (Difco) 5 g; glucose 5 g, growth of all six isolates was observed at initial pH values ranging from 6.0 to approximately 11.0. The effect of pH was studied in more detail with strain BL77/1 in fermenters with continuous control of pH. In YG medium (yeast extract 2.5 g; glucose 5 g; Na_2CO_3 1.1 g; NaCl 1 g; $MgSO_4 \cdot 7H_2O$ 0.06 g) at 25 °C under an aerobic or anaerobic condition the highest growth rates were at pH 8.0 to 10.5, and within this range there were two maxima, at pH 8.5 and 9.5. The highest growth rate occurred in the pH range 8 to 10.5 and the minimum doubling time

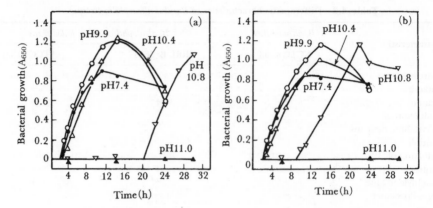

Fig. 4.1 The effect of pH value on growth of alkalitolerant *Bacillus circulans* (Chislett and Kushner, 1961*a*).

(a) Bacteria maintained for 1 year at pH 7.4; (b) bacteria maintained for 1 year at pH 10.7. Bacteria were grown for 20 h on BNA (phosphate-buffered nutrient agar) at pH 7.4(a) or 10.7(b), resuspended in BNB (phosphate-buffered nutrient broth) (pH 7.4) and an inoculum equiv. 0.025 mg dry wt. bacteria in 0.3–0.5 ml broth added to tubes containing 10 ml BNB of the pH values shown. Incubation, shaking, and measurement of absortance conducted as described by Kushner and Lisson (1959).

(Reproduced with permission from Chislett and Kushner, *J. Gen. Microbiol.*, **24**, 189 (1961))

observed was 27 min at approximately 38 °C. Some alkaliphilic bacteria may exhibit two pH optima for growth in complex media.

The optimum pH for growth of ordinary *Bacillus* strains, such as *B. subtilis* and *B. licheniformis*, is around or below pH 7. None of the alkaliphilic bacteria could grow at pH below 6.0 and all indicated optimum pH value for growth above 8. Therefore, bacteria with pH optima for growth in excess of pH 8, usually pH 9 to 11, are defined as alkaliphilic bacteria. This property is the most characteristic feature of these bacteria.

B. Other Microorganisms

Many fungi have a wide pH tolerance. Johnson (1923) reported that *Penicillium variables*, *Fusarium bullatum* and *F. oxysporium* could grow at pH values of 11. He also reported very interesting data. Molds growing in both acidic and alkaline broth changed the pH of the media toward neutral. In acid media, whenever good growth occurred, the pH became neutral. In alkaline media, the pH was reduced. Waksman and Jåffe (1922), in studying actinomycetes, concluded that with media of different hydrogen ion concentrations, the actinomycetes tended to change the media to an optimum. Similar findings were also observed during the cultivation of alkaliphilic bacteria, and further details are presented in the following section.

The blue-green algae are usually most abundant in alkaline habitats (pH 7.5 to 13) (Holm-Hansen, 1968). *Arthrospira plantensis* was found to be present in the highest numbers among 13 different genera of algae present in pH 11, alkaline lakes in Kenya (Jenkin, 1936). Another blue-green alga, *Plectonema nostocorum*, can grow at pH 13, which is probably the highest pH at which life has been recorded. Two green algae, *Chlorella pyrenoidosa* and *C. ellipsoidea*, were also capable of growth under alkaline conditions.

Goto *et al.* (1981) found yeasts which have wide pH tolerance. One of these, belonging to *Exophiala alcalophila*, was proposed as a new species and grows between pH 6 and 10. The present author's research team found alkaliphilic *Aspergillus oryzae* in a soil sample from Kagoshima Prefecture, Japan, which can grow well at pH 9.0 to 9.5. Aono studied several yeasts and yeast-like fungi (433 strains, 296 species, 53 genera) to determine their upper limit of pH for growth (Aono, 1990*a*). Among them, 135 strains of 86 species were found to be capable of growth at pH above 10. These alkali-tolerant strains belonged to 27 genera of yeasts, of which ten genera contained only alkali-tolerant species. Furthermore, phylogenetic relationships of alkali-tolerant yeasts belonging to the genus *Hansenula* were investigated as shown in Fig. 2.11 (Aono, 1992*a*).

Among other organisms, Souza *et al.* (1974) reported that protozoa were found in a pH 11 alkaline spring. Siegel and Giumarro (1966) isolated the fossil-like microorganism *Kakabekia umbellata* from ammonia-rich soil.

C. External pH Change During Cell Growth

Alkaliphilic bacteria can change their environment to a pH value suitable for growth. *Bacillus* sp. No.221, which is a good alkaline protease producer, can grow slowly at neutral pH, changing the pH of the culture broth. It was found that once the pH reached about 9, the bacteria began to grow rapidly and produced a large amount of the same alkaline protease. Ueyama and Horikoshi (unpublished data) isolated an alkaliphilic *Arthrobacter* sp. which utilizes an ε-caprolactam polymer. This microorganism was also capable of changing the broth pH to an optimum (Fig. 4.2). During the course of studies on polygalacturonate lyase by alkaliphilic *Bacillus* RK9, Kelly and Fogarty (1978) reported on the relationship between initial pH values and enzyme production. At initial pH 7.5, neither growth nor enzyme formation was observed until after 30 h at which time the pH began

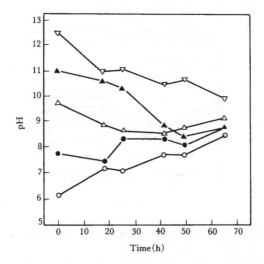

Fig. 4.2 Change in external pH values during cultivation of *Arthrobacter* sp.
(Reproduced with permission from Horikoshi and Akiba, *Alkalophilic Micoroorganisms*, p.36, Springer-Verlag: Japan Scientific Societies Press (1982))

to rise. The highest biomass levels were produced at an initial pH of 9.5, and enzyme production increased up to an initial pH of 9.7. The pH pattern observed in the medium varied depending on the initial pH value. As shown in Fig. 4.3, initial pH values of 10.2 and 10.5 dropped after 30 h and continued to fall until it reached pH 9.6. In the pH range of 9.3 to 9.7, there was a rapid initial decrease in pH and after 24 h the pH rose again in each case. Not only alkaliphilic bacteria but neutrophilic microorganisms can also change their environmental pH values to optimum pH for growth. Johnson (1923) studied the relationships among ions, hydroxyl ions, salt concentrations, and the growth of seven soil molds. During the study, where the molds grew, a change toward neutral and in many cases even over to alkaline conditions occurred, and this change was in proportion to the growth. Such experiments have been conducted by many investigators, and Waksman and

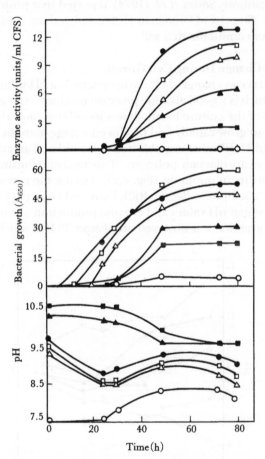

Fig. 4.3   Effect of initial pH on polygalacturonate lyase, biomass, and pH pattern.
○, pH$_i$ 7.5; △, pH$_i$ 9.3; □, pH$_i$ 9.5; ●, pH$_i$ 9.7; ▲, pH$_i$ 10.2; ■, pH$_i$ 10.5. The medium had the following composition, in grams per liter: sucrose, 10.0; bacteriological peptone, 3.0; yeast extract, 3.0; Na$_2$CO$_3$·H$_2$O, 10.0; MnSO$_4$·4H$_2$O, 0.04; MgCl$_2$·6H$_2$O, 0.2; K$_2$HPO$_4$, 1.0.
(Reproduced with permission from Kelly and Fogarty, *Can. J. Microbiol.*, **24**, 1167 (1978))

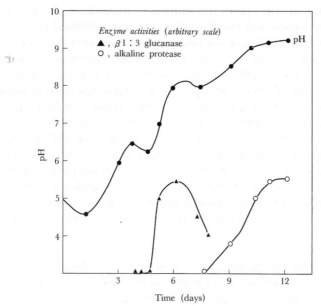

Fig. 4.4 pH and enzyme activities during mixed cultivation of *A. oryzae, B. circulans* and *Bacillus* sp. No. 57.

Jåffe (1922) in studying actinomycetes, concluded that with media of different pH the actinomycetes tend to change the broth to the optimum for growth. The author conducted another experiment, which has not been published. The following microorganisms were used in the experiments. 1) *Aspergillus oryzae*. 2) *Bacillus circulans* IAM 1165, which grows and produces an enzyme that lyzes *A. oryzae* at pH 7-8.3) alkaliphilic *Bacillus* No. A-57, which can grow in the range of pH 8 and 10. These three strains were mix-cultured in Horikoshi-II medium of pH 5 in the absence of 1% $Na_2 CO_3$. The cultivation was carried out for a week with continuous shaking at 30 ℃. *A. oryzae* grew in the medium first. After 3 days incubation, *A. oryzae* began to autolyze and the pH of the broth increased to 6-7. Then *Bacillus circulans* started to grow and produce the lytic enzyme which lyzed *A. oryzae*. The pH of the broth gradually increased to 8. As a result alkaliphilic *Bacillus* A-57 grew well and produced an alkaline protease (Fig. 4.4). This phenomenon is very interesting from the ecological point of view, because it may be the reason why alkaliphilic bacteria can create a microcosmos and live in acidic soil.

4.1.2 Sodium Ion
Another characteristic property of the alkaliphilic *Bacillus* strains is that for many of them sodium ions are an absolute requirement for growth and motility (see section 3.1). The first finding of the sodium requirement was described by Horikoshi and Kurono in a patent application for extracellular production of catalase from alkaliphilic *Bacillus* sp. No. Ku-1 (FERM No.693). The strain isolated could not grow in nutrient broth in the absence of NaCl, but addition of NaCl (1-5%) induced good growth in the nutrient broth, although the pH value of the broth was not changed. In 1973, two scientific papers were published (Kurono and Horikoshi, 1973; Boyer, Ingle and Mercer, 1973), but the sodium

requirement was cited as only one of the cultural characteristics. No one was interested in sodium ions until Kitada and Horikoshi's paper was published (1977). Amino acid uptake into the cells as a function of NaCl was exhibited and they concluded that the presence of NaCl plays an important role in the active transport mechanism of amino acid into the cells. This section discusses the effects of sodium ions on growth, sporulation, germination and flagellar motors. Active transport by sodium is described in section 4.2.

Driessen *et al.* reviewed membrane composition and ion-permeability in extremophiles including alkaliphiles (Driesse *et al.*, 1996). Comparison of the proton and sodium permeability of the membranes of a variety of bacterial and archaeal species that differ in optimal growth temperature reveals that the permeation processes of protons and sodium ions must occur by different mechanisms. Sodium permeability is lower than proton permeability and increases with temperature, but is lipid-independent. Therefore, it appears that for most bacteria the physical properties of the cytoplasmic membrane are optimized to ensure low proton permeability at the respective growth temperature.

A. Growth

Many alkaliphilic *Bacillus* strains cannot grow well in the absence of NaCl in the range of pH 7 to 10 and require sodium ions for growth (Table 4.3). In some of them, K^+ ions can be substituted for Na^+ ions, as shown in Table 3.4.

Table 4.3 Effect of Na^+ ions on growth of alkaliphilic *Bacillus* sp.

Initial pH	NaCl conc.	Biomass (24-h incubation)
Strain No. 8-1		
7.0	None	no growth
	1.5%	0.25 (OD_{610})
9.0	1.5%	0.45 (OD_{610})
Strain C-125		
7.2	None	92 (mg/100 ml)
	0.2 M	303 (mg/100 ml)
10.2	0.1 M	382 (mg/100 ml)
	(as Na_2CO_3)	

Recently, it has been found that very alkaline and saline soda lakes, such as the East Africa Rift Valley lakes, the lakes of Wadi Natrum, some of the lakes in the Kukunda Steppe, Owens Lake, etc., harbor many characteristic alkaliphilic phototrophic bacteria, *e.g.*, *Ectothiorhodospira* together with halophilic archaebacteria *Natronobacterium* and *Natronococcus* (Oren, 1994). Haloalkaliphilic methanogens have also been found in these lakes. For these microorganisms a high concentration of NaCl (1-2 M) is an absolute requirement for life. Haloalkaliphiles are described below.

B. Sporulation

A sodium requirement was also observed in the differentiation process, spore formation and germination. Numerous environmental factors affecting the sporulation of *Bacillus* species have been reported by many investigators. However, only one report on sporulation of an alkaliphilic *Bacillus* strain has been published. Kudo and Horikoshi

Table 4.4 Effect of NaCl concentration on sporulation of *Bacillus* sp. No. 2b-2

NaCl concentration (M)	Final OD_{610}[†]	Growth rate	Percent of sporulation
0	2.9	54	0
0.02	3.7	46	30-40
0.2	4.7	44	70
0.4	4.0	44	30-40
1	2.9	46	1

[†] The final OD and the sporulation rate were measured after 47 h of incubation at 37 °C.
(Reproduced with permission from Kudo and Horikoshi, *Agric. Biol. Chem.*, **43**, 2613 (1979))

(1979) isolated alkaliphilic *Bacillus* sp. No.2b-2 from soil. This strain showed excellent spore yield in alkaline Schaeffer medium. The optimum pH for sporulation is close to that for growth, but the range is narrow. The optimum temperature for sporulation was almost the same as that for growth (34-37 °C). Growth at 45 °C was faster than at 34 °C. Table 4.4 shows the effect of NaCl concentration on sporulation. The optimum concentration was similar to that for growth, but the range was narrow.

C. Germination
1. *Bacillus cereus* and *B. circulans*
The first paper on spore germination of alkaliphilic microorganisms was presented by Chislett and Kushner (1961*a*). They examined the effect of alkaline conditions on the development of the vegetative form from the spore. Three organisms were used in their experiments: *Bacillus cereus* strain Mu-3055 (non-adapted; sensitive to alkali), *B. cereus* strain R (adapted from Mu-3055 and resistant to alkali), and *B.circulans* strain Ru38 (maintained routinely on nutrient agar, pH 7.4, but capable of growth at pH 11.0). The three strains were cultured several times in phosphate-buffered nutrient broth (BNB) (pH 7.4), with incubation at 30 °C, before finally spreading 1 ml of 6-h culture on the surface of 50 ml of phosphate-buffered nutrient agar (BNB) (pH 7.4) containing 5 ppm $MnSO_4$ in a 20-oz bottle. Spores were harvested by washing with distilled water and glass beads. Spore suspensions were added to the BNB medium with pH values from 7.4 to 11.0, to give a final concentration of 10^7 spores/ml. At pH 7.4 both alkali-sensitive and alkali-resistant strains of *B.cereus* showed considerable germination and outgrowth after 2 h at 30 °C or 37 °C. At pH 9.5 spore germination and outgrowth took place much more rapidly and extensively with the alkali-resistant strain than with the alkali-sensitive strain. Spores of the alkali-resistant *B.circulans* strain Ru38 germinated more rapidly and exhibited extensive outgrowth after 2 h at pH 9.5-10.3 than at pH 7.4, but outgrowth was not observed at pH 7.4.

2. Alkaliphilic *Bacillus* sp. No.2b-2 and *Bacillus firmus* OF4
Kudo and Horikoshi (1983*a, b*) reported germination of alkaliphilic *Bacillus* sp. No.2b-2. The results were almost the same as those for *B.circulans* Ru38 obtained by Chislett and Kushner (1961*a*).

In 0.2 M NaCl solution spores of *Bacillus* sp. No.2b-2 germinated very well in the presence of both L-alanine and inosine, but germinated poorly in the presence of only L-alanine. However, no germination was observed in the absence of NaCl (Table 4.5). The

Table 4.5   Germination of *Bacillus* sp. No. 2b-2 spores

Compound added[†1]			Absorbance reduction (%)[†2]
Alanine	+ Inosine	+ NaCl	46
Alanine		+ NaCl	10
	Inosine	+ NaCl	3
Alanine	+ Inosine		0
		NaCl	0

[†1] The final concentration of all germinants was 0.4 mM and the final concentration of NaCl was 0.2 M.

[†2] Germination is expressed as the optical density reduced (%) after 60 min at 37 °C

All experiments were conduced in 0.1M 2-amino-2-methyl-1,3-propanediol (AMPD) buffer (pH 9.7).

(Reproduced with permission from Kudo and Horikoshi, *Agric. Biol. Chem.*, **47**, 666 (1983))

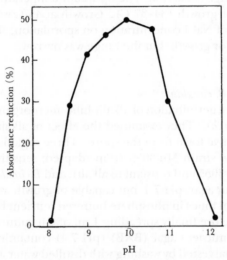

Fig. 4.5   Effect of pH on germination of spores of *Bacillus* sp. No. 2b-2.
(Reproduced with permission from Kudo and Horikoshi, *Agric. Biol. Chem.*, 47, 666 (1983))

optimum temperature for germination was about 37 °C, and germination occurred in the range of pH 8.5 to 11.1. The optimum pH for germination was around 10.0 (Fig 4.5). The optimum concentration of NaCl for the germination was 0.1 to 0.5 M. Other cations such as K^+, NH_4^+, Rb^+, Cs^+, and Ca^{2+} did not show this stimulating effect. Only Li^+ showed weak stimulation. In the absence of Na^+, the loss of heat resistance, acquisition of stainability, and decrease in absorbance was not observed at all even in the presence of germinants such as L-alanine and inosine. However, when Na^+ was added to the medium at the times indicated by arrows the absorbance was decreased immediately without any time lag (Fig.4.6). Although it is too premature to discuss the step stimulated by Na^+, Na^+ may stimulate the uptake of the germinants into the spores.

Quirk (1993) isolated a candidate of *ssp* gene from alkaliphilic *B. firmus* OF4. An 1100-

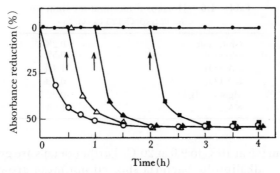

Fig. 4.6 Effect of Na$^+$ on decrease in absorbance.
NaCl was added at the times indicated by arrows to the germination medium (pH 9.7, without NaCl).
(Reproduced with permission from Kudo and Horikoshi, *Agric. Biol. Chem.*, **47**, 667 (1983))

bp DNA fragment cloned from alkaliphilic *B. firmus* OF4 contained an open reading frame deduced to encode a 54-amino-acid, glutamine-rich protein with 35.6% identity to *B. subtilis* small, acid-soluble spore protein-γ [SASP-γ] in a 45-amino acid overlap. This ORF, designated *ssp*A, lacks the lengthy sequence repeat characteristic of previously cloned SASP-γ-encoding genes. Southern analysis under conditions of moderate stringency revealed six bands, suggesting the presence of several genes in the alkaliphile. *Ssp* genes, encoding two types of small, acid-soluble spore proteins (SASP), have been found in several *Bacillus* and *Clostridium* sp. SASP supply amino acids during spore germination, and the α/β help to protect the spore DNA from radiation damage.

3.   Germination of Alkaliphilic Cyanobacteria

Germination of desiccated aged akinetes of alkaliphilic cyanobacteria has been reported (Sili *et al.*, 1994). Morphological and biochemical changes associated with synchronous germination of mature, aged and desiccated akinetes of two alkaliphilic cyanobacteria, *Cyanospira rippkae* and *Cyanospira capsulata*, were described. Akinetes of both strains proved to be highly resistant to desiccation, being able to germinate in the presence of either N$_2$ or nitrate as nitrogen source, with a germination frequency of more than 90% after seven years of storage in a dried state. The first cell division occurred after 8-10 h of incubation, thereafter the germlings of the two strains followed a different pattern of cell differentiation. Akinetes of both species possessed, on a per-cell basis, almost identical amounts of all photosynthetic pigments but, under nitrogen fixing conditions, photosynthetic activity (oxygen evolution) was detected only after new proteins had been synthesized, before a functional heterocyst developed and while total nitrogen remained constant.

4.1.3   Temperature and Nutrition

High temperature for growth is not a characteristic property of alkaliphiles. The highest temperature for bacteria so far reported is 57 °C for *Bacillus* sp. No. 221 (ATCC 21522), which is an alkaline protease producer. Kimura and Horikoshi (1988) isolated several psychrophilic alkaliphiles. One of them, strain 207, which is an aerobic coccus 0.8-1.2 μm in diameter, can grow at tempertures of −5 to 39 °C at pH 8.5. The optimum pH value for

Table 4.6   Horikoshi minimal medium for
Bacillus sp. C-125 (w/v)

Glutamate	0.2%
Glycerol	0.5%
K_2HPO_4	1.4%
KH_2PO_4	0.4%
$MgSO_4 \cdot 7H_2O$	0.02%
NA_2CO_3	1.0%

growth changed from 9.5 at 10 °C to 9.0 at 20 °C. Further details are given in section 7.3.4. However, almost all alkaliphilic bacteria showed optimum growth temperature at 25–45 °C.

During the course of cloning of enzymes Xu *et al.* (1996) isolated a chaperonin gene (gro EL) from the alkaliphilic *Bacillus* sp. strain C-125, and it was cloned in *Escherichia coli.* The gro EL gene encoded a polypeptide of 544 amino acids and was preceded by the incomplete gro ES gene, lacking its 5'-end. The sequence of the derived amino acids was 87.5% identical to that of *B. subtilis,* 85.4 % identical to that of *B. stearothermophilus,* and 60.9% identical to that of *E. coli.* The gro EL protein was expressed in *E. coli.* Purified gro EL protected yeast α-glucosidase from irreversible aggregation at a high temperature and the addition of Mg-ATP was essential for reactivation of the α-glucosidase.

No precise experiment has been reported on nutrient requirements. Takahara, Takasaki and Tanabe (1961) isolated a growth factor of *Bacillus alkaliphilus* nov. sp. No. S-8. Without the addition of the factor, Ala–Ile–Leu–Val–Lys–Glu–Gly, the bacteria could grow, but did not reduce indigo blue. Thus the peptide was necessary to reduce indigo but not for cell growth. No further experiment has been done on this subject. During the development of new host-vector systems, Aono investigated more than 20 strains of alkaliphilic *Bacillus* species and demonstrated that vitamins such as biotin, thiamine and niacin are required for some alkaliphilic *Bacillus* sp. However, no growth factor of peptide has so far been found (Aono, unpublished data). During work on alkaliphily, Kudo *et al.* (1990) found that some alkaliphilic *Bacillus* strains can grow on minimal media; alkaliphilic *Bacillus* sp. No.C-125 (FERM No.7344) grows well on a minimal medium containing glutamate and glycerol, as shown in Table 4.6.

Ikura and Horikoshi (1994) found that the growth of alkaliphilic *Bacillus* no. A-40-2 with nitrate as the nitrogen source was highly stimulated by the addition of 0.1% of certain amino acids, sugars, organic acids, nucleic acids, or Fe^{2+} or Mn^{2+} at concentrations of 10 mM or more to the medium, resulting in maximum growth after 24 h. Other alkaliphilic *Bacillus* strains (A-59, C-125) showed similar results.

A novel facultatively alkaliphilic bacterium that grows on a chemically defined medium containing n-alkanes as the sole carbon source was isolated from soil (Ikeda, Nakajima and Yumoto, 1994). The isolate was obligatory aerobic, non-motile, gram-positive, and formed metachromatic granules. It was not acid-fast and did not form endospores. The cell wall contained *meso*-diaminopimelic acid, arabinose, and galactose; the glycan moiety of the cell wall contained acetyl residues. The bacterium was catalase-positive, oxidase-negative, and the G + C content of DNA was 70.8 mol%. According to these tests, the isolate was assigned to the genus *Corynebacterium.* The bacterium grew well between pH 6.2 to 10.2 and the doubling time in this pH range was 4–6 h. For the growth of the isolate,

Na$^+$ added to the culture medium stimulated growth, but was not indispensable at both pH 7.2 and pH 10.2. In addition to hydrocarbons, the isolate was able to grow on a chemically defined medium containing acetate, glucose, or fructose as the sole carbon source.

4.1.4 Haloalkaliphiles

Tindall, Mills and Grant (1980) and Tindall, Ross and Grant (1984) isolated and characterized haloalkaliphiles from alkaline brines. These isolates require high pH as well as high concentration of NaCl (1-2M) and have a low Mg^{2+} and Ca^{2+} tolerance. They included both rod-shaped and coccoid isolates. Nucleic acid hybridization studies and lipid analysis (De Rosa *et al.*, 1982; Tindall, Ross and Grand, 1984) indicated that they are clearly distinct from halobacteria and halococci. Tindall, Ross and Grand (1984) have proposed the new genera *Natronobacterium* and *Natronococcus* for the rod-shaped and coccoid isolates, respectively. These haloalkaliphiles are widely distributed in soda lakes throughout the world. Shiba, Yamamoto and Horikoshi (1989) also isolated many haloalkaliphiles from Owens Lake in California, U.S.A. The isolate No. M-1213-2 can grow well at pH 10.0 in the presence or absence of a high concentration of saline.

Mwatha and Grant (1993) isolated a novel haloalkaliphilic archaeon from Lake Magadi, an alkaline soda lake in Kenya. Cells of the organism contain large gas vacuoles in the stationary phase of growth, and colonies produced by these archaea are bright pink in appearance. The major polar lipids of these organisms are C$_{20}$C$_{20}$ and C$_{20}$C$_{25}$ derivatives of phosphatidylglycerol phosphate and phosphatidylglycerol, and the organisms contain an unidentified phospholipid as a minor component. The G + C content of the DNA is 62.7 mol%. The name *Natronobacterium vacuolate* sp. nov. has been proposed. The type strain is designated NCIMB 13189.

4.1.5 Anaerobic Alkaliphiles

Alkaliphilic anaerobes have been found in nature; Souza *et al.* (1974) reported an alkaliphilic *Clostridium* sp., then Niimura *et al.* (1987) isolated a facultative anaerobic xylanusing alkaliphile. Alkaliphilic methanogens have also been found from the soda lakes of Kenya (Nakatsugawa and Horikoshi, 1989). Boone *et al.* (1993) isolated and characterized methanohalophiles. Six strains of coccoid, halophilic methanogens were isolated from various salinaria and natural hypersaline environments. These isolates (strains FDF-1, FDF-2, SF-2, Ret-1, SD-1, and Cas-1) grew on media containing methanol and mono-, di-, and trimethylamines as catabolic substrates, but not on media containing dimethyl sulfide, methane thior, H$_2$, formate, or acetate; when cells were provided with H$_2$ in addition to methanol or trimethylamine they grew on the medium containing a methyl substrate but did not catabolize H$_2$. The strain SD-1 was slightly alkaliphilic (fastest growth occurred at pH 7.8).

Li, Mandelco and Wiegel (1993) isolated alkaliphilic, moderately thermophilic anaerobic bacteria from various sewage plants in the United States. The strains, which can grow above pH 10.5 and 55 °C, were motile with two to six peritrichous flagella and formed round to slightly oval terminal spores in terminally distended and slightly enlarged cells. Sporulated cells remained motile. The pH range for growth was between 7.0 and 11.1, with an optimum of around 10.1. At pH 10.1 the temperature range for growth was between 30 and 63 °C, with optimum at 56 °C. The shortest observed doubling time (glucose) was around 16 min at 56°C and pH 10.1. No dissimilatory sulfate reduction was detected. The organism utilized glucose, fructose, sucrose, maltose, and pyruvate but required yeast extract or tryptone for growth. Strain JW-YL-7 (DSM 7308) is designated as the type strain. Li *et al.* (1994) reported *Clostridium thermoalcaliphilum* sp. nov., an anaerobic and thermotolerant facultative alkaliphile.

A strain, JW/YL23-2^T (T = type strain), was isolated from sewage sludge obtained from a sewage plant in Atlanta, GA. At pH 10.1 and 50 °C, the doubling time of this strain was 19 min. Strain JW/YL23-2T, a motile rod-shaped bacterium with 2 to 12 peritrichous flagella, exhibited a negative Gram stain reaction but was gram-type positive as judged by the polymyxin B test. No heat-stable (85 °C, 15 min) endospores were detected. At 50 °C, growth occurred at pH values ranging from 7.0 to 11.0; the optimum pH was 9.6 to 10.1. The temperature range for growth ranged from 27 to 57.5 °C; the optimum temperature was 48 to 51 °C(pH 10.1). *Clostridium thermoalcaliphilum* JW/YL23-2 (= DSM 7309) is the type strain of this new species.

These results strongly suggest that alkaliphilic microorganisms are very similar in kind to neutrophilic microorganisms and can grow under specified extreme conditions.

4.2   Bioenergetics

Alkaliphilic microorganisms, however, have growth optima at pH 9-10.5 and scarcely grow near the neutral pH range.

So a question arises as to how these alkaliphilic microorganisms can grow in such an extreme environment. Is there any difference in the physiological and structural aspects between alkaliphilic and neutrophilic microorganisms?

Since we have isolated many alkaliphilic microorganisms from various sources on alkaline media containing $Na_2CO_3(NaHCO_3)$ or K_2CO_3, extensive bioenergetic and physioligical studies have been carried out by many investigators (Krulwich and Guffanti, 1989). The alkaliphilic microorganisms exhibit remarkable ability to maintain cytoplasmic pH lower than external pH values of 10 to 11. This may lead to secondary bioenergetic problems, *i.e.*, the chemical gradient of protons would be adverse with respect to development of proton motive force (pmf). Thus the energization of active transport and ATP synthesis in alkaliphilic microorganisms is a most attractive problem from the viewpoint of chemiosmotic hypothesis.

4.2.1   Mechanisms Regulating Intracellular pH

A.   Measurements of Intracellular pH

Although actual measurement of intracellular pH (pH$_{in}$) is difficult, several indirect measuring methods have been reported: (1) pH measurement of broken cell lysate or fluids from cells; (2) the use of a visible indicator such as bromothymol blue; (3) measurement of the external and internal distributions of weak acids or bases; (4) the fluorescence method; and (5) high-resolution ^{31}P nuclear magnetic resonance measurements.

The method most applicable to the cells of microorganisms is the weak acid or base distribution techniques (number 3 above). The principle of measurement is based on the ability of weak acids such as 5,5-dimethyl-2,4-oxazolidinedione (DMO) to penetrate membranes in their neutral form. DMO diffuses passively across many biological membranes and the permeability coefficient of the uncharged acid is generally much greater than that of anions. Therefore, DMO permeates cells or vesicles which are internally more alkaline than the surrounding medium. The internal pH of cells or vesicles can be calculated from the distribution by the Henderson-Hasselbach equation. This method was first developed for determining the intracellular pH of muscle and was later applied to the study of pH change in mitochondria and bacteria.

However, if the intracellular pH of the cells is lower than external pH, as expected from pH optima of intracellular enzymes (see p. 95), most of the DMO would be outside the cells. Actually, no DMO uptake by *Bacillus alcalophilus* cells was observed at any external pH (Guffanti *et al.*, 1978).

Rottenberg, Grunwald and Avron (1972) developed a modified method using a weak base, methylamine, instead of DMO. This method is also based on the ability of the base to penetrate membranes in its neutral form, and the ΔpH is calculated assuming that the non-charged amine concentration is equal on both sides due to the permeability of this form. When the intracellular pH is expected to be lower than the external pH, the anion will be more concentrated in the cells. Guffanti *et al.* (1978) used ^{14}C-methylamine to determine intracellular pH values in *B.alcalophilus* by flow dialysis and found that the cytoplasmic pH remained at 9.0-9.5 over a range of external pH values from 9.0 to 11.5. Table 4.7 shows the internal pH values in various microorganisms including acidophiles, neutrophiles, and alkaliphiles at different external pH values. These results imply that internal pH values in acidophiles and alkaliphiles remain nearly neutral under various extracellular pH values. The fluorescent pH indicator 2,7-bis(carboxyethyl)-5(6)carboxyfluorescein, BCECF, can be used to monitor pH_{in} changes in intact cells. Since BCECF fluorescence is free from the binding artifacts and with a pKa near 7, it can provide a convenient and continuous readout of pH_{in} (Davis *et al.*,1987; Brierley *et al.*, 1989; Brierley and Jung, 1990) and be applied to this indicator to measure Na^+/H^+ and K^+/H^+ antiporter activities of isolated heart mitochondria.

Akiba and Horikoshi (1978) reported that the intracellular pH of alkaliphilic *Micrococcus* sp. No. 31-2 may remain nearly neutral even in alkaline environments. They investigated the susceptibility of intracellular α-galactosidase to external pH using toluene-acetone-treated cells in which the membrane permeability was increased. The enzyme of the treated cells was highly sensitive to external pH change and the activity declined in the acid or alkaline regions. The activity of the enzyme in the untreated cells showed no change over a relatively wide pH range (6-9). A similar result was also reported by Ikura and Horikoshi (1979a) using toluene-treated cells of alkaliphilic *Bacillus* sp. No.C-125.

B. Na^+/H^+ Antiport System

As shown in Table 4.7, most bacteria seem to have some mechanism for maintaining pH near normal values from pH 6 to 9. How can the cells of alkaliphilic and acidophilic bacteria maintain such a H^+ ion gradient across their respective membranes?

Most of the obligatory alkaliphilic bacteria show an absolute requirement of Na^+ for growth and are not able to grow on alkaline media without added Na^+. The cells or membrane vesicles from these alkalophilic microorganisms show a typical proton motive force pattern if Na^+ is present (Mandel, Guffanti and Krulwich, 1980; Krulwich *et al.*, 1982). When cells of alkaliphilic *Bacillus firmus* RAB were incubated in the absence of Na^+ at pH 10.5, the cytoplasmic pH immediately rose from less than 9.5 to 10.5, and they lost viability very rapidly, as shown in Table 4.8. On the other hand, cells incubated at pH 10.5 in the presence of Na^+ retained greater viability, and internal pH was maintained below 9.0 (Kitada, Guffanati and Krulwich, 1982). These data indicate the presence of Na^+/H^+ antiporter activity at alkaline pH, catalyzing energy-dependent acidification of the cytoplasmic space.

The importance of a Na^+/H^+ antiporter for pH homeostasis was confirmed in isolat-

Table 4.7 Intracellular pH values in various microorganisms at different external pH values

Microorganism	Optimum pH for growth	Internal pH	External pH	Method
(A) Acidophilic microorganism				
Thermophasma acidophila	2(59 °C)	6.4-6.9	2,4,6	Cell lysis
		6.5	2,4,6	DMO
Bacillus acidocaldarius	4	5.5(55 °C)	3	Fluorescein
		6.5(15 °C)	3	DMO
Sulfolobus acidocaldarius	3(70 °C)	6.3		Cell lysis
Bacillus acidocaldarius	3-4	6.15	2.0	
		5.85	3.0	
				Aspirin method
		5.89	4.0	
		6.31	4.5	
(B) Neutrophilic microorganism				
Streptococcus faecalis		7.2	6.0	
		7.7	7.0	DMO
		8.0-8.2	7.5	
Escherichia coli MRE 600		7.55	6.0-8.0	^{31}P nuclear magnetic resonance
Bacillus megalerium		8.03	7.4	DMO
Escherichia coli ML 308-225		7.8	6.3	
		7.9	6.8	DMO
		8.4	7.8	
		8.8	8.7	
Escherichia coli K12		8.0	6.0	
		7.7	7.0	DMO
		7.7	8.0	
		8.5	9.0	
(C) Alkaliphilic microorganism				
Bacillus alcalophilus	10.5	9.0	9.0	
		9.3	10.0	Methylamine
		9.1	11.0	
Bacillus firmus RAB	10.5	7.7	7.0	DMO
		8.0	9.0	
		9.4	10.5	Methylamine
Bacillus strain YN-2000	9.5	8.2	7.5	
		7.9	8.5	DMO
		8.1	9.5	
		8.4	10.2	Methylamine

Table 4.8 Viability and interacellular pH values of *Bacillus firmus* RAB as a function of the pH and Na$^+$ content of the incubation medium

	pH of the incubation medium					
	10.5		9.0		7.0	
	Intra-cellular pH	Membrane potential mV	Intra-cellular pH	Membrane potential mV	Intra-cellular pH	Membrane potential mV
+ Na$^+$	9.4	− 162	8.0	− 144	7.7	− 41
− Na$^+$	10.5	− 152	8.6	− 142	7.7	−

	pH of the incubation medium					
Incubation time (h)	10.5		9.0		7.0	
	Viable count (%)					
	+ Na$^+$	− Na$^+$	+ Na$^+$	− Na$^+$	+ Na$^+$	− Na$^+$
0	100	100	100	100	100	100
1	86	33	91	60	92	67
2	75	20	83	45	83	58
4	14	0	41	17	74	37

ed membrane vesicles from alkaliphiles. In the presence of added Na$^+$, respiring right-side-out vesicles from *Bacillus firmus* RAB (Krulwich *et al.*, 1982), *Bacillus alcalophilus* (Mandel, Guffanti and Krulwich, 1980) and *Bacillus* sp. No. C-59(Kitada and Horikoshi, 1987) generated a ΔpH (acid in), as demonstrated by methylamine uptake using flow dialysis method. Guffanti *et al.* (1980) and Krulwich (1986) have taken this as evidence for a Na$^+$/H$^+$ antiporter and suggested the occurrence of net proton accumulation on the intravesicular side of the membrane, secondary to respiration-dependent proton translocation and catalyzed by an electrogenic Na$^+$/H$^+$ antiporter.

The presence of a Na$^+$/H$^+$ antiporter has also been observed in experiments using everted membrane vesicles. Everted vesicles from *Bacillus alcalophilus* exhibited Na$^+$-dependent proton extrusion and accumulation of ^{22}Na$^+$ upon energization with NADH (Mandel, Guffanti and Krulwich, 1980) or ATP (Guffanti, 1983).

Garcia, Guffanti and Krulwich (1983) measured Na$^+$ efflux from starved cells which were energized by means of a valinomycin-induced potassium diffusion potential, positive out, (ΔΨ). Their result showed that inactivity of the Na$^+$/H$^+$ antiporter at pH 7.0 was not due to a low ΔΨ generated by respiration at this pH. ^{22}Na$^+$ efflux activity was inhibited by a relatively high internal proton concentration.

Kitada and Horikoshi (1992) studied the kinetic properties of the ΔΨ-dependent Na$^+$/H$^+$ antiporter in right-side-out membrane vesicles of *Bacillus* sp. N-6 strain by examining both Na$^+$ efflux and H$^+$ influx upon imposition of membrane potential (ΔΨ). Imposed ΔΨ (interior negative) increased the Na$^+$ efflux rate (V) linearly and the slope of V versus ΔΨ was higher at pH 9 than at pH 8. Kinetic experiments indicated that the ΔΨ caused pronounced increase in the V_{max} for the Na$^+$ efflux, whereas the K_m values for Na$^+$ were unaffected by the ΔΨ. As the internal H$^+$ concentration increased, the Na$^+$ efflux reaction was inhibited. This inhibition resulted in an increase in the apparent K_m of the Na^+ efflux reaction. The results of H^+ influx experiments showed a good coincidence with those of Na^+ efflux. H^+ influx was enhanced by an increase of $\Delta\Psi$ or internal Na^+ concentration and inhibited by high internal concentration. These results clearly indicate that the Na^+ /H^+ antiport system of this strain operates electrogenically and plays a central role in pH homeostasis at the alkaline pH range. Thus it is quite evident that alkaliphilic bacteria possess a Na^+/H^+ antiporter which is very active in the alkaline range of pH. The crucial role of the antiporter in pH homeostasis in the alkaline range of pH value has been supported by not only extreme alkaliphiles but also by moderate alkaliphiles. From the results of pH-jump experiments, McLaggan, Selwyn and Dawson (1984) suggested that there must be a route that allows rapid entry of Na^+. There is no specific evidence for such a pathway, but a Na^+-coupled solute uptake system can provide one possible pathway of Na^+ entry. The presence of α-aminoisobutyric acid (AIB) allows *Bacillus firmus* RAB to maintain an acidified cytoplasm at a low concentration of Na^+(2.5mM) during the pH shift (Krulwich, Federbush and Guffanti, 1985).

Escherichia coli has three distinct antiporter systems, Ca^{2+}/H^+, Na^+/H^+ and K^+/H^+ antiporters, which function in the extrusion of cations from the cytosol. The K^+/H^+ antiporter system of *E.coli* exhibited a basic pH optimum and catalyzed electroneutral exchange. Thus it is postulated that the function of this K^+/H^+ antiporter is to regulate the cytoplasmic pH. There is no current evidence that a K^+/H^+ antiporter plays a role in pH homeostasis in alkaliphiles in the alkaline range of pH. Mandel, Guffanti and Krulwich (1980) suggested that the K^+/H^+ antiporter could be involved in regulation of cytoplasmic pH in a lower pH range. Koyama and Nosoh (1985) found that a K^+/H^+ antiporter functioned in the alkalization of cytoplasm in the neutral range of pH in studies using a facultative alkaliphile, *Bacillus* YN-2000.

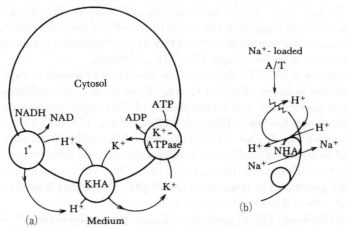

Fig. 4.7 Regulation of intracellular pH by a KHA system (a) (Brey and Rosen, 1980)
and a NHA system (b). (Mandel *et al.*, 1980).
(a) 1°, electrogenic proton pumps; KHA, K^+/H^+ antiporter.
(b) A/T, ascorbate/tetramethyl-*p*-phenylenediamine; NHA, Na^+/H^+
antiporter.
(Reproduced with permission from Horikoshi and Akiba, *Alkalophilic
Microorganisms*, p. 53, Springer-Verlag: Japan Scientific Societies Press
(1982))

Figure 4.7 shows a postulated mechanism of intracellular pH regulation by the K^+/H^+ antiporter or the Na^+/H^+ antiporter by which internal pH can be regulated by cation/H^+ antiporter systems.

4.2.2 Active Transport

The intracellular pH values of alkaliphilic microorganisms are near neutral and there are no significant differences in the pH-activity profiles of intracellular enzymes between alkaliphilic and neutrophilic bacteria. Entry of organic nutrients into microbial organisms must be mediated by membrane transport systems.

Kitada and Horikoshi (1979) have found that the rate of glutamate transport has a limiting effect on growth in a mutant strain which required glutamate for growth as the sole nitrogen source.

It has now been recognized that energization of transport in bacteria which grow at pH extremes, especially at unusually high pH, is of very great interest with respect to the chemiosmotic theory.

A. α-Aminoisobutyric Acid (AIB) Uptake by Cells of Alkaliphilic *Bacillus* Species

α-Aminoisobutyric acid is a very useful compound for transport analysis in whole cells because it is not incorporated into cell materials; nor does it undergo any other mutations.

The typical time course of AIB uptake into cells of alkaliphilic *Bacillus* sp. No. 8-1 in the presence of 150 mM ^{14}C-AIB at pH 9 is shown in Fig. 4.8. The addition of unlabeled AIB to the medium at a concentration of 15mM caused rapid efflux of ^{14}C-AIB from the cells, implying that the accumulated AIB is not chemically modified (Kitada and Horikoshi, 1977; Kitada and Horikoshi, 1980*b*).

The results of the course of AIB uptake into cells as a function of the NaCl concentration during 60 min incubation under neutral (pH 7) and alkaline (pH 9) conditions showed that the AIB uptake was increased by the addition of NaCl up to 0.2 M and the

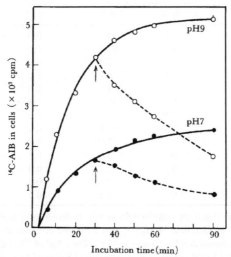

Fig. 4.8 Time course of ^{14}C-AIB uptake and release of ^{14}C-AIB from cells by addition (indicated by arrow) of unlabeled AIB.

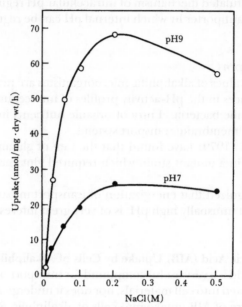

Fig. 4.9 Effect of NaCl concentration on the uptake of ^{14}C-AIB by cells of *Bacillus* sp. No. 8-1.

Table 4.9 Effect of NaCl on uptake of various radioactive amino acids by cells of *Bacillus* sp. No. 8-1

Amino acid	Uptake of amino acid (nmol/mg·dry wt. per 20 min)[†]			
	pH 7 NaCl		pH 9 NaCl	
	0 M	0.2 M	0 M	0.2 M
Gly	3.3	45.1	5.6	56.8
L-Ala	1.7	67.7	5.4	95.3
L-Val	2.1	2.4	0.9	2.4
L-Leu	2.6	6.7	1.2	13.9
L-Ile	1.9	4.9	2.0	6.3
L-Ser	0.7	48.9	1.4	60.5
L-Asp	1.0	6.4	1.2	6.7
L-Asn	0.7	19.2	0.6	44.7
L-Lys	0.3	3.2	1.6	5.3
L-Arg	0.6	3.2	0.8	11.2
L-Met	0.8	2.1	0.9	3.6
L-Pro	0.8	2.0	0.6	4.3
L-Phe	0.8	2.2	0.7	2.4

[†] Cell suspension (0.5 ml) in 0.05 M Tris-hydrochloride buffer containing 0.01 M KCl, 0.05 M MgSO$_4$ and indicated concentrations of NaCl were preincubated for 15 min at 37 °C. Then 5 μl of ^{14}C-labeled amino acid at 0.15 mM concentration was added followed by further incubation for 20 min.

velocity of uptake at pH 9 was about three times faster than at pH 7 (Fig. 4.9). Therefore, the presence of NaCl may play a role in the transport mechanisms of AIB into cells. Other monovalent cations such as K^+, Li^+ and $NH4^+$ could not be substituted for Na^+, and different species of counter-anions for Na^+ did not affect the uptake. The uptake of amino acids other than AIB was also stimulated by the addition of Na^+; this was especially true for glycine, L-alanine, L-serine, and L-asparagine (Table 4.9).

Comparative studies with other alkaliphilic *Bacillus* strains and with *B.subtilis* Pc-I 219 (ATCC 6633), which cannot grow at pH 9, are shown in Table 4.10. The AIB uptake of all alkaliphilic strains was stimulated by the addition of Na^+ at pH 9, whereas no stimulation was observed with *B. subtilis*.

There are many reports of marine bacteria isolated from sea water which require Na^+ in the medium for growth. Koyama, Kiyomiya and Nosoh (1976) also found that alkalophilic *Bacillus* strains require Na^+ for transport of sugars and amino acids.

Because alkaline media used for the isolation of alkaliphilic *Bacillus* strains contained

Table 4.10 Effect of various salts on uptake of AIB by cells of alkaliphilic bacteria and *Bacillus subtilis*

	Uptake of AIB (nmol/mg·dry wt. per h)			
Compound added [†1]	*Bacillus* sp. No. 202-1	*Bacillus* sp. No. K-12-5	*Bacillus* sp. No. A-40-2	*Bacillus subtilis* Pc-I 219
NaCl	69.9	151.2	42.3	47.8
NaNO₃	65.2	143.9	45.4	ND[†2]
Na₂SO₄	60.8	170.0	41.2	ND
KCl	6.4	24.1	1.0	45.6
LiCl	6.6	20.2	3.5	44.8
NH₄Cl	1.6	10.1	0.7	26.8
None	5.6	20.4	0.8	36.7

[†1] At a level of 0.2 M, except for Na₂SO₄, which was added at 0.1 M.
[†2] ND, not determined.

Table 4.11 Effect of various salts on the uptake of AIB by cells of alkaliphilic *Bacillus* strain

	Uptake of AIB (nmol/mg·dry wt. per 30 min)	
Compound added [†]	Medium	
	NaHCO₃	K₂CO₃
NaCl	37.1	66.5
NaNO₃	38.0	59.9
Na₂SO₄	38.9	72.8
KCl	9.3	11.8
LiCl	1.5	2.9
NH₄Cl	0.7	1.4
None	1.2	1.5

[†] At a level of 0.2 M, except for Na₂SO₄, which was added at 0.1 M.

(Reproduced with permission from Kitada and Horikoshi, *J. Biochem.*, **87**, 1280 (1980))

Table 4.12 Kinetic parameters of AIB transport in *Bacillus* sp. No. 8-1 under various conditions

pH	NaCl (M)	K_m (mM)	V_{max} (nmol/mg·dry wt. per min)
7	0.02	0.200	2.0
	0.05	0.111	2.0
	0.20	0.067	2.0
9	0.02	0.167	10.0
	0.02	0.105	10.0
	0.20	0.067	10.0

(Reproduced with permission from Kitada and Horikoshi, *J. Biochem.*, **87**, 1282 (1980))

0.1–0.2 M Na⁺, the composition of the screening media may be responsible for the Na⁺ requirement for growth and AIB uptake as found in marine pseudomonads. To clarify this question alkaliphilic *Bacillu* sp. No. 10A-2 was isolated on an alkaline medium containing 0.5% K_2CO_3 (pH 9.5) (Kitada and Horikoshi, 1980*a*). This strain can grow well in an alkaline medium containing $NaHCO_3$ as well as K_2CO_3, and AIB uptake by the cells grown on $NaHCO_3$ or K_2CO_3 was stimulated by Na⁺, suggesting that the Na⁺ requirement for AIB uptake is one of the special characteristics of alkaliphilic bacteria (Table 4.11).

AIB uptake into the cells of alkaliphilic bacteria was also strongly affected by the pH of the medium (Ohta *et al.*, 1975; Kitada and Horikoshi, 1977; Guffanti *et al.*, 1978). These maximum pH values of alkaliphilic bacteria were in marked contrast to those for the uptake of amino acids by *B.subtilis*, which had a maximum pH of 7. The effects of AIB concentration from 25 to 250mM on the rate of AIB uptake under neutral and alkaline conditions at Na⁺ concentrations between 0.02 and 0.2 M are shown in Table 4.12. When the Na⁺ concentration was raised from 0.02 to 0.2 M the K_m value for AIB transport decreased, although the maximum velocity (V_{max}) was not affected at any pH value. On the other hand, the pH value had a significant effect on maximum velocity.

B. ʟ-Serine Uptake by Membrane Vesicles of Alkaliphilic *Bacillus* Species

Bacteria can take up nutrients actively by means of specific transport systems at the expense of metabolic energy, and the energy required for active transport is derived from either oxidative processes or ATP hydrolysis *via* membrane-bound ATPase (Harold, 1972). By using membrane vesicles the mechanism of energy coupling to active transport has been investigated by many workers, and the relationship between active transport and respiratory oxidation has been clearly established by Mitchell's chemiosmotic hypothesis. Oxidation of electron donors is accompanied by the expulsion of protons into medium, leading to the formation of proton-motive force (Ramos and Kaback, 1977*a*).

Active transport of amino acids into membrane vesicles of alkaliphilic *Bacillus* sp. No.8-1 is dependent on substrate oxidation and the presence of Na⁺ (Kitada and Horikoshi, 1980*b*). The uptake of amino acids was stimulated by the addition of an artificial electron donor system, ascorbate-phenazine methosulfate (PMS), and to a lesser extent by NADH, while succinate, lactate, and α-glycerol phosphate did not stimulate uptake (Table 4.13).

The effect of NADH on serine uptake could be potentiated by the addition of PMS, tetramethyl-*p*-phenylenediamine (TMPD), and cytochrome *c* (cyt.*c*) which could be reduced by NADH chemically (PMS nad TMPD) or enzymatically (cyt.*c*). TMPD and cyt.*c* could also substitute for PMS in the presence of ascorbate. Succinate did not stimulate serine uptake but the addition of PMS caused marked stimulation of the uptake (Table 4.14).

Table 4.13   Effect of various energy sources on L-serine uptake by alkaliphilic *Bacillus* sp. No. 8-1 membrane vesicles

Energy source	Concentration (mM)	Serine uptake[†] (nmol/mg·protein)	
		pH 7	pH 9
Ascorbate + PMS (0.1 mM)	20	1.30	1.80
Ascorbate	20	0.06	1.56
PMS	0.1	0.03	0.02
NADH	20	0.22	0.48
NADPH	20	0.02	0.02
α-Glycerol phosphate	20	0.02	0.03
D-Lactate	20	0.03	0.02
L-Lactate	20	0.04	0.03
Succinate	20	0.03	0.02
Acetate	20	0.03	0.02
Citrate	20	0.03	0.02
D-Glucose	20	0.03	0.02
None		0.03	0.02

[†] Amount accumulated in 10 min.
(Reproduced with permission from Kitada and Horikoshi, *J. Biochem.*, **88**, 1759 (1980))

Table 4.14   Effect of various reducing agents on serine uptake

Substrate (20 mM)	Reducing agent (0.1 mM)	Serine uptake[†1] (nmol/mg·protein)
Absorbate	PMS	1.84
	TMPD [†2]	2.20
	Cytochrome *c*	2.02
	Ferricyanide	0.05
	None	1.50
NADH	PMS	1.80
	TMPD	2.22
	Cytochrome *c*	1.50
	Ferricyanide	0.07
	None	0.48
Succinate	PMS	1.00
	TMPD	0.09
	None	0.06

[†1] Measured at pH 9 for 10 min.
[†2] Tetramethyl-*p*-phenylenediamine dichloride.
(Reproduced with permission from Kitada and Horikoshi, *J. Biochem.*, **88**, 1759 (1980))

Inhibition studies showed that cyanide, sodium azide, and 2-heptyl-4-hydroxyquino-line-N-oxide (HOQNO) inhibited respiration by the vesicles in the presence of NADH, showing that NADH oxidation was cytochrome linked. On the other hand, uncoupling agents such as carbonylcyanide-m-chlorophenylhydrazone and 2,4-dinitrophenol had only a slight effect on NADH oxidation, but almost completely inhibited serine uptake. With the exception of HOQNO, these inhibitors were effective on ascorbate plus PMS-dependent serine uptake.

An absolute requirement for Na^+ was observed for the active transport of amino acids by these membrane vesicles, and the optimum concentration of Na^+ for serine accumulation was found to be 0.05 M. Other monovalent cations could not substitute for Na^+ and different species of counter-anions for Na^+ did not affect the uptake (Table 4.15). Sodium ions stimulated serine transport in the presence of NADH, NADH plus cyt.c, or succinate plus PMS, but had no stimulatory effect on the corresponding dehydrogenase activities.

Figure 4.10 shows the time course of serine uptake into the membrane vesicles in the presence of ascorbate plus TMPD at various pH values (pH 6.8-9). The initial rate and the steady state level of serine accumulation were optimum at pH 8.7. The pH optima of the uptake lay between 8 and 9 for all electron donors. Serine uptake by *E.coli* membrane vesicles was optimum at pH 6-7 in the presence of various electron donors (Lombaridi and Kaback, 1972). The difference in optimum pH in serine uptake between alkaliphilic *Bacillus* and neutrophilic *E. coli* presumably reflects the difference in cell surface of these two strains, and membrane vesicles in the alkaliphilic bacteria still retain their alkaliphilic nature in the transport system.

Prowe *et al.* (1996) published a paper on sodium-coupled energy transduction in the newly isolated anaerobic thermoalkaliphilic LBS3. Strain LBS3 is a novel anaerobic thermoalkaliphilic bacterium that grows optimally at pH 9.5 and 50 °C, and a high concentration of Na^+ ions is required for growth. Right-side-out membrane vesicles exhibited a strict requirement for Na^+ for the uptake of several amino acids. In the case of L-leucine, it was concluded that amino acid uptake occurs in symport with Na^+ ions. Further characteri-

Table 4.15 Effect of various salts on serine uptake by membrane in the presence of ascorbate plus PMS vesicles

Compound (50 mM)	Serine uptake[†] (nmol/mg protein)
NaCl	1.84
NaNO₃	1.66
KCl	0.04
LiCl	0.20
CsCl	0.06
RbCl	0.06
Choline Cl	0.06
NH₄Cl	0.06
None	0.06

[†] Amount accumulated in 10 min at pH 9.
(Reproduced with permission from Kitada and Horikoshi, *J. Biochem.*, **88**, 1761 (1980b))

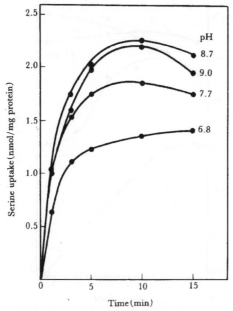

Fig. 4.10 Effect of pH on serine uptake in the presence of ascorbate plus TMPD. (Reproduced with permission from Kitada and Horikoshi, *J. Biochem.*, **88**, 1760 (1980b))

zation of the leucine transport system revealed that its pH and temperature optima closely match the conditions that support the growth of strain LBS3. The ATPase activity associated with inside-out membrane vesicles was found to be stimulated by both Na^+ and Li^+ ions. These data suggest that the primary mechanism of energy transduction in the anaerobic thermoalkaliphilic strain LBS3 is dependent on sodium cycling.

C. Mechanism of Active Transport

It is now well established that a proton electrochemical gradient (pmf) is created by outward translocation of protons coupled to respiration or ATP hydrolysis *via* membrane-bound ATPase. This is composed of a chemical gradient (ΔpH, interior alkaline) and a membrane potential ($\Delta\Psi$, interior negative). The uphill transport of a variety of different substrates (active transport) into the cells occurs as co-transport of protons, which are flowing downhill due to a pmf (Ramos and Kaback, 1977*a*; Ramos and Kaback, 1977*b*; Harold, 1977). Thus the steady state level of the accumulation of transport substrates is determined by the pmf and the stoichiometry between protons and substrates (Rottenberg, Grunwald and Avron, 1972).

Respiring cells of *Bacillus alcalophilus* maintain a constant internal pH value (9-9.5) over a range of external values from 9 to 11.5. Hence ΔpH would be the reverse of the normal direction, that is, interior acid, at external pH values above 9.5, increasing from 36 mV at pH 10.0 to 151mV at pH 11.5. Therefore, although the membrane potential, $\Delta\Psi$, increases from −84 mV at pH 9.0 to -152 mV at pH 11.5, pmf decreases with increasing external pH, and pmf is only −80 mV at pH 10.5, at which AIB transport is maximum (Fig. 4.11).

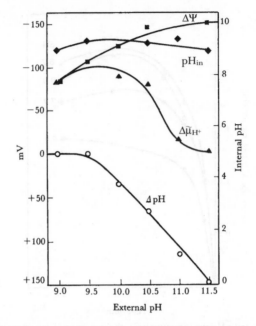

Fig. 4.11 Effect of external pH on internal pH, ΔpH, ΔΨ, and Δμ̃$_{H^+}$ in alkaliphilic
Bacillus sp.
(Reproduced with permission from Guffanti *et al.*, *J. Biol. Chem.*, **253**, 709
(1978))

How can the alkaliphilic bacteria retain a high transport activity in that highly alkaline pH range? The observation that alkaliphilic bacteria require Na$^+$ for active transport
and possess a Na$^+$/H$^+$ antiporter system to keep internal pH lower than the external pH
may account for this (Mandel, Guffanti and Krulwich, 1980). The Na$^+$/H$^+$ antiporter catalyzes the exchange of external protons for internal sodium and that of *B. alcalophilus* is
found to be ΔΨ-dependent, reflecting an electrogenic transport of Na$^+$/H$^+$ of less than 1.
By this means, an electrogenic Na$^+$/H$^+$ antiporter would be expected to convert the pmf
to an electrochemical sodium gradient, which drives substrate transport through a Na$^+$
/substrate symport mechanism. Thus, the alkaliphiles solve the problem of accumulating
solute in the presence of a low pmf by using Na$^+$ as a coupling ion for solute uptake.

Guffanti *et al.* (1978) suggested that AIB transport occurred by electrogenic symport
with Na$^+$ in response to the ΔΨ. AIB transport driven by a pmf was also observed in membrane vesicles of alkaliphilic *Bacillus* sp. No. 8-1. Imposition of a sodium gradient by the
addition of NaCl to the reaction mixture containing membrane vesicles and ^{14}C-AIB
caused a transient transport of AIB into the membrane vesicles (Fig. 4.12). A greater accumulation of Na$^+$ gradient-stimulated amino acids in membrane vesicles was observed when
an interior negative membrane potential was imposed by the addition of permeant anions
(SCN$^-$) or by electrogenic potassium diffusion potential catalyzed by valinomycin. These
results supported the theory that energization of amino acid transport involved a sodium
gradient and ΔΨ (interior negative).

A facultatively alkaliphilic *Bacillus* strain YN-2000 grows well between pH 6.8 and 10,
and the available size of pmf is high around neutral pH. The energy coupling system for

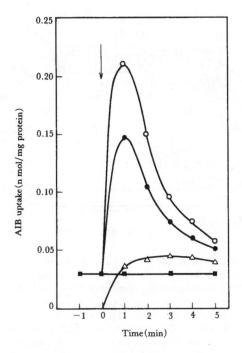

Fig. 4.12   Transient uptake of ^{14}C-AIB by the addition of Na$^+$.
Membrane vesicles were preincubated with ^{14}C-AIB for 3 min at 30 °C, then
NaCl (●), NaSCN (○), or KCl (■) was added at a final concentration of 0.05
M at the time indicated by the arrow. A control experiment (△) was carried
out by preincubating the membrane vesicles with 0.05 M NaCl for 5 min
before the addition of ^{14}C-AIB.
(Reproduced with permission from Kitada and Horikoshi, *J. Biochem.*, **87**,
1282 (1980))

glutamate transport and flagellar motors in this strain, however, was fixed as the Na$^+$-dri-
ven type, and a switch of the energy source for the membrane functions from the H$^+$-dri-
ven type around neutral to the Na$^+$-driven type in alkaline was not observed. Therefore,
it is evident that a facultatively alkaliphilic *Bacillus* strain YN-2000 has essentially the same
energy coupling system as obligatory alkaliphilic Bacilli (Sugiyama, Matsukura and Imae,
1985; Sugiyama *et al.*, 1986). Krulwich (1986) presented a model of an alkaliphile cell
including the Na$^+$ cycle composed of Na$^+$/H$^+$ antiporter and Na$^+$/solute symporters (Fig.
4.13).

4.2.3   ATPase and ATP Synthesis
The synthesis of ATP is energized by electrochemical proton gradients (inside positive, inside alkaline) *via* a membrane-bound ATPase. Alkaliphilic bacteria in which proton gradients exist in the reverse of the normal direction (inside acid) are considered to
be unfavorable for ATP synthesis, but intracellular ATP levels in alkaliphilic bacteria are
no lower than those in neutrophilic bacteria. ATPase can function not only in the ATP

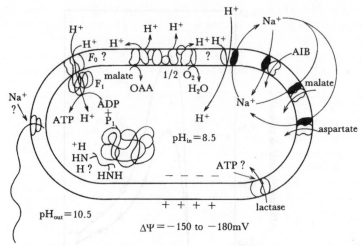

Fig. 4.13 Schematic representation of an alkaliphile cell. Features of the model include: typical cytoplasmic pH and $\Delta\Psi$ values during growth on malate at pH 10.5; a proton-pumping respiratory chain; a Na^+/H^+ antiporter catalyzing an electrogenic exchange of H_{out}^+ (perhaps utilizing some protons that are transported along a localized pathway?) for Na_{in}^+; Na^+ solute symporters that complete the Na^+ cycle; an F_1F_0 = ATPase involved in ATP synthesis and, perhaps, utilizing some localized proton pathway; some basic molecules shown as protecting the chromosome, representative of unidentified, basic buffering components of the cytoplasm; involvement of Na^+ in flagellar rotation; and an ATP-dependent lactose transport system. The Na^+-translocating porters are shown with a defined Na^+ site; that site is probably not, according to the most recent data, a distinct subunit. Some regulatory or common synthetic step may link the Na^+-translocating porters, or a pleiotropic nonalkaliphilic phenotype may be produced by mutational generation of a Na^+ leak.

(Reproduced with permission from Krulwich, *J. Membr. Biol.*, **89**, 118 (1986))

synthesis of oxidative phosphorylation but also as an ATP-driven H^+-pump in the reverse direction, and generates a proton motive force (pmf), which is utilized for active transport. Therefore, ATPase is a very important enzyme for energy-coupling reaction in alkaliphilic bacteria.

Koyama, Koshiya and Nohso (1980) purified ATPase from an alkaliphilic *Bacillus* isolated from indigo leaves. The enzyme had a molecular weight of 410,000 and was composed of five subunits, α, β, γ, δ, and ε, with molecular weights of 60,000, 58,000, 34,000, 14,000, and 11,000, respectively. The enzyme was activated by Mg^{2+} and Ca^{2+}. The pH optima of this enzyme with 0.1 and 0.2 mM Mg^{2+} were 9 and 6, and those with 1 and 10 mM Ca^{2+} were 8–9 and 7, respectively. The V_{max} and K_m values of the enzyme with ATP in the presence of 10 mM Ca^{2+} or 0.6 mM Mg^{2+} at pH 7.2 were 17 or 0.5 units/mg protein and 1.2 or 0.3 mM, respectively.

Koyama (1996) purified ATPase from the anaerobic alkaliphile *Amphibacillus xylanus*. The membrane-bound ATPase of the facultatively anaerobic alkaliphile *Amphibacillus xylanus* strain Ep01 was solubilized with a non-ionic detergent, decanoyl *N*-methylglucamide, and purified. The purified enzyme contained five polypeptides with molecular masses of approximately 78, 59, 40, 22 and 13 kDa. The enzyme that was specifically activated by NH^{4+} and Na^+ utilized only triphosphates of nucleosides as a substrate and exhibited the highest hydrolyzing activity for ATP. The polypeptides with molecular masses of 78 and 59 kDa cross-reacted with an antiserum against the catalytic units (subunits A and B) of V-type ATPase from *Enterococcus hirae*. Based on the results, it is concluded that the unique *Amphibacillus xylanus* ATPase belongs to V-type ATPase.

The ATPase was extracted from everted membrane vesicles of alkaliphilic *Bacillus firmus* RAB by low ionic strength treatment and purified to homogeneity by hydrophobic interaction chromatography and sucrose density gradient centrifugation. The ATPase had the characteristic F_1-ATPase subunit structure, with molecular weight values of 51,500(a), 48,900(b),34,400(g),23,300(d), and 14,500(e). Methanol and octyl glucoside elevated the Mg ATPase activity of the purified F_1 more than 150-fold. It is concluded that in alkaliphilic bacteria, ATP hydrolysis and ATP synthesis are catalyzed by an F_1F_0-ATPase rather than a Na^+ ATPase, as shown in Table 4.16, although the F_0 has not yet been characterized (Hicks and Krulwich, 1986).

As already mentioned, most alkaliphiles exhibited very low pmf values at optimum external pHs for growth (pH 9.5-11), because an acidified cytoplasmic pH was maintained through the operation of a Na^+/H^+ antiporter; -80 mV at pH 10.5 and -15 mV at pH 11.0 in cells of *B.alcalophilus* (Guffanti *et al.*, 1978). Guffanti, Bornstein and Krulwich (1981) studied an oxidative phosphorylation by membrane vesicles from *B.alcalophilus* and found a large quantitative discrepancy between the pmf generated upon addition of an electron donor and the amount of ATP formed. The pmf of K^+-loaded vesicles energized with ascorbate/TMPD at pH 9.0 was -125 mV and that at pH 10.5 was -40 mV. At steady state, ADP and Pi-loaded vesicles whose internal pH was buffered to 9.0 exhibited ATP synthesis of 3.2 nmol/mg protein at pH 9.0 and 1.6 nmol/mg protein at pH 10.5. Therefore, the phosphate potential (ΔG_p) was calculated to be 12 kcal/mol (-252 mV) and 11 kcal/mol (-231 mV), respectively. Thus, theoretical mV equivalents of the ΔG_p values were considerably greater than the observed pmf. A question then arises as to how a proton-translocating ATPase can synthesize ATP at such a low pmf value. The possibility that ATP synthesis uses an electrochemical gradient of sodium as in other bioenergetic

Table 4.16   A comparison of the enzymatic properties of ATPase between two alkaliphilic *Bacillus* sp.

Property	Alkaliphilic *Bacillus*	*Bacillus firmus* RAB
Molecular weight	410,000	
Subunit	$\alpha, \beta, \gamma, \delta, \varepsilon$	$\alpha, \beta, \gamma, \delta, \varepsilon$
Optimum pH	6.0 (Mg^{2+})	8.5 (Mg^{2+})
	7.0 (Ca^{2+})	8.0 (Ca^{2+})
K_m, mM	1.2 (Ca^{2+})	1.9 (Ca^{2+})
	0.3 (Mg^{2+})	1.6 (Mg^{2+})
V_{max}, unit/mg protein	17(Ca^{2+})	0.23 (Ca^{2+})
	0.5 (Mg^{2+})	0.13 (Mg^{2+})
Activator	Mg^{2+}, Ca^{2+}	Mg^{2+}, Ca^{2+}, Ethanol Octyl–glucoside

(Reproduced with permission from Koyama *et al.*, *Arch. Biochem. Biophys.*, **199**, 103-109 (1980); Hicks and Krulwich, *J. Biol. Chem.*, **261**, 12896-12902 (1986))

processes is eliminated by the following findings of no special role for Na^+ in ATP synthesis (Guffanti, Bornstein and Krulwich, 1981; Guffanti, Chiu and Krulwich, 1985; Krulwich and Guffanti, 1983): (1) A DCCD-sensitive inward translocation of protons during ATP synthesis was demonstrated, (2) no stimulation of ATP synthesis by added Na^+ or K^+ could be demonstrated, and (3) DCCD inhibited ATP synthesis.

Krulwich and her colleagues have studied oxidative phosphorylation by alkaliphilic *Bacillus* species from the viewpoint of challenging the chemiosmotic hypothesis (Guffanti *et al.*, 1984). They suggested that the bulk transmembrane gradients of proton are not the relevant ones, but that intramembranal gradients or localized gradients give rise to the observed discrepancies between the ΔG_p and pmf. This localized gradient may be the form of energy which is directly coupled to ATP synthesis. When large valinomycin-mediated K^+ diffusion potentials were imposed across starved cells of *Bacillus firmus* OF4 from pH 7.5 to pH 10.5, there was a sharp decrease in the rate of ATP synthesis as the external pH increased above pH 8. The rate of ATP synthesis fell to zero by pH 9.2–9.4, although electrogenic Na^+/H^+ antiporter and Na^+/AIB symport proceeded at a substantial rate throughout. On the other hand, when synthesis was energized by an electron donor, cells synthesized ATP at a rapid rate up to pH 10.5. This shows that the proton transfers that occur during respiration-dependent oxidative phosphorylation at pH 10.5 may depend upon specific complexes. This idea was supported by results showing that cells grown at pH 7.5 had one-third the levels of the *caa*3-type terminal oxidase and supported only low rates of ATP synthesis at pH 10.5 even though energy dependent symport and antiport rates are comparable with those in pH 10.5 grown cells. The model in Fig. 4.14 shows the oxidative phosphorylation by alkaliphilic *Bacillus* that involves a nonchemiosmotic direct intramembrane transfer of protons from specific respiratory chain complexes to the F_0 sector of ATPase (Guffanti and Krulwich, 1992).

Recently, Gilmour and Krulwich (1997) constructed a mutant of *Bacillus firmus* OF4 having a disrupted *cta* operon and purified a novel cytochrome *bd*. The *cta* operon encoding the oxidase was disrupted by the insertion of a spectinomycin resistance cassette. The mutant could not oxidize ascorbate in the presence of N,N,N',N'-tetramethyl-p-phenylenediamine (TMPD). Absorption spectra of membranes confirmed the loss of the enzyme and indicated the presence of cytochrome *bd*-type terminal oxidase. The mutant could grow on glucose at pH 10.5 but was unable to grow on malate or other nonfermentative carbon sources, despite the cytochrome *bd*. Cytochrome *bd* consisting two subunits was purified from the mutant. From properties including subunit size, the cytochrome *bd* is different from its counterparts in *E. coli* and *Azotobacter vinelandii*. Construction of mutants having different genetic properties is one of the most effective methods to understand "alkaliphily."

Guffanti and Krulwich (1994) have studied the effect of the presence of cell wall surface potential on ATP synthesis in ADP-Pi loaded membrane vesicles upon energization with respiration, and proposed three models for oxidative phosphorylation involving proton flow localized near, on or in the cell membrane.

4.2.4 Respiration
Respiration is one of the most fundamental physiological activities of living cells. The results of pH-respiratory activity profiles of alkaliphilic *Bacillus* isolated from an indigo ball indicated that oxygen consumption of this strain was maximum at pH 9-10 when amino

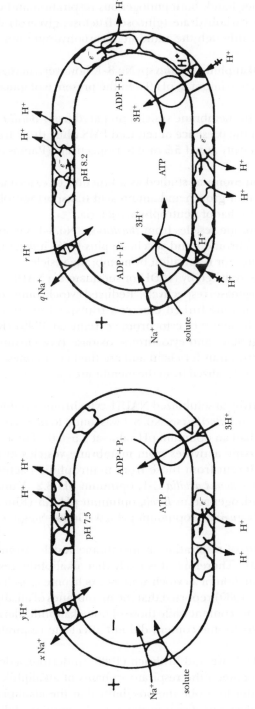

Fig. 4.14 A model for oxidative phosphorylation by alkaliphilic *B. firmus* OF4. In the diagram, several Na^+/H^+ antiporters, catalyzing electroneutral or electrogenic antiport, are represented by one antiporter with overall stoichiometry in which $x > y$ and $q > r$, but the aggregate antiport stoichiometry is different at pH 7.5 than at much more alkaline pH values. (Reproduced with permission from Guffanti and Krulwich, *J. Biol. Chem.*, **267**, 9587 (1992))

acids such as casamino acid, L-alanine, L-serine, and L-glutamic acid were used as substrates (Ohta *et al.*, 1975). On the other hand, both endogenous respiration and oxygen consumption upon the addition of carbohydrate (glucose, fructose, glycerol) did not vary over the pH range from 7 to 11, although the addition of carbohydrate stimulated respiration.

Respiration by the cells of alkaliphilic *Bacillus* sp. No.8-1, however, increased with the increase in external pH and was maximum at pH 9.0 in the presence of glucose or succinate.

The rates of oxygen uptake by membrane vesicles prepared from *Bacillus* sp. No. 8-1 were also maximum at pH 9.0 in the presence of reduced PMS while those from *E. coli* or *B. subtilis* exhibited maximum activity at pH 5.5 or 6.8, respectively (Ramos and Kaback, 1977*a*).

Thus when the rate of respiration was studied as a function of external pH, a relationship similar to that observed for growth and amino acid transport was obtained and found to be in marked contrast to that of neutrophilic bacteria.

Amino acid uptake by membrane vesicles from *Bacillus* sp. No.8-1 was stimulated by the addition of NADH plus cytochrome *c* and succinate plus PMS, indicating that they were also able to serve as substrates for respiration by membrane vesicles.

Cytochrome *c* or PMS can be reduced by NADH or succinate *via* NADH-cytochrome *c* reductase or succinate dehydrogenase, respectively. Reduced cytochrome *c* or PMS then enter electrons into the cytochrome-linked electron transport chain and reduce cytochrome a/a_3 (reduced cytochrome *c*) or cytochrome *c* (reduced PMS) (Konings and Freese, 1972). These dehydrogenases and cytochrome oxidase (cytochrome a/a_3) are located at the portal of the electron transfer chain and are the key enzymes in respiratory activity. These enzymes can be solubilized from the membrane with treatment by Triton X-100 and/or deoxycholate.

The optimum pH for the activity of solubilized NADH-cytochrome *c* reductase or succinate dehydrogenase prepared from *Bacillus* sp. No. 8-1 was found to be 8.5 or 8.0, respectively. The cytochrome oxidase had an optimum pH value at 5.5-6.0. These results indicate that the pH optima of enzyme activities from membrane vesicles of alkaliphilic *Bacillus* sp. No. 8-1 are not different from those from neutrophilic bacteria: NADH-cytochrome *c* reductase in *Pseudomonas arvilla* C-1, optimum pH 8.2 (Yamaguchi and Fujisawa, 1978), succinate dehydrogenase in *E.coli*, optimum pH 7.5 (Kim and Bregg, 1971), cytochrome oxidase in *P.aeruginosa*, optimum pH 6.0 (Gudat, Singh and Whaton, 1973).

As mentioned above, the rate of respiration by membrane vesicles from *Bacillus* sp. No. 8-1 was maximum at pH 9.0. Therefore, it is likely that alkaliphilic respiration by reduced PMS depends on the function of a cytochrome *c*-cytochrome a/a_3 fragment.

Lewis, Belkina and Krulwich (1980) reported that the membranes of alkaliphilic bacilli contained high cytochrome heme content while those of nonalkaliphilic derivatives contained much less heme. But cytochrome content did not affect the respiratory rates of whole cells (Table 4.17).

Lewis *et al.* (1981) and Kitada, Lewis and Krulwich (1983) undertook a detailed comparison of midpoint potentials between the respiratory chains of alkaliphiles and their nonalkaliphilic derivatives in order to probe the possibility that the alkaliphiles possess adaptations that facilitate particularly effective transduction. A summary of the species in membranes of *B.alcalophilus* and *B.firmus* RAB is shown in Fig. 4.15. The presence of quantitatively more cytochromes with a complex assortment of *b*-type cytochromes in the wild type membranes could represent an adaptation of the special bioenergetic costs of life at very alkaline pH. Measurements of respiration-dependent proton translocation by *B.firmus* RAB and its nonalkaliphilic mutant (RABN) in the presence of K^+ and valinomycin showed that the alkaliphile (RAB) had much higher H^+/O ratio stoichiometry (9-13 at pH 9) than did the mutant (4 at pH 7). Both the wild type and the mutant exhibited H^+/O ratio near 9 at pH 7 (Lewis, Kaback and Krulwich, 1982; Lewis *et al.*, 1983). These results are consistent with suggestions that the alkaliphilic respiratory chain is especially well adapted for effective energy transduction at alkaline but not neutral pH.

Table 4.17   Cytochrome contents of membranes from alkaliphilic bacteria and their nonalkaliphilic mutant derivatives

Bacterial strain	Concentration of cytochrome (nmol/mg·membrane protein)					
	a-type		*b*-type		*c*-type	
	Difference spectra	Heme extraction	Difference spectra	Heme extraction	Difference spectra	Heme extraction
Bacillus alcalophilus	0.51	1.3	2.2	1.7	2.6	2.5
B. alcalophilus KM23	0.23	—	0.54	—	0.50	—
B. firmus RAB	0.57	1.7	2.7	1.1	3.8	2.7
B. firmus RABN	0.43	—	0.50	—	0.53	—

(Reproduced with permission from Lewis *et al.*, *Arch. Biochem. Biophys.*, **95**, 857-863 (1980))

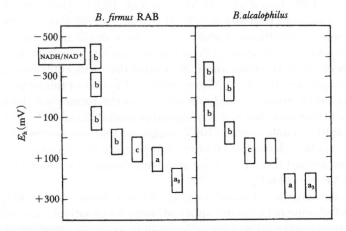

Fig. 4.15   Summary of the respiratory chain components identified on the basis of redox potentiometry in membranes from *Bacillus alcalophilus* and *Bacillus firmus* RAB at pH 9.0. The data are taken from Lewis *et al.* (1981) and Kitada *et al.* (1983). The blocks representing each component indicate the potential range over which the component becomes 9-91% oxidized or reduced. The lateral positioning is arbitrary.

(Reproduced with permission from Krulwich, *J. Membr. Biol.*, **89**, 115 (1986))

Both soluble and membrane-bound cytochrome *c* have been distinguished in *B.firmus* RAB. The membrane-bound cytochrome *c* was copurified with the cytochrome oxidase, which has been characterized as possibly having two major polypeptide subunits, 2 moles of copper and 2 moles of heme (Kitada and Krulwich, 1984). A soluble cytochrome *c* and cytochrome *b* were purified from alkaliphilic *B.firmus* RAB. Above pH 8.3, cytochrome *c* exhibited a pH-dependent decrease in midpoint potential while at both pH 7 and 8.3, midpoint potential was +66 mV. The soluble cytochrome *b* exhibited a pH-independent midpoint potential of +28 mV (Davidson *et al.*, 1988).

From the alkaliphilic *Bacillus firmus* OF4, Gilmour and Krulwich (1996) partially purified and characterized the succinate dehydrogenase complex. The partially purified cytochrome *b* is actually composed of three different *b*-type cytochromes which can be separated and purified by a combination of ion-exchange, hydroxyapatite and gel filtration chromatographies. Two of the cytochromes were CO-reactive but lacked the characteristic multisubunit composition of known terminal oxidases. Neither purified cytochrome catalyzed quinol or ferrocytochrome *c* oxidation. The third, non-CO-reactive cytochrome *b* was associated with substantial succinate dehydrogenase activity and was purified as a three-subunit succinate dehydrogenase complex with high specific activity (17.7 mμ mol/min/mg). Limited N-terminal sequence of each subunit demonstrated marked similarity to the complex from *Bacillus subtilis*. The cytochrome *b* of the alkaliphile enzyme was reduced about 50% by succinate compared to the level of reduction achieved by dithionite. The enzyme reacted with both napthoquinones and benzoquinones. The results presented indicate that *Bacillus firmus* OF4 contains a succinate dehydrogenase complex with properties very similar to those of the enzyme from *Bacillus subtilis*, but does not contain a cytochrome *o*-type terminal oxidase under the growth conditions studied.

A facultative alkaliphilic *Bacillus* YN-2000 possesses a novel aco_3-type cytochrome oxidase (Qureshi *et al.*, 1990; Orii *et al*, 1991; Yumoto *et al.*, 1993). The purified enzyme was composed of three subunits with MrS 50,000, 41,000 and 22,000 and contains 1 molecule each of cytochrome *a*, cytochrome *c* and cytochrome o_3 in the minimal structural unit. The heme *o* was assumed to react directly with molecular oxygen while the heme *a* moiety acted like an electron reservoir and may control the intracellular electron transfer through the main pathway (a-c-O_2). The membrane fraction of this strain was found to contain two *c*-type cytochromes, *c*-553 and *c*-552, and the former was the physiological electron donor and had a very low redox midpoint potential (95 mV) (Yumoto, Fukumori and Yamanaka, 1991). Yumoto *et al.* (1993) suggested that this low midpoint potential may be necessary to allow electrons to flow from the outside (cytochrome *c* or cytochrome *a* moiety) to the inner part (cytochrome o_3) of the membrane, overcoming the larger electrical potential ($\Delta\Psi = -188$ mV).

Studies on bioenergetics of *Bacillus* sp. C-125 have been started. Respiration was increased around alkaline pH near the upper pH-limit for growth of the organism (Aono *et al.*, 1996). Oxygen uptake rates of the cells grown in a complex medium at pH 7-9 and 9.5-10 were 0.98-1.4 and 2.4 mμ mol O atom/min/mg cell protein, respectively. Membrane vesicles from the cells grown at pH 7-9 and 9.9 incorporated O_2 at rates of 1.1-1.4 and 2.5 mμ mol O atom/min/mg envelope protein, respectively, using exogenous NADH as an electron donor. In the presence of menadione as an exogenous electron acceptor, the membrane vesicles from the cells grown at pH 7-8.5 and 9.9 oxidized NADH at rates of 1.4-1.7 and 6.3 mμ mol NADH/min/mg envelope protein, respectively. Levels of respiratory and NADH-oxidation activities of the organism are dependent on the growth pH, and higher than those reported previously in alkaliphilic *Bacillus* sp.

4.3   Intracellular Enzymes

Properties of intracellular enzymes of alkaliphilic microorganisms were very attractive when the present author began to study the physiology of alkaliphiles in the 1970's. How did extreme environments affect intracellular enzymes? Were there critical differences between intracellular enzymes of neutrophiles and alkaliphiles? Many intracellular enzymes had been isolated and purified, and their properties investigated. Some enzymes exhibited relatively higher pH optima than neutrophilic bacteria. However, there appeared to be nothing exceptional in intracellular enzymes. Cell-free protein synthesis systems showed maximal activity at between pH 8 and 8.5, but this was only 0.5 pH units higher than that observed for neutrophilic *Bacillus subtilis*. This is apparently due to the ability of alkaliphiles to regulate intracellular pH towards neutrality.

4.3.1   α-Galactosidases

The first work on these enzymes was done by Akiba and Horikoshi (1976*a,b*; 1978). Two kinds of α-galactosidase-producing bacteria, *Micrococcus* sp. No. 31-2 and *Bacillus* sp. No. 7-5, were isolated from soil on a modified Horikoshi-I medium (1% raffinose was substituted for glucose). Alkaliphilic *Micrococcus* sp. No. 31-2 induced a cytoplasmic α-galactosidase while alkaliphilic *Bacillus* sp. No. 7-5 produced an extracellular α-galactosidase constitutively. The properties of these enzymes resembled each other, as shown in Table 4.18. The optimum pH range of these enzymes was higher than that of yeast, mold and plant seeds.

Then Ikura and Horikoshi (1987*c*) isolated a group of α-galactosidase-producing bacteria which belong to the genus *Flavobacterium*, although they grew better in an alkaline medium than in a conventional neutral medium. The isolation medium contained D-galactose, 15.0g; polypeptone, 5.0g; yeast extract, 5.0g; $MgSO_4 \cdot 7H_2O$, 0.2g; $MnSO_4 \cdot 4/5H_2O$, 0.02g; Na_2CO_3, 10.0g. The colonies appearing on the solid medium were cultured in the same medium for 24 h with shaking. The α-galactosidase activity in the tolu-

Table 4.18   Comparison of the enzymatic properties of α-galactosidase between alkaliphilic and neutrophilic microorganisms

Property	*Micrococcus* sp. No. 31-2	*Bacillus* sp. No. 7-5	*Mortierella vinacea*
Molecular weight	367,000	312,000	
Optimum pH	7.5	6.5	4.0-6.0
pH stability range	7.5-8.0	6.0-8.5	7.0-11.0
Optimum temperature, °C	40	40	
*K*m ONPG [†1], mM	0.47	1.0	0.36 ± 0.014
Meliboise, mM	1.5	7.9	0.39 ± 0.029
Raffinose, mM	12.6	24.1	1.83 ± 0.13

[†1] *o*-Nitrophenyl-α-D-galactoside.
(Reproduced with permission from Horikoshi and Akiba, *Alkalophilic Microorganisms*, p. 55, Springer-Verlag : Japan Scientific Societies Press (1982))

enized cells was assayed using p-nitrophenly-α-D-galactoside as the substrate. Seven strains were isolated and all grew better at the initial pH of 10.2 than at pH 7.2. More enzyme was produced at pH 10.2 than at pH 6.7, and the optimum pH for enzyme action was in the range of 7.0–7.5. Addition of 0.5% glycine and 1.5% galactose in the medium (w/v) stimulated enzyme synthesis tremendously in all strains. Addition of disaccharides raffinose or starch (except melibiose) did not increase enzyme production.

4.3.2 β-Galactosidases

β-Galactosidase was produced inducibly by alkaliphilic *Bacillus* sp. No. C-125 not only in alkaline medium but also in neutral medium. However, the induction of the enzymes was much faster at pH 10.2 than at pH 7.2 (Ikura and Horikoshi, 1979*a*). The enzymes produced in media of different pH values possessed the same enzymatic properties (Ikura and Horikoshi, 1979*b*). The molecular weight of the enzyme was about 185,000. The enzyme was most active at pH 6.5 and stable over the pH range 5.5 to 9.0, suggesting that the enzymatic properties of alkaliphilic *Bacillus* sp. No. C-125 β-galactosidase are essentially similar to those of neutrophilic *Bacillus* species.

It was reported (Ikura and Horikishi, 1988) that in the presence of Mn^{2+}, glycine, glycine derivative, amino acid analogous, or D-amino acids, release of β-galactosidase by

Table 4.19 Effect of cation on production and release of β-galactosidase in the presence of glycine ethyl ester

Cation added	Growth (OD_{600})	β-galactosidase activity (units/mL)	
		Extracellular	Total
None	6.6	0.2	1.1
Glycine ethyl ester, 1.5%	13.7	4.2	8.2
Glycine ethyl ester, 1.5%, plus :			
Min^{2+}, 100 µM	22.0	16.2	27.2
Ca^{2+}			
100 µM	15.8	4.8	9.0
5 mM	14.9	5.2	8.8
Ba^{2+}			
100 µM	12.1	4.0	7.9
5 mM	7.7	1.8	4.4
Zn^{2+}, 100 µM	13.0	3.2	6.4
Co^{2+}, 100µM	14.2	4.3	7.4
Fe^{2+}, 100 µM	13.3	4.6	7.8
Ni^{2+}, 100 µM	13.3	4.7	8.2
Cu^{2+}, 100 µM	10.9	2.4	4.7
Cd^{2+}, 100 µM	3.2	0.6	1.7
Sn^{2+}, 100 µM	11.1	2.0	4.2
Pb^{2+}, 100 µM	10.4	3.0	6.1
Mo^{2+}, 100 µM	13.9	4.7	8.1

Note: β-Galactosidase activity was measured after 24 h cultivation. Mn^{2+}, Ca^{2+}, Ba^{2+}, Zn^{2+}, Co^{2+}, Cd^{2+}, Sn^{2+}, and Pb^{2+} were added as chlorate compounds: Fe^{2+}, Ni^{2+}, and Cu^{2+} as sulfate compounds; and Mo^{2+} was added as $Na_2MoO_2 \cdot 2H_2O$.
(Reproduced with permission from Ikura and Horikoshi, *Can. J. Microbiol.*, **34**, 1179 (1988))

alkaliphilic *Bacillus* No. C-125 in the culture broth was observed. The addition of more than 100 μM Mn^{2+} was indispensable for increased enzyme production and β-galactosidase was released even in a nutrient-rich medium. When both Mn^{2+} and 2.0% glycine ethyl ester were added simultaneously, 65% of the total activity was found extracellularly (Table 4.19). Protease production was stimulated by Mn^{2+} but repressed by glycine ethylester. The simultaneous addition of Mn^{2+} and 2.0% glycine ethylester repressed protease activity strongly, indicating that this method is advantageous for practical application to extracellular production of intracellular proteins.

Choi *et al.* (1995) extracted a β-galactosidase from alkaliphilic and thermophilic *Bacillus* sp. TA-11. The enzyme was purified 20-fold over the crude extract by ion-exchange and gel-filtration chromatography with a recovery of 32%. The molecular mass of the native enzyme was estimated to be 200 kDa. SDS/PAGE revealed three protein bands of 62, 40 and 34 kDa. Maximum enzyme activity was observed at pH 6.0 and 40°C, but the enzyme was stable over the pH range of 6-12 and below 55°C. The K_m and V_{max}, values for *o*-nitrophenyl β-D-galactopyranoside were 13.5 mM and 251 mm M/min per mg of protein. The enzyme activity was significantly inhibited by EDTA, bivalent metal ions (Zn^{2+}, Hg^{2+} and Cd^{2+}) and galactose.

4.3.3   D-Xylose Isomerase

D-Xylose isomerase from *Bacillus* sp. KX-6 has been purified and characterized (Kwon, Kitada and Horikoshi, 1987). Kitada, Dobashi and Horikoshi (1989) later isolated another D-xylose isomerase from thermophilic alkaliphile, *Bacillus* TX-3. The growth medium of strain KX-6 contains (grams per liter): D-xylose, 5.0; polypeptone, 5.0; yeast extract, 5.0; K_2HPO_4, 1.0; $MnCl_2 \cdot 4H_2O$, 0.1; $CoSO_4 \cdot 7H_2O$, 0.05; and Na_2CO_3, 10.0. An alkaliphilic *Bacillus* sp. KX-6 was selected as a D-Xylose isomerase producer out of 150 colonies. The enzymatic activity in the culture was measured at pH 7.5 by a conventional method. The isolate was similar to *Bacillus circulans* in taxonomic characteristics except for its growth pH. The enzyme extracted from the cultured cells overnight using a French press was purified by ammonium sulfate fractionation (45-90% saturation) and DEAE-Sepharose CL-6B chromatography followed by Sephacryl S-200 gel filtration. The enzyme preparation had strong activity towards D-xylose but weak activity on D-glucose and D-arabinose. The other enzymatic properties are summarized in Table 4.20.

Table 4.20   Some enzymatic properties of D-xylose isomerase

Optimum pH and temperature: 7.0-9.0; 60°C
Stable pH : 6.5-11.0
Molecular weight: 120,000 with two subunits of 58,000
pI: 4.5
Sedimentation constant: 9.35 S
Activator: 1 mM Co^{2+} and Mn^{2+}
Inhibitor: 1 M Hg^{2+}, Ag^{2+}, and Cu^{2+}
Substrate specificity: D-xylose ($K_m = 0.076$ M)

4.3.4 Dehydrogenases

A. Dehydrogenases from *Corynebacteria*

Facultative alkaliphilic bacteria *Corynebacterium* sp. No. 93-1 and No. 150-1 were isolated from soil on alkaline media containing 1% Na_2CO_3 (Kobayashi and Horikoshi, 1980*d*; Kobayashi and Horikoshi, 1980*a*; Kobayashi and Horikoshi, 1980*b*; Kobyashi, Ueyama and Horikoshi 1980; Kobayashi and Horikoshi, 1980*c*; Kobayashi, Ueyama and Horikoshi, 1982). The cell-free extract of strain No. 93-1 contained three kinds of dehydrogenase: (1) NAD-dependent maltose dehydrogenase, (2) NAD-dependent D-glucose dehydrogenase, and (3) NADP-dependent maltose dehydrogenase. The cell-free extract of strain No. 150-1 also contained two kinds of dehydrogenase: (1) NAD-dependent maltose dehydrogenase, and (2) NAD-dependent D-glucose dehydrogenase. The enzymatic properties of all five dehydrogenases from the two strains are summarized in Table 4.21. (Kobayashi, Ueyama and Horikoshi, 1982).

1. NAD-dependent D-glucose dehydrogenase

Glucose dehydrogenase is known in various microorganisms. The enzyme has been classified into two types, particle type and soluble type. All the dehydrogenases from alkalophilic *Corynebacterium* species were located in soluble fractions obtained by centrifugation at $100,000 \times g$. The NAD-dependent D-glucose dehydrogenase of strain No. 93-1 (Kobayashi and Horikoshi, 1980*a*) was different from those so far reported in its lack of ability to oxidize 2-deoxy-D-glucose and *N*-acetylglucosamine. The D-glucose dehydrogenase of strain No. 93-1 was inhibited by sulfhydryl group inhibitors, Hg^{2+} and Ag^+. The D-

Table 4.21 Comparison of the properties of NAD-dependent maltose dehydrogenase (NAD–MDH), NAD-dependent D-glucose dehydrogenase (NAD–GDH), and NADP-dependent maltose dehydrogenase (NADP–MDH) produced by *Corynebacterium* sp. No. 150-1 and No. 93-1

	NAD–MDH		NAD–GDH		NADP–MDH
	Strain No. 150-1	Strain No. 93-1	Strain No. 150-1	Strain No. 93-1	Strain No. 93-1
Molecular weight by gel filtration	36,000	40,000	54,000	55,000	45,000
SDS Polyacrylamide gel electrophoresis	36,000	38,000	48,000	51,000	48,000
Isoelectric point	4.70	4.50	4.40	4.05	4.48
Optimal pH	10.8	10.2	10.0	9.8	10.2
K_m value (mM)					
D-Glucose	35	23	59	51	—
D-Galactose	4.7	3.0	—	—	—
D-Xylose	—	—	148	125	150
Maltose	3.4	2.1	—	—	250
Lactose	0.54	0.49	9.8	—	—
Cellobiose	17	13	9.8	11	—
Gentiobiose	—	—	5.6	3.2	—
Maltotriose	1.3	1.4	—	—	270
NAD (NADP)	0.15	0.15	0.67	0.44	(1.5)

glucose dehydrogenase of both strains No. 93-1 and No. 150-1 oxidize well, reducing di-saccharides containing β-linked D-glucose residues such as cellobiose and gentiobiose.

2. NAD-dependent maltose dehydrogenase

NAD-dependent maltose dehydrogenase of strain No. 93-1 is composed of a single peptide chain and does not have a subunit structure (Kobayashi and Horikoshi, 1980d; Kobayashi and Horikoshi, 1980c). This is unique in comparison with other previously reported dehydrogenases, which have subunits. The substrate specificity is very wide and different from the enzymes so far reported. The enzyme from strain No. 93-1 oxidizes lactose, maltose, cellobiose and maltotriose, as shown in Table 4.10. In addition, the enzyme also showed oxidative activity on maltotetraose and maltopentaose. The enzyme was completely inhibited by sulfhydryl group inhibitors Hg^{2+} and Ag^{2+}, suggesting the contribution of a sulfhydryl group to the enzyme activity. NAD-dependent maltose dehydrogenase of strain No. 150-1 closely resembled that of strain No. 93-1.

3. NADP-dependent maltose dehydrogenase

The enzyme of strain No. 93-1 (Kobayashi and Horikoshi, 1980b) is composed of a single peptide chain with no subunit structures, the same as NAD-dependent maltose dehydrogenase and NAD-dependent glucose dehydrogenase. The enzyme shows relatively high oxidative activity on D-xylose, maltose and maltotriose, and to a lesser extent on maltotetraose and maltopentaose, the same as the NAD-dependent maltose dehydrogenase of this strain. Sulfhydryl group inhibitors inhibit the enzyme activity.

In order to clarify the distribution of these dehydrogenases in coryneform bacteria, 26 strains of *Corynebacterium, Curtobacterium, Brevibacterium* and *Cellulomonas* which have been reported to produce acids from maltose were examined for sugar dehydrogenase activity. NAD-dependent maltose dehydrogenase and NAD-dependent D-glucose dehydrogenase were detected in only two of the 26 strains, but the activity of these two dehydrogenases was less than one-fifth that of alkaliphilic strains No. 93-1 and No. 150-1. These results lead to the conclusion that these types of dehydrogenase are distributed specifically in alkaliphilic *Corynebacterium*.

4. NADP-linked isocitrate dehydrogenase

Shikata *et al.* (1988) reported a very labile NADP-linked isocitrate dehydrogenase from a strain of alkaliphilic *Bacillus* sp. KSM-1050. The bacterium was isolated from a soil sample and is the potential producer of an alkaline carboxymethyl cellulase, although no microbial properties have been reported yet. The enzyme was purified by a simple method, with an overall yield of over 76% of the original activity. The organism was grown in a 500-ml flask containing 100 ml of an alkaline medium composed of (w/v): 0.8% nutrient broth, 0.5% NaCl, 0.1% KH_2PO_4, 0.005% $CaCl_2$, 0.001% $MnSO_4 \cdot 4\text{-}6H_2O$, 0.001% $FeSO_4 \cdot 7H_2O$, 0.5% $MgSO_4 \cdot 7H_2O$, and 0.5% Na_2CO_3 at 30 °C on a reciprocal shaker for 40 h. Cells were harvested by centrifugation, washed twice in chilled buffer A (12 mM Tris-HCl buffer, pH 7.8) containing 5 mM $MnCl_2$, 1mM Tri-sodium citrate, 5 mM 2-mercaptoethanol and 10% (v/v) glycerol, then frozen at −20 °C. Frozen cells were thawed, suspended in buffer A, and sonicated. The supernatant fluid was used as a crude extract for the following purification process. The crude extract was passed through a DEAE-Toyopearl 650 S column and the enzyme was eluted by applying a gradient of sodium chloride from 0 to 0.5 M. Active fractions were collected and dialyzed. Then the dialysate was loaded onto a column of Affi-Gel blue equilibrated with buffer A and dehydrogenase was eluted with NaCl(gradient 0-0.5M). The purified enzyme was free from any activity

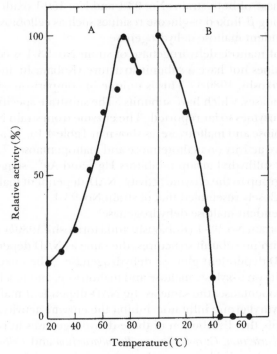

Fig. 4.16 Effect of temperature.
(A) Optimum temperature
(B) Thermal stability

against citrate, *cis*-aconitate and α-ketoglutarate. The molecular weight on Sephadex G-200 was around 90,000, and by SDS-PAGE it was about 44,000. Its isoelectric point was pH 4.7. The enzyme required Mn^{2+} for the reaction and for stability. The optimum pH for the action was in the range of 7.8 to 8.4 at 30 °C; the optimum temperature at pH 8.0 was 75 °C. However, in the absence of the substrate the enzyme was very thermolabile, as shown in Fig. 4.16. This enzyme was inhibited by NADPH, glyceraldehyde 3-phosphate, 3-phosphoglycerate, phosphoenolpyruvate, *cis*-aconitate, α-keto-glutarate, and oxaloacetate. In addition, it was subject to a concerted inhibition by a combination of glyoxylate and oxaloacetate, and to a cumulative inhibition by nucleoside triphosphates. The authors concluded that the regulation of activity of isocitrate dehydrogenase in the alkaliphilic *Bacillus* strain was analogous to that in neutrophilic bacteria. The intracellular pH value must be maintained at neutral by a H^+/Na^+ antiporter system (see section 4.2).

B. NADH Dehydrogenase

Xu *et al.* (1991) isolated NADH dehydrogenase from an alkaliphilic *Bacillus* strain YN-1, and cloned it in *E. coli*. The cloned DNA fragment contained an open reading frame of 1,557 nucleotides encoding a polypeptide composed of 519 amino acid residues. The predicted amino acid sequence was consistent with the partial amino acid sequences including the N-terminal and C-terminal sequences reported. Sequence comparison with other flavoenzymes revealed high homology.

C. Glutamate dehydrogenases

Two types of glutamate dehydrogenases have been isolated from alkaliphiles. Jahns (1996) isolated and purified an NAD-specific glutamate dehydrogenase from the alkaliphile *Amphibacillus xylanus* DSM 6626. The enzyme was highly specific for the coenzyme NAD(H) and unusually resistant towards variation of pH, chaotropic agents, organic solvents, and was stable at elevated temperatures, retaining 50% activity after 120 min incubation at 85 °C.

Another type of NADP-specific glutamate dehydrogenase was isolated from alkaliphilic *Bacillus* sp. KSM-635 (Koike *et al.*, 1996). An NADP-specific glutamate dehydrogenase from alkaliphilic *Bacillus* sp. KSM-635 was purified 5840-fold to homogeneity by a several-step procedure involving Red-Toyopearl affinity chromatography. The native protein, with an isoelectric point of pH 4.87, had a molecular mass of approximately 315 kDa consisting of six identical subunits each with a molecular mass of 52 kDa. The pH optima for the aminating and deaminating reactions were 7.5 and 8.5, respectively. The optimum temperature was around 60 °C for both. The enzyme could be stored without appreciable loss of enzyme activity at 5 °C for half a year in phosphate buffer (pH 7.0) containing 2 mM 2-mercaptoethanol.

4.3.5   Restriction Enzymes

Two site-specific deoxyribonucleases (DNases) that cleave DNA strands at unique sites have been studied. Shibata *et al.* (1976) reported on a restriction enzyme in alkaliphilic *Bacillus* sp. No. 170, which was found to be a penicillinase producer. The nucleotide sequence recognized by the enzyme was

$$5'\text{-CTGCAG-}3'$$
$$3'\text{-GACGTC-}5'$$

which was also recognized by endonuclease *Pst* I of *Providenca stuartii* 164.

Furthermore, Moriya *et al.* (1992) reported the isolation of a type II restriction enzyme, *Bsp* O4 I, from alkaliphilic *Bacillus* sp. O-4 (ATCC21536). The enzyme was an isoschizomer of *Puv* II, demonstrating the six-base aplindromic sequence of $5'\text{-CAGCTG-}3'$ and cleaves between G and C to produce a blunt-ended cleavage product.

4.3.6   RNA Polymerases

Sporulation and germination are typical differentiation processes of bacteria. *Bacillus subtilis* has been studied biochemically and genetically for many years. Several types of evidence suggest that a specific change in transcription level causes dramatic differentiation. In particular, sigma or sigmalike subunits are responsible for the specificity of transcription in the differentiation process.

Many alkaliphilic *Bacillus* strains have been isolated in the author's laboratory, but no information on differentiation of these bacteria was obtained. Kudo and Horikoshi (1978*a*, *b*) studied the RNA polymerase from vegetative cells and spores of alkaliphilic *Bacillus* sp. No. 2b-2, which is the best spore former, to ascertain whether any changes occurred during the differentiation process. The molecular weight of $\beta\beta'$, σ and α was 165,000, 97,000(?) and 42,000, respectively. These subunits were essentially identical to those of *Bacillus subtilis*. The σ factor was the same as σ-43 in *B. subtilis*. Another σ factor was also found in *Bacillus* sp. No. 38-2, which corresponds to σ-30.

4.3.7 Protein Synthesizing System

Only one paper has been published on a cell-free protein synthesizing system using alkaliphilic *Bacillus* sp. No. A-59 and No. C-125 (Ikura and Horikoshi, 1978). Microorganisms grown in Horikoshi-I medium were collected by centrifugation ($6,000 \times g$) and washed twice with 0.01 M Tris-HCl buffer (pH 7.5) containing 0.01 M MgCl$_2$, 6 mM 2-mercaptoethanol, and 0.06 M KCl (TMM buffer).

The enzymes were extracted with TMM buffer equivalent to twice the volume of cells ground with two to three times its weight in alumina in a prechilled mortar. The extract was centrifuged twice at $30,000 \times g$ for 30 min. The upper two-thirds of the supernatnat was dialyzed against TMM buffer for 3 h. From this supernatant fluid, ribosomes were precipitated by centrifugation at $105,000 \times g$ for 2 h, washed once with TMM buffer, and finally resuspended in TMM buffer. The supernatant was centrifuged at $105,000 \times g$ for 2.5 h and the upper two-thirds was used as the S-100 fraction. tRNA was prepared from the cells by the conventional method.

The optimum pH for protein synthesis directed by poly-U or endogenous mRNA was about 8.5, which was about 0.5 higher than that of *B.subtilis* (Fig. 4.17). As the reference, the protein synthesizing system of *B.subtilis* was tested under the same condition. To confirm this result, protein synthesis directed by endogenous mRNA was performed and the

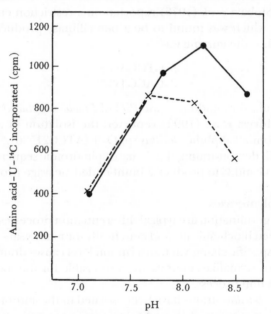

Fig. 4.17 Effect of pH on incorporation of amino acid mixture into protein directed by endogenous mRNA.
Bacillus sp. No. A-59 system contained 0.2 mg ribosomal protein and 0.35 mg S-100. The *Bacillus subtilis* system contained 0.32 mg of ribosomal protein and 0.45 mg S-100. Zero time values were subtracted. Other conditions are described in the text. ●—●, *Bacillus* sp. No. A-59, ×---×, *Bacillus subtilis*.
(Reproduced with permission from Ikura and Horikoshi, *Agric. Biol. Chem.*, **42**, 755 (1978))

optimum pH was also slightly higher than that of *B. subtilis* (about pH 8.5).

Ribosomes of alkaliphilic *Bacillus* sp. No. A-59 and No. C-125 were of the 70S type, and no difference was observed in their thermal denaturation curves. Phenylalanyl-tRNA synthetase activity at different pH values also indicated no remarkable difference between alkaliphilic *Bacillus* and *B. subtilis* Marburg 168. Phenylalanine incorporation was tested in a series of homogeneous and heterogeneous combinations. The results are shown in Table 4.22. In the systems containing alkaliphilic *Bacillus* ribosomes or S-100, phenylalanine incorporation at pH 8.4 was higher than at pH 7.5, although the activity of heterogeneous systems was lower than that of homogenous systems. In conclusion, alkaliphilic *Bacillus* sp. No. A-59 grows well under highly alkaline conditions, but the protein synthesizing mechanism is essentially the same as that of *B. subtilis*.

No direct method has been established for determining the internal pH. The pH optimum of the protein synthesizing mechanism strongly suggests that the internal pH value may be 8 to 8.5, not 10. Heterogeneous combination also supports this possibility.

The results clearly indicate that the differences between alkaliphilic *Bacillus* and neutrophilic *Bacillus* exist in the cell surface and not within the cells.

Table 4.22 Phenylalanine incorporation in homogeneous and heterogeneous systems

Ribosomes	S-100	Phenylalanine incorporated (nmol)		Ratio between
		pH 7.5	pH 8.4	pH 8.4 to 7.5
Bacillus A-59	*B.* A-59	5.82	9.37	1.61
—	*B.* C-125	6.12	8.65	1.41
—	*B. subtilis*	5.84	6.37	1.09
B. C-125	*B.* A-59	5.17	6.62	1.28
—	*B.* C-125	5.32	8.12	1.53
—	*B. subtilis*	5.63	6.46	1.15
B. subtilis	*B.* A-59	6.31	7.15	1.13
—	*B.* C-125	6.71	8.34	1.24
—	*B. subtilis*	9.65	6.59	0.68

The reaction mixture contained 0.2 mg ribosomal protein and 0.3 mg of S-100 protein; the other conditions are described in the text. Blanks (− poly U) are subtracted.
(Reproduced with permission by Ikura and Horikoshi, *Agric. Biol. Chem*, **42**, 756 (1978))

4.4   Polyamines

Polyamines are widely distributed in plants and animals, including algae and microorganisms. In the past decade, several new polyamines have been found from thermophilic bacteria. These new polyamines appear to play essential roles in protein synthesis in these bacteria. Polyamines also respond to environmental stresses and cellular polyamine composition changes upon change in external conditions.

4.4.1 Polyamines

The first report on polyamines from an alkaliphilic *Bacillus* strain was published by Yonezawa and Horikoshi (1978). An alkaliphilic strain, *Bacillus* sp. No. Y-25, which can grow at pH 7.0, was used to compare the contents of polyamine isolated from the cells cultivated at pH 10.0 and pH 7. As shown n Fig. 4.18, six ninhydrin-positive peaks were detected from the samples. Three of these peaks were cadaverine, spermine, and spermidine. The ratio of spermine and spermidine showed a dramatic change between pH 7 and 10. Spermidine was the major polyamine in alkaliphilic *Bacillus* sp. No. Y-25 cultivated at pH 10.0, and only a trace amount of spermine was detected. However, spermine was the major component in pH 7 cells. Recently, Hamana *et al.* (1989) confirmed our results, using a different alkaliphilic strain. Of three polyamines, putrescine, spermidine, and spermine, spermidine was the major polyamine in alkaliphilic bacilli and no spermine was detected. Furthermore, the amount of spermidine in *Bacillus alcalophilus* was reported to account for more than 90% of total polyamines (Chen and Cheng, 1988). The dissociation constants (pK) of the polyamines are in the range of 8.5 to 10.0. Each nitrogen of spermidine is protonated at pH 7.0 ($N^+H_3-(CH_2)_3-N^+H_2-(CH_2)_4-N^+H_3$), but only 30% of the nitrogen is protonated at slightly alkaline pH in cytoplasm of alkaliphiles. Therefore, the physiological function of spermidine in alkaliphiles is different from that of other extremophiles such as thermophiles. One reason for this may be that the cell membrane is stabilized by spermidine, which is a more basic compound than spermine.

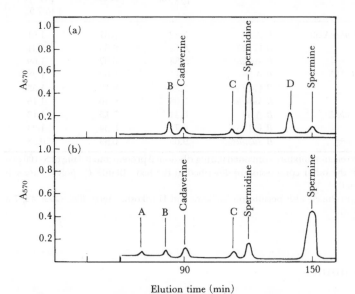

Fig. 4.18 Polyamines in alkaliphilic *Bacillus* sp. No. Y-25.
Polyamines extracted were analyzed with an automatic amino acid analyzer.
(a) The elution profile of polyamine from cells cultured at pH 10. (b) From pH 7 cells.
(Reproduced with permission from Horikoshi and Akiba, *Alkalophilic Microorganisms*, p. 91, Springer-Verlag: Japan Scientific Societies Press (1982))

It is known that ornithine is a precursor of polyamines in bacteria, and ornithine decarboxylase is a key enzyme; however, this enzyme was not detected in *B.alcalophilus*. Another pathway of polyamine synthesis is from arginine. This pathway in *B.alcalophilus* was revealed by Chen and Cheng (1988). They incubated the strain with a medium containing (^{3}H) arginine and found more than 90% of the radioactivity recovered in spermidine.

Recently, Hamasaki *et al.* (1993) reported distribution of polyamines in various alkaliphilic *Bacillus* strains. These results are shown in Table 4.23. During the course of their study, large amounts of 2-phenylethylamine were isolated from the culture broth of alkaliphilic *Bacillus* sp. YN-2000 described on page 288.

Table 4.23   Polyamine and 2-phenylethylamine contents of various alkaliphilic *Bacillus* spp.

Strain [†1]	Amount detected (nmol/g [wt/wt])[†2]			
	Put	Spd	Spm	2-PEA
Bacillus sp. YN-2000	25	2,100	65	800
Bacillus sp. YN-1	tr	109	4	ND
B. alkalophilus	tr	248	107	ND
B. firmus RAB RA-1	23	1,223	307	ND
B. firmus OF-4	29	2,167	467	ND

[†1] Strains YN-2000 and YN-1 were provided by Y. Nosoh, Iwaki-Meisei University, and *B. alkalophilus* (ATCC2764), *B. firmum* RAB RA-1 (Sm^r), and a homologous strain OF-4 were provided by Y. Imae, Nagoya University.
[†2] ND, not detected.
(Reproduced with permission from Hamasaki *et al.*, *Appl. Environ. Microbiol.*, **59**, 2720 (1993))

4.4.2   Polyamine Oxidases

Although we have not studied polyamine oxidase of alkaliphilic *Bacillus* strains, an alkaline-tolerant fungus, *Penicillium* sp. No. PO-1, which was isolated on an alkaline medium of pH 10.0, produced polyamine oxidase extracellularly (Kobayashi and Horikoshi, 1981; Kobayashi and Horikoshi, 1982). Polyamine oxidase catalyzes the oxidation of spermine and spermidine to form putrescine, 3-aminopropinaldehyde, and H_2O_2, as shown in Fig. 4.19. Polyamine oxidases are widely distributed in fungi, including *Penicillium* and *Aspergillus* species, but most of them are known to be intracellular enzymes.

Polyamine oxidase of *Penicillium* sp. No. PO-1 was inducible by the addition of spermine and spermidine, but these are expensive as inducers. As shown in Table 4.24, the most effective and cheapest inducer was 1,3-propanediamine, which had an optimum concentration of 0.2%. Thus the best medium for enzyme production was as follows: 0.2% 1,3-propanediamine, 0.1% glucose, 0.1% K_2HPO_4, and 0.05% $MgSO_4 \cdot 7H_2O$, at pH 5.5.

The enzyme was easily purified from culture fluid by chromatofocusing between pH 3 and 5 with a yield of over 90%. The properties of the enzyme are shown in Table 4.25. The enzyme protein consisted of two subunits, each of which had a molecular weight of about 64,000. The enzyme contained flavin adenine nucleotides and its activity was not

$$NH_2(CH_2)_3NH(CH_2)_4NH(CH_2)_3NH_2$$

Spermine

$$\downarrow \quad -H_2O+O_2$$

$$NH_2(CH_2)_3NH(CH_2)_4NH_2+NH_2(CH_2)_2CHO+H_2O_2$$

Spermidine   3-Aminopropion aldehyde

$$\downarrow \quad -H_2O+O_2$$

$$NH_2(CH_2)_4NH_2+NH_2(CH_2)_2CHO+H_2O_2$$

Putrescine   3-Aminopropion aldehyde

Fig. 4.19   Oxidation of spermine by polyamine oxidase.
(Reproduced with permission from Horikoshi and Akiba, *Alkalophilic Microorganisms*, p.140, Springer-Verlag: Japan Scientific Societies Press (1982))

Table 4.24   Effect of amines on the production of polyamino oxidase by *Penicillium* sp. No. PO-1

Amine	Activity (units/ml)
Ethylamine	0
Methylamine	0
n-Propylamine	0
n-Butylamine	10.1
n-Heptylamine	0
Benzylamine	0
n-Amylamine	8.1
Histamine	0
Tyramine	0
Tryptamine	0
Ethylendiamine	0
3,3'-Diaminodipropylamine	0
1,3-Propanediamine	36.7
1,4-Butanediamine (Putrescine)	10.6
1,5-Pentanediamine (Cadaverine)	15.6
1,6-Hexanediamine	7.5
1,7-Heptanediamine	0
1,8-Octanediamine	0
1,9-Nonanediamine	5.6
1,10-Decanediamine	0
Spermidine	11.2
Spermine	10.7

(Reproduced with permission from Horikoshi and Akiba, *Alkalophilic Microorganisms*, p.140, Springer-Verlag : Japan Scientific Societies Press (1982))

Table 4.25   Properties of polyamine oxidase from *Penicillium* sp. No. PO-1

Property	Polyamine oxidase
Optimum pH	4.0
Optimum temperature, °C	45
Molecular weight	135,000
Isoelectric point	4.5
Sedimentation constant	7.6
Stable pH[†1]	4.0-6.0
Stable temperature[†2], °C	Up to 45.
Spermine	
K_m, µM	5.3
V_{max}, µM/mg/min	13.3
Spermidine	
K_m, µM	8.2
V_{max}, µM/mg/min	16.4

[†1] At 37°C for 30 min.
[†2] At pH 4 for 10 min.

(Reproduced with permission from Horikoshi and Akiba, *Alkalophilic Microorganisms*, p.142, Springer-Verlag : Japan Scientific Societies Press (1982))

inhibited by thiol reagents such as *p*-chloromercuribenzoate, monoiodoacetate, heavy metals or metal chelating agents, but completely inhibited by phenylhydrazine.

The enzyme showed specific activity on spermine, spermidine, and its *N*-acetyl derivatives (*N*-and N^8-acetylspermidine) but no activity on mono- and diamines such as tyramine and putrescine.

Although the fungus we isolated was not truly alkaliphilic but alkaline-tolerant, it is very interesting from an industrial point of view that the polyamine oxidase, which had been thought to be a typical intracellular enzyme, was produced as an extracellular enzyme.

5

Molecular Biology

Apart from the isolation of mutants, little is known about the genetics and molecular biology of alkaliphiles. No detailed genetic map has been constructed except for an incomplete map of *Bacillus* sp. No. C-125 (Sutherland *et al.*, 1993*a*).

It is perhaps inevitable that the techniques of molecular biology, developed using other organisms, have been applied to alkaliphiles (mainly bacilli) and a significant number of genes from alkaliphiles have been cloned into alkaliphilic or neutrophilic hosts. This may have advantages such as increased levels of expression of the required enzyme in the host, alterations in the properties of the enzyme and excretion of normal intracellular enzymes into the medium. Many investigations on genes of extracellular enzymes and their expressions have been conducted, and the details are described in Chapter 7.

5.1   Alkaliphilic Microorganisms as DNA Sources

5.1.1   Secretion Vector

The author and his coworkers have tried to produce *Escherichia coli* strains which could secrete gene products from the cells. *E. coli* is widely used in genetic engineering experiments because it has been extensively studied and much is known about its genetics and biochemistry. Unfortunately, with the exception of a few proteins such as colicin (Zhang Faro and Zubay, 1985), cloacin (De Graaf and Oudega, 1986) and hemolysin (Mackman *et al.*, 1986), *E. coli* does not secrete gene products from the cell. If *E. coli* could be modified to secrete recombinant DNA products, it would be of considerable interest from an industrial point of view.

One of the most important processes in the fermentation industry is the extracellular production of proteins. The reasons are as follows: (1) If gene products remain in the cells, the products cannot exceed the maximum volume of cells. Also, the process of secretion allows for production from continuous culture, and substances which have an inhibitory effect to microbial metabolism may be produced as extracellular products from cells. (2) Usually, the number of proteins secreted from a cell is not so large, so that purification processes are relatively simple and can be used in fermentation industries.

A.   Cell Surface of Bacteria

Bacteria are classified into two groups, gram-positive and gram-negative, according to the nature of the cell surface. Gram-positive bacteria have a plasma membrane and a pep-

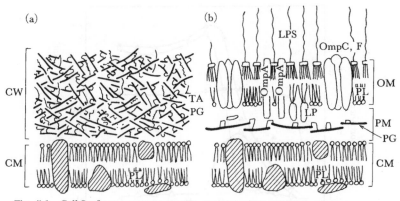

Fig. 5.1   Cell Surfaces
(a) Gram-positive bacteria, (b) Gram-negative bacteria
CM, Cytoplasmic membrane; OM, Outer membrane; PM, Periplasm;
PG, Peptidoglycan; TA, Teichoic acid; LPS, Lipopolysaccharide; PL,
Phospholipid

tidoglycan layer in the cell surface, while gram-negative bacteria have a plasma membrane (inner membrane), peptidoglycan and an outer membrane (Fig. 5.1). The plasma membrane is a phospholipid bilayer containing approximately 300 different proteins, many of which are involved in respiration, electron transport, nutrient uptake and membrane biogenesis. The outer membrane in gram-negative bacteria is a glycolipid-phospholipid bilayer, with the polysaccharide groups of the glycolipid being exposed on the outer surface of the cell. The periplasm contains approximately 100 proteins involved in nutrient uptake and catabolism. Protein molecules have signal peptides which allow for their secretion into either the culture medium in the case of gram-positive bacteria or into the periplasmic space in the case of gram-negative bacteria. In the case of gram-negative bacteria such as *E. coli*, the outer membrane acts as a barrier and the proteins are trapped in the periplasmic space.

B.   Changing *E. coli* to Become Permeable

A penicillinase gene from alkaliphilic *Bacillus* sp. No.170 was cloned in *E. coli* HB101 by using pM9. Plasmid-borne penicillinase was found to be produced in the culture medium (Kudo, Kato and Horikoshi, 1983), as shown in Fig. 5.2. The alkaliphilic *Bacillus* sp. No. 170, a penicillinase producer, was digested with *Hin*d III or *Eco*RI restriction enzymes and shotgun cloned in pMB9 by conventional means. Two plasmids, pEAP1 and pEAP2, were obtained from the transformants. The cleavage maps of pEAP1 and pEAP2 are shown in Fig. 5.3. The 2.4-kb *Hin*d III fragment containing the penicillinase gene was located in the middle of the 4.5-kb *Eco*RI fragment. The plasmid-encoded penicillinase was immunologically crossed with penicillinase III of alkaliphilic *Bacillus* sp. No. 170.

E. coli carrying plasmids were aerobically grown in LB-broth for 20 h at 37 °C and enzymatic activities determined. Most of the penicillinase produced by these cells was detected in the culture medium. Less than 15% of the total activity was observed in the periplasmic and cellular fractions. However, almost all the β-lactamase produced by *E. coli* HB101 (pBR322) was trapped in the periplasmic space. *E. coli* HB101 carrying pEAP2 was inoc-

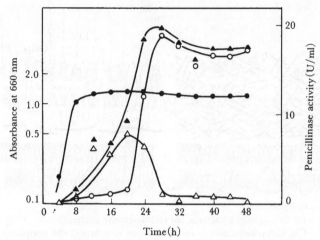

Fig. 5.2 Bacterial growth and penicillinase production by *Escherichia coli* HB101 (pEAP2). *E. coli* HB101 (pEAP2) was inoculated in LB-broth containing 0.2% glycerol and cultured at 37 °C on a rotary shaker. Bacterial growth (absorbance at 660 nm, ●) and penicillinase activities (extracellular, ○; intracellular, △; and total, ▲) were determined.

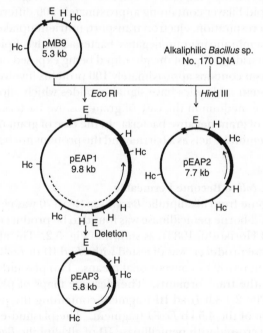

Fig. 5.3 Construction of plasmids pEAP1, pEAP2 and pEAP3 containing the *Bacillus* penicillinase gene. (——), vector DNA; (——), *Bacillus* DNA. Arrow denotes direction of transcription of the *Bacillus penicillinase* gene. E, *Eco* RI; H, *Hind* III; Hc, *Hinc* II.

(Reproduced with permission from Kobayashi *et al., J. Bacteriol.,* **166**, 729 (1986))

Table 5.1 Distribution of enzymes in *E. coli* HB101 carrying plasmids

Plasmids		Extracellular	Periplasmic	Cellular	Total
None	Protein	0.04 (4%)	0.07 (6%)	0.97 (90%)	1.08 (100%)
	APase	0.02 (2%)	1.05 (79%)	0.26 (20%)	1.33 (100%)
	β-gal	0.03 (2%)	0.00 (0%)	1.20 (98%)	1.23 (100%)
pMB9	Protein	0.01 (1%)	0.03 (4%)	0.75 (95%)	0.79 (100%)
	APase	0.01 (2%)	0.32 (67%)	0.15 (31%)	0.48 (100%)
	β-gal	0.00 (0%)	0.00 (0%)	0.80 (100%)	0.80 (100%)
pEAP2	Protein	0.18 (21%)	0.12 (14%)	0.55 (65%)	0.85 (100%)
	PCase	10.70 (83%)	0.40 (3%)	1.80 (14%)	12.90 (100%)
	APase	0.29 (58%)	0.15 (30%)	0.06 (12%)	0.50 (100%)
	β-gal	0.09 (10%)	0.03 (3%)	0.78 (87%)	0.90 (100%)

E. coli strains were aerobically grown in the LB-broth for 20 h at 37 °C. Enzymatic activities of alkaline phosphatase (APase) and β-galactosidase (β-gal) are expressed as absorbance at 420 nm. Penicillinase (PCase) activity is expressed as units per milliliter of broth. Protein concentration is expressed as milligrams in 1 ml of broth.
(Reproduced with permission from Kudo, Kato and Horikoshi, *J. Bacteriol.*, **156**, 951 (1983))

ulated in 500-ml flasks containing 100 ml of LB-broth with 0.2% glycerol and cultured at 37 °C. As shown in Fig. 5.2, the bacteria reached maximum cell concentration at 16 h, and no lysis of the cells was observed up to 48 h (viable counts $3\text{-}2 \times 10^9$). The extracellular penicillinase activity increased at about 24 h and reached maximum at 28 h. No intracellular penicillinase was observed after 28-h cultivation. The distribution of β-galactosidase and alkaline phosphatase activities were also investigated (Table 5.1). Essentially, neither protein nor enzymatic activity was detected in the culture broth of *E. coli* HB101 or *E. coli* HB101 (pMB9). About two thirds of the alkaline phosphatase was observed in the periplasmic space and almost all of the β-galactosidase was in the cellular fraction. On the other hand, in *E. coli* HB101 (pEA2) it is striking that 21% of the total protein, 58% of the alkaline phosphatase and 83% of the penicillinase were found in the culture broth. About 87% of the activity of β-galactosidase, a typical intracellular enzyme, was detected in the cellular fraction and not in the periplasmic fraction. These results suggest that the outer membrane of *E. coli* was changed by the introduction of pEAP2 into the cells, because a periplasmic enzyme, alkaline phosphatase, was released from the periplasmic space.

The fragment contains the penicillinase gene and an open reading frame of the *kil* gene with a promoters. Our results showed that the activation of the *kil* gene of pMB9 by a promoter (*Ex* promoter) in the inserted fragment increased outer membrane permeability (Kobayashi, Kato *et al.*, 1986). Insertional mutation in the −35 region of the *Ex* promoter inhibited secretion. High resolution S1 nuclease mapping indicated that the *kil* gene was activated by the *Ex* promoter, which had a TAT TAT sequence at −10 and TTGA-TA at −35 (Fig. 5.4). The penicillinase itself was not responsible for this secretion.

C. Role of the *kil* Gene

Penicillinase was not responsible for protein secretion. The *Ex* promoter is not very strong, being almost the same as the tetracycline promoter of pMB9. The *kil* gene and *Ex* promoter are necessary for the excretion of penicillinase from the cells. The *kil* gene of

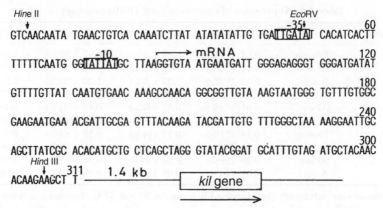

Fig. 5.4 Nucleotide sequence of the *Ex* promoter isolated from alkaliphilic *Bacillus* sp.
170. The putative -35 and -10 are boxed. The location and direction of the
kil gene are indicated. The space between the *Hin*d III site and the *kil* gene
is approximately 260 kb in pEAP37.
(Reproduced with permission from Kobayashi *et al.*, *J. Bacteriol.*, **166**, 730
(1986))

ColEl is required for mitomycin-induced lethality and release of colicin from the cells.
However, under our experimental conditions, *kil* gene did not kill the *E. coli* cells but
instead made the outer membrane permeable. Aono (1988*a*) succeeded in isolating the
kil gene proteins from the *E. coli* cell envelope. One of the peptides (M_r 4800), which was
probably a precursor peptide, was detected in the inner-membrane fraction from the
organism when envelope proteins were subjected to differential solubilization. The other
(M_r 3500), which was a mature peptide, was detected in the outer-membrane fraction of
the organism. The mature peptide was only detected in the envelope of cells releasing
the penicillinase transiently accumulated in the periplasm into the culture medium. As
shown in Fig. 5.5, the *kil* gene exhibits partial homology with the gene for Braun's lipoprotein detected in the outer membrane of *E. coli*, not only at the level of the DNA sequence
but also at the level of the amino acid sequence. Therefore, the product of the *kil* gene
may act as a perturbator of the cell membrane.

Aono (1988*b*; 1989*b*) found that the extracellular production of alkaliphilic *Bacillus*
penicillinase by *E. coli* HB101 carrying pEAP31 was affected by several parameters such as
concentration of carbohydrates and NaCl, pH values of culture broth, culture temperature and aerobic conditions. He noted that the most critical condition is culture temperature, and no secretion was observed at temperatures lower than 26 °C. Cultivation at various temperatures (22 to 32 °C) revealed that the penicillinase accumulated in the
periplasmic space of *E. coli* at 22 °C was released from the cells at 32 °C (Table 5.2). The
following conditions for the production of extracellular penicillinase by *E. coli* HB101 carrying pEAP31 were recommended. The organism should be inoculated in 200 to 300 ml
of LG-broth [10 g Bactotryptone (Difco), 5g Bacto yeast extract (Difco), 2g glycerol, 1g
glucose and 10 g NaCl in 1000 ml of deionized water] in a 500-ml volume cultivation flask
and then shaken at 30 °C on a high speed reciprocal shaker (172 oscillation/min with 3.2-
cm strokes). Selection of cultivation conditions is the most important factor in producing extracellular enzymes. *E. coli* HB101 carrying pEAP31 grown at 30 °C released the

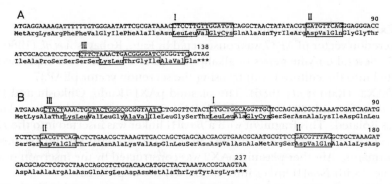

Fig. 5.5 Comparison of the nucleotide and the derived amino acid sequences of the *kil* gene (A) and the lipoprotein gene (B). Homologous regions I, II, and III, are boxed.
(Reproduced with permission from Kobayashi *et al., J. Bacteriol.*, **166**, 731 (1986))

Table 5.2 Distribution of enzymes in *E. coli* HB101 carrying pEAP 31 at various temperatures

Cultivation Temper -ature	Total PCase (U/ml)	Distribution of PCase(%)				Release from cells			Growth A660
		Extra	Intra	Peri	Cyto	APase (%)	β-gal (%)	Protein (μg/ml)	
16°C	73	1	99	87	12	1	ND	36	5.3
20	174	1	99	87	12	2	ND	36	5.7
22	181	2	98			2	ND		5.5
24	206	5	95	88	7	4	2	51	5.7
26	206	17	83			11	ND		5.6
28	183	66	34	31	4	42	1	186	4.4
30	155	81	19	17	2	55	14	339	3.4
32	148	95	5	3	2	85	26	468	2.3
36	102	94	6	4	2	73	37	470	2.3
38	88	75	25	20	5	66	11	250	3.0
40	55	85	15	10	5	67	7	273	3.2
43	<1					43	ND	44	2.6

outer membrane proteins, lipopolysaccharide and phosphatidylethanolamine, as well as penicillinase into the culture medium. Aono analyzed the fatty acid content in *E. coli* HB101 carrying or not carrying pEAP31 but observed no difference. Recently, Aono reported that the penicillinase that accumulated in particular subcellular fractions of *E. coli* grown under different conditions was purified and characterized. Periplasmic or extracellular penicillinase (24 kDa) was mature protein, indicating that the putative precursor (27 kDa) was processed at the correct amino acid residue, probably by signal peptidase I. Cytoplasmic penicillinase contained two unusual proteins (25 kDa) produced by proteolytic cleavage of the precursor within its signal sequence (Aono, 1992*b*).

D. Extracellular Production of Microbial Enzymes from *E. coli*

An excretion vector pEAP37 was constructed by Kato, Kobayashi *et al.* (1986) as shown in Fig. 5.6. Several enzyme genes of alkaliphilic *Bacillus* strains were expressed in *E. coli* and secreted into the culture broth by using the secretion vector pEAP37.

(1) p7AX2: (Kato *et al.*, 1986): The plasmid pAXI (Kudo, Ohkoshi and Horikoshi, 1985) was digested with *Bgl* II and a 4.0-kb DNA fragment containing the xylanase gene (xyl-L) was isolated. This fragment with a *Bam*HI linker was inserted into the *Hinc* II site of pEAP37. p7AX2 was isolated from the Cm-resistant, Ap-sensitive and xylanase-producing transformants. Another plasmid pAX2 was constructed by the insertion of the 4.0-kb DNA fragment with *Bam*HI linker into the *Bam*HI site of pBR322.

(2) p7NK1 and p7FK1: Plasmids pNK1 (Sashihara, Kudo and Horikoshi, 1984) and

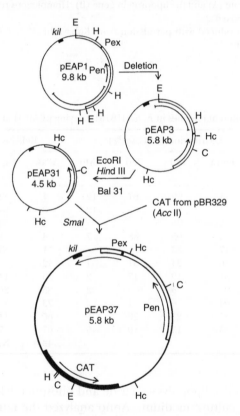

Fig. 5.6 Construction of pEAP37. The thin line represents the DNA region derived from pMB9. The open bar represents the DNA fragment containing the penicillinase gene from an alkaliphilic *Bacillus* sp. and the black bar represents the DNA fragment containing the chloramphenicol-resistant gene (CAT) derived from pBR329. The location of the *kil* gene and *Ex* promoter are indicated by a black box. The dotted lines in pEAP1 and pEAP3 indicate the deletion regions. Abbreviations: C, *Cla* I; S, *Sma* I.
(Reproduced with permission from Kato *et al.*, *FEMS Microbiol. Lett.*, **36**, 32 (1986))

Table 5.3   Distribution of the xylanases and cellulases in E. coli

Plasmid	Extracellular fraction	Periplasmic fraction	Cellular fraction	Total activity
pAX2	23 (4.8)	147 (30.4)	315 (64.8)	485
p7AX2	135 (62.1)	8 (3.6)	74 (34.3)	217
pNK1	12 (9.2)	60 (44.0)	64 (46.8)	136
p7NK1	282 (66.7)	92 (21.8)	49 (11.5)	423
pFK1	7 (6.1)	88 (74.3)	23 (19.6)	118
p7FK1	93 (60.2)	16 (10.6)	45 (29.2)	154
pAX1	10 (1.6)	290 (47.9)	350 (50.5)	605
pXP102	120 (88.8)	10 (7.4)	5 (3.8)	135

Note: The activities (units) of the enzymes are presented in each column.
The percentages of the total activities are shown in parentheses.
(Reproduced with permission from Kato et al., FEMS Microbiol. Lett., **36**, 33 (1986))

pFK1 (Horikoshi et al., 1984; Fukumori, Kudo and Horikoshi, 1985; Fukumori, Kudo et al., 1986; Fukumori et al., 1986) containing cellulase genes of different alkaliphilic Bacillus strains have been found in our laboratory. A 2.0-kbp Hind III fragment from pNK1 and a 3.0-kb Hind III-Hinc II fragment from pFK1, each containing cellulase genes, were isolated. These fragments were filled in and ligated with Hinc II-digested pEAP37. Plasmids p7NK1 and p7FK1 were constructed (Kato et al., 1986; Kato et al., 1986).

E. coli HB101 carrying each constructed plasmid was cultured aerobically in LB-broth at 37 °C. After 24 h, the cells were harvested by centrifugation and fractionated into extracellular, periplasmic and cellular fractions. Most of the xylanase and cellulase activities were detected in the extracellular fractions (Table 5.3). In control experiments using E. coli carrying pBR322 with inserts coding for xylanase and cellulase, the enzymatic activities remained in the periplasmic and cellular fractions.

(3) Georganta et al. (1991) investigated the expression level of CGTase gene of alkaliphilic Bacillus No. 38-2 in various hosts. E. coli carrying pEAP85-CGT excreted 70% of total CGTase (500 U/ml) into the culture broth. This amount was higher than that produced by the parent strain. Bacillus subtilis carrying pGK-1 produced a very low level of CGTase (less than 0.1 U). These results were only from laboratory scale experiments and a stock strain was used. In Japan, genetically engineered alkaliphilic Bacillus No. 38-2 has not been used in industrial scale plants. Mutagenesis of the strain and optimization of the culture conditions made the enzyme production much higher than 500 U/ml.

E.   Extracellular Production of Human Proteins
1.   Human growth hormone (hGH) (Kato et al., 1987)

A Dra I fragment containing the promoter and the structure gene isolated from pEAP2 was digested by Rsa I and connected with a Hind III linker. The fragment (PS fragment) thus obtained contained the promoter and the whole signal peptide sequence of penicillinase. This fragment was ligated with hGH gene, inserted into the Hind III site of pBR322 and introduced into E. coli. The transformant (Ap^r, Cm^s) was cultured in LB-broth for 24 h at 37 °C. The distribution of the hGH produced was analyzed by radioim-

munoassay. More than 90% of the hGH was detected in the periplasmic space of *E. coli.*
The fragment having the PS fragment and hGH gene was inserted into the *Hinc* II site of
the secretion vector pEAP37 and introduced into *E. coli.* The transformant was cultured
by the method described above. About 60–70% of total hGH activity was detected in the
culture broth. Western blotting indicated that the protein produced by the plasmid was
immunologically the same as that of the authentic hGH sample. The yield of hGH in the
culture broth was about 60–70 mg/l.

  2. Production of the Fc fragment in the culture broth

  The Fc fragment (Fig. 5.7) is a protein fraction crystallized from the papain
hydrolyzate of immunoglobulin. Its molecular weight is 50,000 and it has two intramole-
cular disulfide bonds. The Fc fragment has several biological activities. The DNA fragment of the Fc structure gene was ligated with a DNA fragment coding the penicillinase
signal peptide isolated from alkaliphilic *Bacillus* sp.No.170 and inserted into the secretion
vector. After plasmid was introduced into *E. coli* HB101 and cultured in the LB-broth for

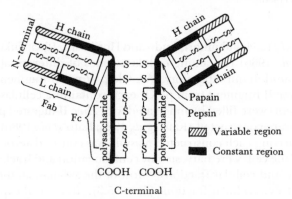

Fig. 5.7 Human immunoglobulin.

Time (hrs)		6	1 2	1 8	2 4	3 0
h IgG - Fc	E					
	P					
	C					
Alkaline phosphatase (%)	E	0	0	3 6	8 3	8 2
	P	7 3	3 8	5 0	4	4
	C	2 7	6 2	1 4	1 3	1 4
β - galactosidase (%)	E	0	0	2	1 3	1 7
	P	3	0	2	3	9
	C	9 7	1 0 0	9 6	8 4	7 4

Fig. 5.8 Production of hIgG-Fc in the culture broth.
(Reproduced with permission from Nakamura *et al., Appl. Microbiol.
Biotech.,* **28**, 55 (1988))

24 h, more than 60% of the gene products was detected in the culture broth (Fig. 5.8). Without the signal sequence, no gene product was detected in either the periplasmic space or the culture broth (Kitai *et al.*, 1988). In conclusion we believe that the secretion vector pEAP37 may prove to be useful industrially for extracellular production of proteins from genetically modified *E. coli.*

5.1.2 Promoters

Promoters of *Bacillus* strains have been studied from the industrial point of view for the hyperproduction of extracellular enzymes, such as alkaline proteases and alkaline cellulases.

A.   *Ex* Promoter of Alkaliphilic *Bacillus* sp. No.170

As described in section 5.1.1C, Kobayashi *et al.* (1986) isolated the *Ex* promoter, which had a TATTAT sequence at -10 and TTGATA at -35 (Fig. 5.4). This promoter is not so strong, but this was the first finding of a promoter of alkaliphiles. In order to develop a more useful system for extracellular protein production from *E. coli*, Murakami *et al.* (1989) have constructed the new excretion vectors, pEAP82-1, 82-2, and 82-3, which are derivatives of the secretion vector pEAP37. These vectors have, respectively, a single, double, and triple *Ex* promoter upstream from a penicillinase structural gene; *E. coli* HB101 carrying pEAP82-2 or 82-3 produced, respectively, about two or three times as much penicillinase protein than that produced by *E. coli* carrying pEAP82-1, and 70% to 80% of the protein was excreted into the culture medium. The *E. coli* carrying pEAP82-3 was cultivated at various temperatures and the optimum for extracellular penicillinase production was found to be 30–33 °C. Using this multipromoter excretion system the amount of extracellular production of human growth hormone was increased several fold as observed with penicillinase excretion.

B.   Promoter of *Bacillus* sp. KSM-635

During studies of industrial production of alkaline cellulase for laundry detergents, Ozaki *et al.* (1990) isolated a gene for alkaline cellulase from the alkaliphilic *Bacillus* sp. KSM-635. There was an open reading frame (ORF) of 2823 bp, which encoded 941 amino acid residues. It was noteworthy that two putative ribosome binding sites and a sigma 43-type, promoter-like sequence were found upstream from an initiation codon in the ORF. Such a multiple promoter system may produce a large amount of alkaline cellulase in an industrial scale plant.

C.   A Conditionally Regulated DNA Fragment

Alkaliphilic *Bacillus* strains sporulate well at pH 10, which is also the optimum pH for germination of the spores. Almost all extracellular enzymes are fully produced at the stationary phase. For alkaline cyclodextrin glycosyltransferase production, the alkaliphilic *Bacillus* strain No. 38-2 must be grown for at least 48-72 h. Why are these extracellular enzymes produced at particular times? One approach for analyzing the regulation mechanisms of production of extracellular enzyme is to clone elements that control developmental gene expression using expression probe vectors. One of these expression vectors is pGR71 (Goldfarb, Rodoriguez and Doi, 1982). In *B. subtilis*, expression of the chloramphenicol acetyltransferase (CAT) gene carried by pGR71 requires insertion of *Bacillus* promoters and ribosomal binding sites into the *Hin*d III site immediately upstream from the CAT gene. In order to screen for insertionally activated and developmentally induced gene expression, Kudo *et al.* (1985) cloned fragments and examined their CAT activities during the vegetative and sporulation phases of growth. Fragments which were controlled by temperature, pH or NaCl concentration in the pGR71 CAT expression system were also isolated.

Table 5.4 Strains and vectors used

Strain or plasmid	Genotype or property	Source
Alkaliphilic *Bacillus* sp. No. 38-2	Produces an alkaline amylase	Author's laboratory
B. subtilis		
NIG1121	*met his spo*[†]	Y. Sadaie
NIG1131	*met his spo OA*34	Y. Sadaie
NIG1132	*met his spo OB*36	Y. Sadaie
NIG1133	*met his spo OC*7	Y. Sadaie
NIG1134	*met his spo OD*8	Y. Sadaie
NIG1139	*met his spo OE*81	Y. Sadaie
NIG1140	*met his spo OF*221	Y. Sadaie
NIG1135	*met his spo OG*14	Y. Sadaie
NIG1136	*met his spo OH*17	Y. Sadaie
NIG1137	*met his spo OJ*87	Y. Sadaie
NIG1138	*met his spo OK*141	Y. Sadaie
IS17	*trp* C2 *phe* A1 *spo OE*11	BGSC[†]
IS22	*trp* C2 *rpo* B2 *spo OH*17	BGSC
IS24	*trp* C2 *phe* A1 *spo OH*81	BGSC
1012	*lec* A8 *met* B5 *hsr M*1	H. Honda
pGR71	Km^r	D.S. Goldfarb

[†] BGSC, *Bacillus* Genetic Stock Center, Ohio State University, Columbus.

Bacterial strains and vectors	Genotypes or properties
Alkaliphilic *Bacillus* No. 221	Produces an alkaline protease
Alkaliphilic *Bacillus* No. 38-1	Produces an alkaline amylase
E. coli CV512	*leu* A371
E. coli HB101	*leu* B, *pro, thi, rec* A, *hsd* R, *hsd* M, *str* A
E. coli CV520	*leu* C171
E. coli CV526	*leu* D101
B. subtilis 1012	*leu* A8, *met* B5
B. subtilis GU134	*leu* B6, *trp* C2
B. subtilis GU741	*leu* C7, *trp* C2
B. subtilis GU229	*ilv* B2, *trp* C2
B. subtilis DB-1	Wild type strain
Cl. butyricum M588	Used as gastrointestinal tonic
Plasmid pBR322	Amp^r, Tc^r
Plasmid pGR71	Km^r

1. Media and strains used

The following media were used throughout the experiments: Medium $2 \times$ SSG (modified Schaeffer medium) containing 16 g of nutrient broth (Difco Laboratories), 0.5 g of $MgSO_4 \cdot 7H_2O$ and 2 g of KCl; after autoclaving, $Ca(NO_3)_2$, $MnCl_2$, Fe_2SO_4 and glucose were added to concentrations of 1 mM, 0.1 mM, 1 µM and 0.1%, respectively. Alkaliphilic *Bacillus* No. 38-2 (ATCC 21783) was grown aerobically at 37 °C to early log phase in Horikoshi-II medium for DNA extraction. Other strains and plasmids used are listed in Table 5.4.

2. Construction of recombinant plasmids

Chromosomal DNA of alkaliphilic *Bacillus* No.38-2 and plasmid DNAs were completely digested with *Hind* III at 37 °C. After digestion, 1 µg of pGR71 and 3 µg of chromosomal DNA were ligated with T_4 ligase overnight at room temperature. Competent *B. subtilis* 1012 cells were transformed with the ligation mixture described above. After 90 min of incubation to allow expression of the pGR71 kanamycin resistance the cells were plated onto LB plates containing kanamycin (20 µg/ml). To analyze the chloramphenicol acetyltransferase (CAT) activity profile, *B. subtilis* carrying the pGR71 derivatives was cultivated in the $2 \times$ SSG sporulation medium containing 10 µg/ml kanamycin. The culture was sampled at various stages of growth and sporulation, and the cell extracts were prepared by sonication for CAT enzyme activity. As shown in Fig. 5.9, the CAT activity profile of pGR71-5 indicated that CAT expression from the plasmid was induced after the cessation of vegetative growth and that CAT specific activity increased until late in sporulation. The expression of CAT activity of pGR71-5 was strongly suppressed by the addition of glucose in the $2 \times$ SSG medium. This expression was dramatically decreased in *spoOE* and *spoOH* mutants, which had a very low level of CAT activity, as shown in Fig. 5.10. This loss of activity was a function of the host cell genotype and not due to plasmid rearrange-

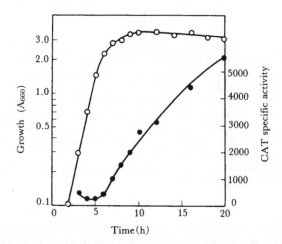

Fig. 5.9 CAT activity as a function of growth and sporulation in *B. subtilis* 1012 harboring plasmid pGR71-5. Symbols: ○, cell growth; ●, CAT specific activity. One unit of CAT specific activity was defined as 1 nmol of DTNB reduced per min per mg of total soluble protein. A_{660}, absorbance at 660nm.
(Reproduced with permission from Kudo, Ohkoshi and Horikoshi, *J. Bacteriol.*, **161**, 159 (1985))

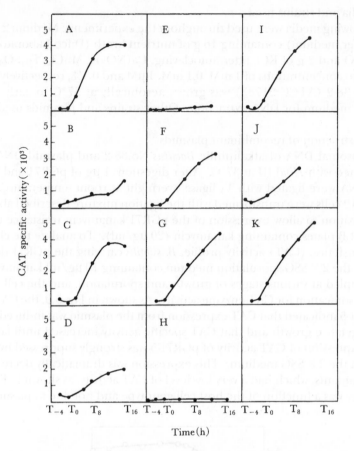

Time(h)

Fig. 5.10   Effect of various *spoO* mutations on CAT expression. *B. subtilis* cells carrying pGR71-5 were inoculated into 500ml of medium 2 × SSG and cul-
tured at 37 °C on a rotary shaker. CAT activity was measured at different
times during the growth and sporulation phases.
(A) Strain NIG1131 (*spoOA*). (B) Strain NIG1132 (*spoOB*). (C) Strain
NIG1133 (*spoOC*). (D) Strain NIG1134 (*spoOD*). (E) Strain NIG1139
(*spoOE*). (F) Strain NIG1140 (*spoOF*). (G) Strain NIG1135 (*spoOG*). (H)
Strain NIG1136 (*spoOF*). (I) Strain NIG1137 (*spoOJ*). (J) Strain NIG1138
(*spoOK*). (K) Strain NIG1121 (*spo*+).
(Reproduced with permission from Kudo, Ohkoshi and Horikoshi, *J.
Bacteriol.*, **161**, 160 (1985))

ment. When pGR71-5 was isolated from *spoOE* or *spoOH* strains carrying the plasmid and
reintroduced into the *spo*+ strains, the high CAT activity characteristic of the original
pGR71-5 isolate was restored. The nucleotide sequence of the entire 50-bp fragment was
determined and the site of regulated transcription initiation located by high resolution S1
nuclease mapping of the *in vivo* transcript (Fig. 5.11). From the results, the promoter
sequences were deduced to be 5'CGAATCATGA3' at −10 and 5' AGGAATC3' at
−35. This transcript was not detected in either *spoOE* or *spoOH* mutants, indicating that

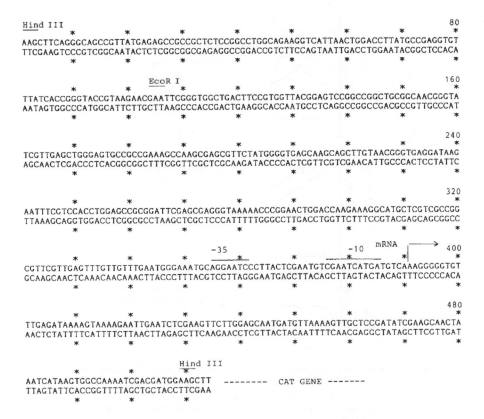

Fig. 5.11 DNA sequence of the pGR71-5 *Hin*d III fragment. The arrow indicates the probable transcription initiation site. The putative -35 and -10 regions are indicated.
(Reproduced with permission from Kudo, Ohkoshi and Horikoshi, *J. Bacteriol.*, **161**, 161 (1985))

these gene products control the developmentally regulated CAT expression at the transcription level.

In conclusion, the *Hin*d III fragment of pGR71-5 was developmentally regulated in the sporulation medium of *B.subtilis* and controlled by the *spoOE* and *spoOH* loci. The authors reported that the promoter obtained from alkaliphilic *Bacillus* resembled the promoters recognized by SigH (sigma-30) of *B.subtilis* having 5'GAATTNNT3' at −10 and 5'GCAGGANTT3' at −35 and that this promoter was under glucose and *spoO* control in a heterologous system.

D. DNA Fragments Controlled by Temperature, pH or NaCl Concentration in pGR71 CAT Expression

Alkaliphilic *Bacillus* DNA banks made by strains No. 38-2, No.17-1 and No.221, cloned in the expression probe plasmid pGR71, have been screened for the presence of genetic elements regulated by temperature, pH or NaCl concentration (Yoshitake *et al.*, 1984). Several kinds of regulatory elements whose expression was controlled by temperature, pH

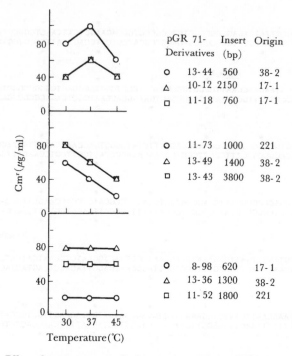

Fig. 5.12 Effect of temperature on Cm^r gene expression in pGR71 derivatives.
The Cm^r was determined as described in the text. 38-2, 17-1 and 221 indicate alkaliphilic *Bacillus* sp. No. 38-2, 17-1 and 221, respectively.
(Reproduced with permission from Yoshitake *et al.*, *Agric. Biol. Chem.*, **48**, 2624 (1984))

or NaCl concentration were obtained. As shown in Fig. 5.12, in *B.subtilis* the 1-kb *Hind* III fragment (pGR71-73) of alkaliphilic *Bacillus* sp. No. 221 was shown to decrease CAT activity at higher temperatures. However, CAT activity of *B.subtilis* carrying plasmid pGR71-13-36 (1.3-kb *Hind*III fragment of alkaliphilic *Bacillus* sp. No. 38-2) was not changed by temperature.

The experiments reported are very primitive, but these studies paved ways of expression of genes under extreme conditions such as high pH value, temperature and pressure. (Kato *et al.*, 1994). Although the experiment of Kato *et al.* has not been done on alkaliphiles, hydrostatic pressure may regulate the expression of genes in microorganisms. Recombinant plasmids with the chloramphenicol acetyltransferase (CAT) structural gene behind several kinds of promoters were tested for expression in *Escherichia coli* during growth at atmospheric pressure (0.1 MPa) and at high pressure (30 MPa). Expression of the CAT gene from the *lac* promoter was remarkably activated (approx. 78-fold) by high pressure in the absence of the inducer isopropyl-β-D-thiogalactopyranoside (IPTG). The stimulation of the CAT activity by the *lac* promoter at high pressure did not simply result from an increased plasmid copy number.

5.2   Host-Vector Systems of Alkaliphiles

There is no genetic information on alkaliphiles before 1983. The first recombinant DNA experiment in this field was the cloning and expression of leucine genes of alkaliphilic *Bacillus* strains in *E. coli*. Since then, our interest has focused on genetic engineering of alkaliphilic bacteria to study alkaliphily. One of the characteristic properties of alkaliphiles is pH homeostasis in the cell. Intracellular pH values are maintained around 7-8, although pH values of the culture media are 9.5-11.0. This pH difference is caused by the cell surface. In order to analyze the genetic properties of alkaliphilic *Bacillus* strains we had to make host-vector systems. If suitable host-vector systems of alkaliphilic *Bacillus* strains are established, more information on alkaliphily can be obtained.

We carried out tedious experiments, such as investigation of the longevity of alkaliphiles in nature, toxicity, etc., to receive approval of the Japanese government. We had to wait three years to start recombinant DNA experiments on alkaliphilic *Bacillus* strains. Our first task was the construction of host-vector systems, for which we mapped out a strategy. Host *Bacillus* strains must have the following properties: good growth in synthetic media, good protoplast formation and regeneration, and high frequency of transformation. If possible, the host should be *rec* minus. Vectors must be easy to introduce into hosts and be genetically stable.

5.2.1   Selection of Host Strains

The following alkaliphilic strains were used in this study. Strain No. 2b-2 is a strict alkaliphile that belongs to group 1. Strain Nos. A-59 and C-125 are facultative alkaliphiles and belong to group 2. Strain No. M-29 is a facultative alkaliphile belonging to group 3. All of them grow very well at pH 10.3 after 18 h incubation at 37 °C either in liquid or on solid media.

Strain Nos. 2b-2 and M-29 could not grow well in Horikoshi-I or -II medium in the absence of yeast extract. However, *Bacillus* sp. Nos. A-59 and C-125 grew well in the absence of yeast extract. Kudo *et al.* (1990) made a synthetic medium containing 0.2% glutamate and 0.5% glycerol for these strains. In 1992, Aono developed a novel minimal medium containing citric acid, glucose and ammonium sulfate for No. C-125 (Aono *et al.*, 1992). Alkaliphilic *Bacillus* strain No. A-59 (ATCC21591) and C-125 (FERM. No. 7344) can grow well in both media in a range of pH 7.5 to 11. Further details on the minimal media are given in Chapter 2.

5.2.2   Stability of Antibiotics in Horikoshi-I Medium

In the field of microbial genetics, many antibiotics have been used as resistant markers in order to concentrate or screen mutants having antibiotic-resistant genes. However, under alkaline environments several antibiotics so far tested are very unstable and decompose after 2-3 days incubation at 37 °C. Furthermore, many alkaliphilic microorganisms can raise the pH values of the media during cultivation, even if they are inoculated into conventional neutral media.

Stability of the five most popular antibiotics in microbial genetics was tested in media of varying pH values (Usami, Kudo and Horikoshi, 1990). We have many alkaliphilic *Bacillus* strains which can grow in the range of pH 7.0 to 11, and seven strains, including

Table 5.5   Growth at various pH values in the presence of antibiotics

Antibiotic	pH values			
	7.6	8.5	9.5	10.2
Tetracyline				
50 μg/ml	−	−		+
100	−	−	+	+
250	−	−	+	+
Ampicillin				
50 mg/μl	+	+	+	+
100	+	+	+	+
250	−	−	+	+
Chloramphenicol				
20mg/μl	−	−	−	−
40	−	−	−	−
100	−	−	−	−
200	−	−	−	−
Kanamycin				
40	−	+	+	+
100	−	+	+	+
200	−	+	+	+

Symbols : − no growth, + growth. Media used : nutrient agar for pH 7.6, others were modified Horikoshi-I medium. Growth was observed after 3days incubation at 37 ℃.

Bacillus sp. No. A-59 (ATCC21591) and No. C-125 (FERM. No. 7344), were tested. One of them, alkaliphilic *Bacillus* sp. No. 13 (ATCC 31006), which is a producer of alkaline amylase (see p. 162), is shown in Table 5.5 as an example. Our results indicated that chloramphenicol can be used as a resistance marker in the field of genetics of alkaliphilic *Bacillus* strains. Neither tetracycline nor ampicillin could be used.

Nutrient requirements and chloramphenicol-resistance are useful markers for the alkaliphilic host.

5.2.3   Preparation of Mutants

Mutants of alkaliphilic strains can be readily obtained by conventional mutagenesis at pH 7.5-8, *e.g.*, with nitrosoguanidine or ethyl methanesulfonate, or spontaneous mutagenesis. (Kudo *et al.*, 1990; Aono and Ohtani, 1990). Auxotrophic mutants were concentrated by the penicillin-screening method (1-3 mg/ml) at pH 8, then selected by the replica method using a synthetic alkaline medium. A spontaneous streptomycin-resistant mutant was isolated from one of the auxotrophic mutants on a neutral complex medium containing streptomycin (1 mg/1). Details are given in Chapter 6.

5.2.4   Preparation of Stable Protoplasts, Transformation and Regeneration

Alkaliphilic *Bacillus* strains are shaped in the form of rods by rigid layers of the Alγ-type of peptidoglycan, which can be readily hydrolyzed with egg white lysozyme. The first protoplast transformation of alkaliphilic *Bacillus* sp. by plasmid DNA was reported by Usami, Kudo and Horikoshi (1990). An overnight culture of the host strain, alkaliphilic *Bacillus* sp. No. A-59 (ATCC21591) (leu⁻, Cm^S) was diluted 20-fold with 50 ml of Horikoshi-II medium and grown at 37 °C with shaking at an early log phase of growth ($A_{660} = 0.4$). The cells were harvested and resuspended in 5 ml of ASMMP buffer. Lysozyme was added to a final concentration of 500 mg/ml and the suspension was incu-

Table 5.6 Effect of molecular weights of PEG on transformation efficiency.

PEG	Average mol wt	No. of transformants (per µg DNA)
None	—	0
PEG 200	200	5.0×10^3
PEG 1000	1000	3.6×10^5
PEG 2000	2000	4.0×10^5
PEG 4000	3000	7.2×10^5
PEG 8000	7000	1.1×10^6
PEG 20000	20000	4.8×10^5

The final concentration of each PEG used for transformation was 22.5%. pHW1 DNA prepared from alkaliphilic *Bacillus* sp. A-59 was used as the donor plasmid.
(Reproduced with permission from Usami, Kudo and Horikoshi, *Starch/Stärke*, **46**, 231 (1990))

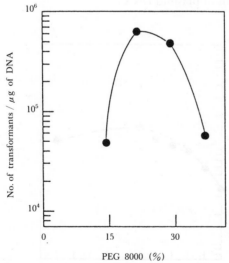

Fig. 5.13 Effect of concentration of PEG8000 on transformation efficiency. The concentration indicated in the figure is the final concentration in the reaction mixture. pHWI DNA prepared from *Bacillus* sp. A-59 was used as the donor plasmid.
(Reproduced with permission from Usami *et al.*, *Starch/Stärke*, **46**, 231 (1990))

bated at 37 °C with gentle shaking. ASMM buffer consisted of 0.54 M sucrose, 0.02 M maleate and 0.02 M MgCl$_2$ (pH 7.2). ASMMP medium was prepared by mixing equal volumes of 4 × strength Penassay broth (Difco) and 2 × strength ASMM. Protoplasts were collected by centrifugation at 2,600 × g for 20 min, washed once with ASMMP buffer, and resuspended in 5 ml of the same buffer. Plasmid pHWl (0.1–1 mg) was mixed with 0.5 ml of the protoplast suspension, then 1.5 ml of 30% polyethylene glycol (PEG) 8000 solution (Table 5.6 and Fig. 5.13) was added. After 5 to 10 min incubation, 5 ml of ASMMP buffer was added, and the protoplasts were collected by centrifugation. The collected protoplasts were resuspended in 1 ml of ASMMP containing 0.3% NaCl and incubated for 15 h at 30 °C. Transformants were selected on a modified DM-3 regeneration plate containing 20 µg/ml of chloramphenicol. The modified DM-3 regeneration medium (pH 7.3) consisted of 0.8% agar, 13.5% sodium succinate, 0.5% casamino acid (Difco), 0.5% yeast extract (Difco), 0.35% K$_2$HPO$_4$, 0.15% KH$_2$PO$_4$, 0.5% glucose, 0.41% MgCl$_2$, 0.05% filter-sterilized bovine serum albumin and 1% NaCl. Under optimized transformation conditions, the transformation efficiency was approximately 2×10^6 transformants per µg of plasmid DNA (Fig. 5.14). The use of a slightly higher pH for buffer and medium and a lower concentration of PEG with a longer incubation period resulted in higher frequency of transformation.

A modified procedure for *Bacillus* sp. C-125 was reported by Kudo *et al.* (1990). Protoplasts were prepared in ASMMP buffer, consisting of 0.5 M sucrose, 0.02 M maleate, and 0.02 M MgCl$_2$ in double strength Penassay broth (pH 7.5), and polyethylene glycol # 8000 (final concentration, 22.5%) was used to introduce the plasmid DNA into the protoplasts. Modified DM-3 protoplast regeneration medium (pH 7.3) described above was used. Under these conditions, the transformation efficiency was approximately 5×10^6 transformants per milligram of plasmid DNA.

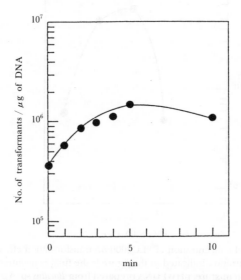

Fig. 5.14 Effect of incubation time with PEG on transformation frequency. The final concentration of PEG used for transformation was 22.5%.
(Reproduced with permission from Usami, *Starch/Stärke*, **46**, 232 (1990))

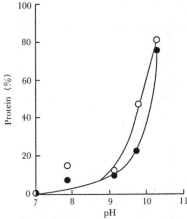

Fig. 5.15   Release of protein from protoplasts exposed to alkaline pH.
Protoplasts were prepared from alkaliphilic *Bacillus* sp. C-125 (●) or neutrophilic *Bacillus subtilis* GSY1026 (○). The protoplasts were incubated at
various pH values at 30 °C for 20 min. Initial A_{660} of the suspensions was
about 1. The pH values shown are those immediately after the protoplast
was suspended in the buffer.
(Reproduced with permission from Aono, Ito and Horikoshi, *Biochem. J.*,
285, 101 (1992))

Aono, Ito and Horikoshi (1992) extensively studied the stability of the protoplast
membrane of *Bacillus* strain C-125 at varying pH values and reported an improved regeneration method of protoplasts (Aono *et al.*, 1993*b*). Protoplasts prepared from the strain
with lysozyme regenerated cell walls at neutral pH, but not at pH above 8.5. As shown in
Fig. 5.15, cellular protein was released from the protoplasts exposed to various pH values
for a short period. The amount of protein released into the medium was dependent on
the pH of the suspension. Alkaline pH-dependent liberation of protein was also found in
the protoplasts prepared from neutrophilic *B. subtilis* GSY1026. The liberation was almost
identical for the protoplasts of the two organisms. Liberation of protein was not found at
pH 7–9. About 80% of the cellular protein was liberated from both protoplasts exposed
to pH 10.2 within 20 min. The colony-forming ability decreased rapidly in the protoplasts
exposed to high alkaline pH. The number of colonies formed by protoplasts decreased
to 0.08% of the initial number after only 3 min incubation at pH 10.6. These results indicated that the protoplasts exposed to alkaline pH rapidly burst and lost the ability to regenerate cell walls (Fig. 5.16). The alkali unstability was similar to that of protoplasts from
neutrophilic *B.subtilis* 168. The membrane vesicles were also labile at alkaline pH. The
acidic wall components of strain C-125 may contribute to stabilization of the cytoplasmic
membrane of cells growing at alkaline pH, probably by shielding the membrane from
direct exposure to an alkaline environment.

Furthermore, regeneration of protoplasts from the four alkaliphilic strains of *Bacillus*
sp. described in section 5.2.1 was also studied. These strains were aerobically grown at
37 °C in the following alkaline medium: glucose, 0.2%; polypeptone, 0.5%; yeast extract,
0.5%; citric acid, 0.034%; K_2HPO_4, 1.37%; KH_2PO_4, 0.59%; $MgSO_4\cdot7H_2O$, 0.005%;
Na_2CO_3, 1.06%(w/v). Cells in the exponential phase of growth were collected by cen-

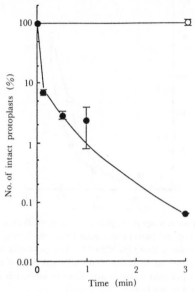

Fig. 5.16 Death of protoplasts exposed to alkaline pH.
The protoplasts were prepared from strain C-125-002 and incubated in
SMMP medium of pH 10.6 (●) or pH 7.2 (○) at 30 °C. Periodically the suspension was diluted with SMMP medium of pH 7.2 and spread on the neutral regeneration medium. Colonies formed at 37 °C by 4 days were counted. The examination was carried out in duplicate.
(Reproduced with permission from Aono, Ito and Horikoshi, *Biochem. J.,*
285, 101 (1992))

trifugation and washed once with double-strength Penassay broth (Difco Laboratories)
containing 0.5 M sucrose, 20 mM $MgCl_2$, and 20 mM maleic acid, pH 7.0 (SMMP medium) and resuspended in the SMMP medium to a concentration of A_{660} 15-20. One-hun-
dredth volume of 1% lysozyme solution was added to the cell suspension described above.
The protoplasts were collected by centrifugation at $1,000 \times g$ for 30 min at 10 °C and
washed twice with the SMMP medium.

The protoplasts of the alkaliphilic strains were spread onto a regeneration medium
containing 5 g of yeast extract, 5 g of casamino acids, 20 g of glucose, 30 mM $MgCl_2$, 1.25
mM $CaCl_2$, 0.5 M monosodium succinate, 4 g of bovine serum albumin, 10 g of agar and
30 mM Tris in 1 liter of deionized water. The pH of the medium was adjusted with NaOH
or HCl. After incubation of the protoplasts at 37 °C for 2-4 days, colonies formed on the
regeneration medium.

The protoplast from strain A-59 showed the highest regeneration frequency among
the four strains tested, as shown in Fig. 5.17. The colony formation was extremely low at
alkaline pH. The strain C-125 was strikingly dependent on the regeneration pH: At pH
6.7, 30% of the protoplasts were regenerated, but only 0.04% of the protoplasts formed
colonies at pH 7.0. Once the protoplasts were regenerated, optimum pH for growth was
pH 9 to 10 exhibiting characteristics of alkaliphilic *Bacillus* strains.

These results clearly indicate that protoplasts of the alkaliphilic *Bacillus* strains form

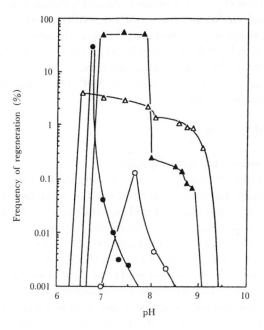

Fig. 5.17   Colony formation by protoplasts prepared from alkaliphilic strains of
Bacillus spp.
Protoplasts were prepared as described in the text from the alkaliphilic
strain 2B-2 (○), A-59 (▲), C-125 (●), or M-29 (△). Protoplasts in the suspension were microscopically counted with a Thoma's hematocytometer.
The protoplast suspension was spread on the regeneration medium adjusted to various pHs with NaOH or HCl. Resulting pH of the surface of the
medium was measured with a flat type of commercial glass electrode immediately before use. These pHs are indicated in the figure. The protoplasts
were incubated at 37 °C for 4 days. Colonies formed on the regeneration
medium were counted every day. A graph shows the ratio of the number of
colonies found by 4 days to that of the protoplasts spread on the medium in
duplicate examinations.
(Reproduced with permission from Aono, Ito and Horikoshi, *Biosci. Biotech.
Biochem.*, **57**, 1597 (1993))

colonies and regenerate cell walls at neutral pH. The efficiency was somewhat lower at
the alkaline pH optimum for growth of strains. The protoplasts were not tolerant of the
alkaline pH that was optimum for the growth of the strains.

5.2.5   Cell Fusion of Alkaliphiles

A procedure for genetic recombination after cell fusion of protoplasts from the facultative alkaliphile *Bacillus* sp. C-125 was developed by Aono *et al.* (1994). Protoplasts were
prepared from two auxotrophic and antibiotic-resistant strains (Met⁻ Nal^r and Thr⁻ Str^r,
respectively) of the facultative alkaliphile *Bacillus* sp. C-125 by treatment with lysozyme.
Equal volumes of two parental protoplasts were mixed and collected by centrifugation followed by the addition of polyethylene glycol (# 4000). The protoplasts fused effectively

in the presence of polyethylene glycol. Fusants obtained between two parental protoplasts were regenerated on solid medium of the modified DM-3 containing the two antibiotics nalidixic acid (Nal) and streptomycin (Str) after 5 days incubation at 37 °C. Parental protoplasts were regenerated at high frequency (43–97%) on non-selective medium but not on selective medium. The Nal[r] Str[r] fusants had the form of bacilli. Met and Thr markers segregated among the fusants with a predominantly Met[+] Thr[+] phenotype. The exfusants seemed to restore the parental ploidy.

5.2.6   Selection of Vectors

A.   Screening of Plasmids from Alkaliphiles
No information on plasmids in alkaliphilic microorganisms had been reported prior to 1983. Plasmids of alkaliphilic bacteria are required to make new host-vector systems. Plasmid screening was performed using a rapid-isolation method with about 200 *Bacillus* strains from our stock culture collection of alkaliphilic bacteria. Plasmids were found in two strains of alkaliphilic *Bacillus* sp. No. H331 and No. 13 (Usami *et al.*, 1983). In *Bacillus* sp. No. H331 two plasmids, pAB3311 (16 kb) and pAB3312 (6 kb), were observed in an electron micrograph. The molecular size of pAB13 in *Bacillus* sp. No.13 was estimated to be 7.9 kb from the molecular size of the fragments produced by restriction endonucleases and from the electron micrograph. The restriction map of pAB13 is shown in Fig. 5.18. Although these plasmids are still cryptic, the estimated molecular sizes, stabilities and restriction maps indicate that pAB13 is a most suitable candidate for use as a vector.

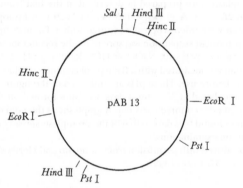

Fig. 5.18   Restriction map of pAB13.
(Reproduced with permission from Usami. *et al.*, *Agric. Biol. Chem.*, **47**, 2102 (1983))

Further work on this is not in progress because a plasmid vector pHW1 constructed for *Bacillus subtilis* is available for alkaliphilic *Bacillus* sp. C-125 (FERM. No.7344) and *Bacillus* sp. A-59 (ATCC21591). The plasmid pHW1 contains a chloramphenicol acetyltransferase gene as a marker. This antibiotic is the only one bound to be stable under alkaline conditions (Horinouchi and Weisblum, 1982).

B.   Phage Vectors

Horikoshi and Yonezawa (1978) isolated a phage from nature which infects alkaliphilic *Bacillus* sp. Y-25. The phage was assayed for plaque-forming units by the double-layer method. Turbid plaques (1–2 mm in diameter) were observed after 24 h of incubation at 30 °C. An electron micrograph of the phage showed that the phage has an icosahedral head of about 100 nm in diameter attached to a tail 210 nm long and 8 to 10 nm wide. The phage grew well at pH 9 to 11, the highest titer being observed at pH 10.5. Since then no one has worked on alkaliphilic phages. Conventional phages of *Bacillus subtilis* may infect *Bacillus* sp. C-125 or A-59 under neutral conditions such as pH 7–7.5. These phage vectors may be useful for further work.

5.2.7   Summary and Conclusion

Two hosts and one vector, pHW1, have been discussed this section. Transformation efficiency of both hosts is approximately 10^6. Regeneration of protoplasts depends on the pH value of the regeneration medium. DM-3 of pH 6.7–7.3 must be used in the case of strain C-125. Strain A-59 regenerates the cell wall in the pH range of 7 to 8. However, the same strain flocculated in conventional alkaline or neutral liquid media during shaking culture. As a result we used alkaliphilic *Bacillus* sp. C-125 (FERM. No. 7344) as the host in further genetic experiments.

5.3   Leucine Gene of Alkaliphilic *Bacillus* Strains

It is well known that genes specific to leucine biosynthesis are clustered on the chromosomes of *Salmonella typhimurium*, *Escherichia coli* and *Bacillus subtilis*. It was of interest to compare the leucine genes of alkaliphilic *Bacillus* strains and those of *E. coli*.

The first cloning of leucine genes was carried out using alkaliphilic *Bacillus* sp. No. 221, a producer of alkaline protease (Honda *et al.*, 1984). Alkaliphilic *Bacillus* sp. No. 221 was grown aerobically to the early stationary phase at 37 °C in Horikoshi-II medium. Chromosomal DNA of alkaliphilic *Bacillus* sp. No. 221 was hydrolyzed with *Hind* III. *Hind* III fragments of chromosomal DNA were inserted into the *Hind* III site of pBR322 and ligated with T_4 ligase. *E. coli* HB101 [*leuB*; β-isopropylmalate (β-IPM) dehydrogenase-less mutant] was used as the host.

Among ten recombinant plasmids, the smallest was purified and designated pHK101. The plasmid consisted of pBR322 and *Hind* III fragments of 4.0 kb and 3.2 kb. As shown in Table 5.7, the cloned fragment contains a leucine gene cluster. The 4.0 kb *Hind* III fragment could substitute for the α-IPM synthetase gene of *E. coli*, and the 3.2 kb *Hind* III

Table 5.7   Transfomation of *E. coli* with recombinant plasmid pHK101

Strains used	Number of Leu$^+$ Amp^r transformant per 10 µg DNA
	pHK101
CV512 *leu*A	4.0×10^4
HB101 *leu*B	3.0×10^7
CV520 *leu*C	6.2×10^4
CV526 *leu*D	1.2×10^5

fragment could substitute for the α-IPM isomerase gene of *E. coli*. Both fragments are necessary to substitute for the β-IPM dehydrogenase gene. Expression of the β-IPM dehydrogenase gene of alkaliphilic *Bacillus* sp. No. 221 was repressed by the addition of leucine into the culture medium of *B. subtilis* carrying the plasmid pHK111, which was constructed from pHK101 and pGR71. The 7.2-kb fragment could not substitute for the *iluB* gene of *B. subtilis*.

These results indicate that the 7.2-kb fragment contains the leucine gene cluster and its regulatory region. Therefore, the gross structure of the leucine cluster in alkaliphilic *Bacillus* sp. No. 221 is very similar, if not identical, to the gross structure of the leucine cluster in other bacteria.

Later Kato *et al.* (1984) cloned β-isopropylmalate dehydrogenase gene from alkaliphilic *Bacillus* sp. No. 38-2, a strain which produces cyclodextrin glycosyltransferase (CGTase). Chromosomal DNA of alkaliphilic *Bacillus* sp. No. 38-2 was prepared and the recombinant DNA of pBR322 and *Hind*III fragment containing leucine gene was constructed in the same manner as described above. A plasmid pHK202 containing a 1.1-Kbp *Hind* III-*Hinc* II fragment could substitute for the *leuB* phenotype. Therefore it is clear that the 1.1-kb *Hind* III *Hinc* II fragment contained the β-IPM dehydrogenase gene. These results strongly indicate that several intracellular enzymes such as leucine-synthesizing enzymes are essentially the same as those of neutrophilic microorganisms such as *Bacillus subtilis* and *Escherichia coli*.

6

Alkaliphily

How alkaliphiles adapt to their alkaline environments is one of the most interesting and challenging topics facing microbiologists.

6.1 Alkaliphily of *Bacillus* sp. C-125

6.1.1 Physical Map of Chromosomal DNA

To aid our studies on alkaliphiles, we decided to construct a physical map for the alkaliphilic *Bacillus* sp. C-125 (FERM7344) chromosome (Sutherland *et al.*, 1993*a*). This strain was chosen as a model alkaliphile since 1) it exhibits good growth between pH 7.0 and 11, and requires sodium ions for growth; 2) it grows well on minimal medium, allowing cloning of genes by a selective marker and complementation approach; 3) an efficient transformation system for cloning *Bacillus* sp. C-125 is available; 4) several genes have been cloned from this strain and used as markers on its chromosome map; and 5) this strain is not closely related to *Bacillus firmus* of Krulwich's group, but has been assigned by Aono to *Bacillus lentus* group 3 (see section 2.3.1A) (Aono, 1995).

A. Preparation of Insert and Digestion of Chromosomal DNA

Cells of *Bacillus* sp. C-125 were harvested from a 10-ml stationary-phase culture. The cells were washed once in TE buffer (10 mM Tris, pH 8.0, 1 mM EDTA) before being resuspended in 0.25 ml TE buffer. The following were then added to the cell suspension: 20 μl 10 mg/ml RNAase A, 200 μl 0.1 mg/ml lysozyme, 1 ml 1.5% (w/v) in Cert agarose solution (made in TE buffer). The solution was vortexed briefly before being poured into a single well of a 24-well Falcon tissue culture plate. The plate was kept at −20 °C for 2 min. The solidified gel was cut into 8–10 pieces and these pieces were transferred to 2 ml of a 0.5M EDTA (pH 9.5), 1% *N*-lauroylsarcosine solution containing 2 mg proteinase K/ml. The plugs were incubated at 50 °C for 24 h. They were then washed 5 times with TE buffer at room temperature for 15 min, once with 0.1 mM PMSF for 1 h at room temperature and 3 times with TE buffer. Plugs prepared as described above were stored indefinitely in TE buffer at 4 °C until required for digestion. For digestion, a portion of a DNA plug (25–50 μl in volume) was incubated twice in 200 μl of the appropriate digestion buffer at room temperature for 1 h and then in 100 μl digestion buffer containing 5–20 U restriction endonuclease at the appropriate temperature for 16 h. Partial digestion was achieved by lowering the concentration of enzyme. For lading, the DNA plug reaction mixture was incubated at 65 °C for 10 min and then loaded directly onto the gel prior to submerging the latter in the electrophoresis buffer.

DNA fragments were resolved in a Pharmacia-LKB Pulsaphor using a hexagonal array of electrodes. The 1% (w/v) agarose gels were subjected to electrophoresis at 14 °C in TBE buffer (0.05 M Tris-HCl, 0.05 M boric acid, 2mM Na$_2$EDTA 2H$_2$O of pH 8.0) at 160V for 24 h with a pulse time of 5 to 55 s, varied according to the optimum size of separation desired.

B. An *Asc*I Physical Map

Restriction endonucleases which recognize an 8bp sequence were tested for their ability to digest the chromosome. *Pac*I enzyme with the recognition sequence 5'-TTAATTAA-3' generated over 50 fragments of less than 50 kb. *Sfi* I enzyme (with the recognition sequence 5'-FFCCNNNNGGCC-3') generated approximately 20 fragments ranging in size from 20 kb to 700 kb. However, smearing made it hard to identify fragments and their sizes accurately. The two remaining enzymes, *Asc*I and *Sse*83871, generated 17 and 18 resolvable fragments, respectively. The mean total genome size of *Bacillus* sp. C-125 is estimated to be 3.7 Mb. As the *Asc*I enzyme gave consistently better digestion than the *Sse* 83871 enzyme, we decided to make a physical map of the *Asc*I sites on the chromosome.

Unfortunately, we could not isolate all the necessary linking clones required to determine the positions of every *Asc*I site on the chromosome. To obtain more information and to add detail to the map, clones of *Bacillus* sp. C-125 and *B. subtilis* DNA, and the 20kb *Asc*I fragment were used as hybridization probe (Table 6.1).

Table 6.1 Probes

Probe	Size (kb)	Gene product/phenotype
C-125 *unc* H, A, G	2.2	F$_1$ ATPase α, β & γ subunits
C-125 hag	2.7	Flagellin
C-125 pALK	0.5	Alkaline resistance
C-125 23S *rrn*	3.0	23S rRNA
C-125 *sec* Y	2.0	Secretion of proteins
C-125 *xyl* (A)	2.6	Xylanase resistance (alkaline)
C-125 *xyl* (N)	0.9	Xylanase (neutral)
*B. subtilis trp*S	0.9	Tryptophanyl tRNA synthase
*B. subtilis spo*IIG	0.8	Stage II sporulation
B. subtilis pNEXT39	3.4	*B. subtilis Not*I-linking clone
B. subtilis pNEXT41	2.9	*B. subtilis Not*I-linking clone
C-125 pLINK H1	2.9	C-125 *Asc*I-linking clone
C-125 pLINK H4	4.0	C-125 *Asc*I-linking clone
C-125 pLINK H5	1.5	C-125 *Asc*I-linking clone
C-125 pLINK H10	1.4	C-125 *Asc*I-linking clone
C-125 pLINK H20	3.7	C-125 *Asc*I-linking clone
C-125 pLINK H23	0.7	C-125 *Asc*I-linking clone
C-125 pLINK E6A	1.5	C-125 *Asc*I-linking clone

(Reproduced with permission from Sutherland *et al.*, *J. Gen. Microbiol.*, **139**, 664 (1993*a*))

C. Putative Chromosome Map

Southern blot analysis of single, double and partial digests of the chromosomal DNA allowed a putative chromosome map of *Bacillus* sp. C-125 to be constructed (Fig. 6.1). The allotted positions of cloned genes on the putative chromosome map of the strain compare favorably with their positions on the genetic and physical maps of *B.subtilis*. The corresponding starting point of the strain can be tentatively assigned to the 9A, or possibly the 12A, fragment. It is noteworthy that using the 6A DNA fragment as a probe, Sutherland *et al.* (1993*b*) isolated the F_1ATPase genes of *B.subtilis* in the 306–315° region of the genetic map. The similarity of the putative *Bacillus* sp. C-125 genetic map to that of *B. subtilis* is encouraging, and suggests that gene order on the chromosomes of closely related organisms is conserved.

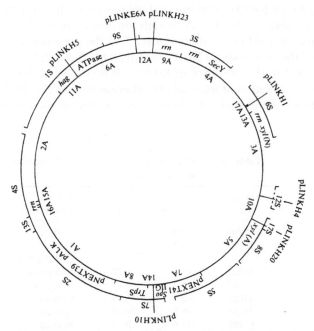

Fig. 6.1 Putative *Asc*I physical map of the *Bacillus* sp. strain C-125 chromosome. The correct orientations of the 13A and 17A doublet, the 15A and 16A doublet, the *rrn* and *xyl*(N) on 3A or 6S, the *rrn* and *secY* on 4A or 3S, and the pALK and pNEXT39 on 1A or 2S are unknown.
(Reproduced with permission from Sutherland *et al.*, *J. Gen. Microbiol.*, **139**, 667 (1993))

6.1.2 Alkaliphily and Cell Surface

Two series of experiments have been conducted by the author's colleagues. Kudo *et al.* (1990) focused on the cell membrane of *Bacillus* sp. C-125, mainly on H^+/Na^+ antiporters that regulate intracellular pH values. Aono *et al.* (1992) analyzed the cell walls of *Bacillus* sp. C-125. During the course of their studies, it has been reported that cell walls as well as cell the membrane are responsible for pH homeostasis.

A. Isolation and Properties of Alkali-sensitive Mutants

Kudo *et al.* (1990) formed an alkali-sensitive mutant of alkaliphilic *Bacillus* sp. C-125 by mutagenesis. This strain was selected for study since it grows well in Horikoshi minimal medium (see p. 6) over the pH range of 7 to 11.5 at 37 °C. Alkaliphilic *Bacillus* sp. strain C-125 (Trp$^-$ Ura$^-$ Cm^S) was treated with nitrosoguanidine in 25 mM glycine–NaOH–NaCl buffer of pH 8.5 and plated on Horikoshi-II medium (pH 7.5) containing 5g/l NaCl instead of sodium carbonate. Colonies that appeared on the plates were transferred onto Horikoshi-II medium (pH 10.3). After 16 h of incubation at 37 °C, seven alkali-sensitive mutants that could not grow at pH 10.5 but grew well at pH 7.5 were obtained from 7 × 10^4 colonies. Two mutants, 18224 and 38154, were selected (Table 6.2). The two mutants showed different properties from each other.

One of the mutants, No. 38154, was unable to sustain low internal pH in the presence of either Na$_2$CO$_3$ or K$_2$CO$_3$. The internal pH was 10.4, which was the same as that for all strains in the presence of K$_2$CO$_3$. Although the other mutant, No. 18224, cannot grow at pH >9, the internal pH of the mutant was 8.7 in the presence of Na$_2$CO$_3$. This value is close to that of the parent strain No. C-125 (pH 8.6). No. 18224 mutant cannot maintain a low internal pH in the presence of K$_2$CO$_3$. It is suggested that mutant No.38154 was defective in the regulation of internal pH, whereas mutant No. 18224 apparently showed normal regulation of internal pH values: Na$^+$ ion is also assumed to play an important role in pH homeostasis

Table 6.2 Alkali-sensitive mutants of alkaliphilic *Bacillus* sp. strain C-125

Strain No.	Growth capable pH	Internal pH[†]	Cell morphology
12797	<8.5	8.55	Curling
13797	<8.5	8.64	Curling, elongation
18224	<8.5	8.72	Elongation
19363	<8.5	8.92	Elongation
47629	<8.5	8.64	Curling
38154	<8.5	10.40	Normal
C-125	<11.0	8.64	Normal

[†] Measured by incorporation of [^{14}C] methylamine into cells at pH 10.5.
(Reproduced with permission from Hashimoto *et al.*, *Biosci. Biotechnol. Biochem.*, **58**, 2090 (1994))

B. Molecular Cloning of DNA Fragments Conferring Alkaliphily

The DNA fragment of the parental *Bacillus* strain C-125 that restores alkaliphily to the alkali-sensitive mutant strain No. 38154 was found in a 2.0-kb DNA fragment, which was cloned in a recombinant plasmid pALK2 (Fig. 6.2). A downstream 0.4-kb DNA fragment cloned in another recombinant plasmid, pALKl, restores alkaliphily to another alkali-sensitive mutant strain, No. 18824.

Nucleotide sequence analysis and restriction mapping indicated about 1.6 kb of com-

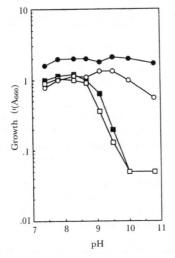

Fig. 6.2 Effect of pH on growth of alkali-sensitive mutant 38154 and its derivatives.
pH was adjusted by the addition of NaOH, Na_2CO_3, $NaHCO_3$, or NaCl.
Overnight cultures were added to fresh Horikoshi-II medium at an
absorbance at 660nm (A_{660}) of 0.01. Each point represents cell growth after
7h at 37°C. Parental strain C-125 (●), mutant 38154 (■), mutant 38154 carrying pHW1 (□), and mutant 38154 carrying pALK2 (○) were cultivated.
(Reproduced with permission from Kudo *et al., J. Bacteriol.*, **172**, 7283 (1990))

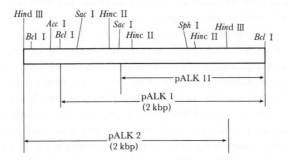

Fig. 6.3 Restriction maps of plasmids pALK1, pALK11, and pALK2. Restriction sites
and their coordinates are indicated.
(Reproduced with permission from Kudo *et al., J. Bacteriol.*, **172**, 7283 (1990))

mon sequence between pALK1 and pALK2 (Fig. 6.3). Plasmid pALK1 conferred alkaliphily on mutant 18224 but not on mutant 31154. Mutant 38154, which could not regulate its internal pH value in alkaline media, recovered pH homeostasis after the introduction of pALK2 (Table 6.3) but not after the introduction of pALK1. These results indicate that there are at least two factors involved in alkaliphily located in a closed linked
region of chromosomal DNA, although we do not know the precise functions of these two
DNA fragments. These two DNA fragments hybridized with DNAs isolated from other
alkaliphilic *Bacillus* strains, such as strains 221, A-59 and 17-1, but not with DNA of *B.sub-*

Table 6.3 Intracellular pH of mutants 18224 and 38154 and of transformants[†]

Strain	Intracellular pH
C-125	8.64
18224	8.72
38154	10.40
38154 (pALK2)	8.80

[†] Cells were grown to stationary phase at pH 8, washed with 20 mM Tris buffer (pH 8), and suspended in 100 mM carbonate buffer (pH 10.4) prepared with Na_2CO_3. The cell suspensions were incubated for 10 min at 25 °C with vigorous aeration in the presence of 2.2 μM [^{14}C]-methylamine. Samples were removed after 10 min and filtered for determination of ΔpH. Control experiments in which 20 μM gramicidin was included in the incubation mixture were conducted for each assay. Assuming that ΔpH = 0 in the presence of gramicidin, internal pH was calculated from the difference in the presence and absence of gramicidin.
(Reproduced with permission from Kudo *et al.*, *J. Bacteriol.*, **172**, 7283 (1990))

tilis. These results strongly suggest that at least two genetic loci that are required for alkaliphily are clustered on the chromosome. This was the first report of the recovery of alkaliphily in an alkali-sensitive mutant by the insertion of a specific gene sequence.

The nucleotide sequence of the fragment exhibited four open reading frames (−4). The transformant was able to maintain an intracellular pH that was lower than that of external pH and contained an electrogenic Na^+/H^+ antiporter driven only by ΔΨ (membrane potential, interior negative) (Kitada *et al.*, 1994). By subcloning the fragment, they demonstrated that a 0.25-kb DNA region is responsible for the recovery. Direct sequencing of the mutant's corresponding region revealed a G to A substitution. The mutation resulted in an amino acid substitution from Gly-393 to Arg of the putative ORF1 product, which was deduced to be an 804-amino acid polypeptide with a molecular weight of 89070. Membrane vesicles prepared from 38154 did not show membrane potential (ΔΨ)-driven Na^+/H^+ antiporter activity. Antiporter activity was resumed by introducing a parental DNA fragment which led to recovery of the mutant's alkaliphily. These results indicate that the mutation in 38154 affects, either directly or indirectly, the electrogenic Na^+/H^+ antiporter activity. This is the first report of a DNA fragment responsible for an Na^+/H^+ antiporter system in the alkaliphily of alkaliphilic microorganisms (Hamamoto *et al.*, 1994; Hashimoto *et al.*, 1994; Seto *et al.*, 1995).

C. Alkaliphily and Cell Surface

Aono *et al.* (1992) analyzed the acidic components constructing the cell wall of facultative alkaliphile *Bacillus* sp. C-125. Cell wall-defective mutants were previously isolated from the strain. During the course of studies, several alkali-hypersensitive mutants, whose cell walls did not appear to be defective, were isolated. A threonine auxotrophic C-125-001 was first isolated from wild strain C-125 treated with 1% ethylmethanesulfonate (EMS) with Aono's alkaline synthetic medium (see p. 6). A conventional penicillin-screening method was used for the C-125-001 cells mutagenized with 1% EMS. The alkali-sensitive

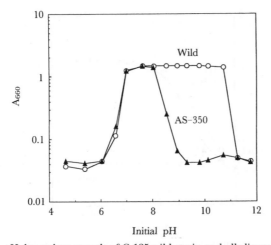

Fig. 6.4 pH-dependent growth of C-125 wild strain and alkali-sensitive mutant AS-350.

Medium was adjusted to various pHs indicated in the figure by addition of NaCl, NaHCO₃, Na₂CO₃, and NaOH. Both strains were grown in neutral medium (pH 7.3) at 37 °C for 12h. Each culture was diluted 100-fold with fresh media of various pH. All the subcultures were aerobically incubated at 37 °C. Absorbance at 660 nm of each culture was recorded after 8 h.

mutants that grew on a neutral complex medium (pH 7.2) but not on an alkaline medium (pH 10.3) were selected by the replica method. About 0.5% of the colonies grown at pH 7.2 showed somewhat poorer growth at pH 10.3 than the parent strain. Our collection of mutants isolated while looking for cell wall-defective mutants contained three types of alkali-sensitive mutants in pH-dependent growth. Among these, a type 1 mutant (AS-399) was duly alkali-sensitive. The upper limit of pH for growth was about 10, and that of the parent strain C-125 was about 11. The pH-dependent growth of these mutants appeared to be similar to those of mutants 18224 and 38154. Type 2 mutants (AS-187, -292 and -350) were more sensitive than mutant AS-399 (Fig. 6.4). The upper limits were almost identical to that of neutrophilic *Bacillus subtilis* GSY1026 used as a reference strain. A type 3 mutant (AS-409) was extremely alkali-sensitive, and grew only at around pH 7.5. Phenotypes of these mutants may become neutrophilic. However, none of the mutants in this study grew at pH below 6.5. The lower limit of pH for growth of the mutants was identical to that of the parent strain. Therefore, these mutants did not phenotypically become neutrophiles. Measurement of 9-aminoacridine incorporation in cells of the mutants indicated that the mutants were defective in the regulation of internal pH in an alkaline environment.

D. Molecular Biology of Cell Surface Mutants

An alkali-hypersensitive mutant AS-350 was used as a host for protoplast transformation. A gene that complemented the mutation in AS-350 and restored the alkaliphilic growth of the mutant was cloned from the parent strain (Aono *et al.*, 1993). Of approximately 1,000 chloramphenicol-resistant transformants, two grew on the alkaline medium. Plasmids harbored by these two transformants were found to be the same in size and

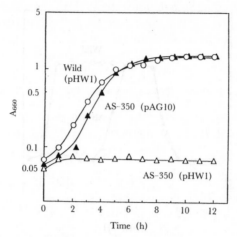

Fig. 6.5 Restoration of alkaliphily in mutant AS-350 transformed with pAG10.
C-125 wild strain carrying pHW1, mutant AS-350 carrying pHW1 or pAG10
was aerobically grown in alkaline medium (pH 10.3) containing chloramphenicol (10 μg/ml) at 37 °C. Absorbance at 660nm (A_{660}) was measured
periodically.

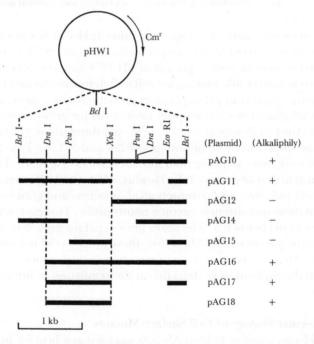

Fig. 6.6 Deletion derivatives from pAG10.
Each plasmid from pAG11 to 18 is a deletion derivative from pAG10. Mutant
AS-350 carrying each plasmid was examined on the alkaline agar plate (pH
10.3) for whether alkaliphily would be restored (+) or not (−). Cm^r, chloramphenicol-resistant gene.

restriction map. The plasmid, designated pAG10, grew in the alkaline medium as well as the parent strain C-125 (Fig. 6.5). The upper and lower limits of pH for growth of AS-350 carrying pHG10 was the same as those of the parent strain. The plasmid pAG10 restored only mutant AS-350 to alkaliphily. The pH range for growth of AS-350 carrying pAG10 was not expanded in comparison to that of the parent strain. Other type 2 mutants (AS-187 and -292) were also restored to alkaliphilic growth with pAG10. On the other hand, a type 1 mutant did not change to alkaliphilic. A type 3 mutant (AS-409) was partially restored to alkaliphily with the plasmid. The plasmid pAG10 contained an approximately 2.7-kb foreign DNA in the *Bcl*I side of the vector pHW1. Various deletion plasmids were prepared. As shown in Fig. 6.6, the mutation in the AS-350 should be complemented with a 1.0-kb *Dra*I-*Xbz*I region of the plasmid.

The restriction map of pAG10 is different from that of pALK1 or pALK2. These results strongly suggest that all the genes responsible for alkaliphily are not clustered on the same loci.

E. Role of the Cell Wall in Alkaline Environments

As shown in section 3.2.2, the cell wall of *B.lentus* C-125 contains peptidoglycan, teichuronic acid (TUA) and teichuronopeptide (TUP). The peptidoglycan is an A1γ-type that is identical to that in the cell wall of the neutrophilic *Bacillus subtilis*. The TUA consists of galacturonic acid, glucuronic acid and N-acetyl-D-2-amino-2,6-dideoxygalactyose (fucosamine) in a molar ratio of 1:1:1. TUP is a polymer in which polyglutamic acid binds covalently to polyglucuronic acid (PGU) and has a 1:4.5 molar ratio of glucuronic acid to L-glutamic acid (Aono, 1985; Aono, 1987; Aono, 1989*a*). The quantities of acidic polymers in the cell wall of strain C-125 increase when cells are grown at alkaline pH (Aono and Horikoshi, 1983; Aono, 1985). Mutants defective in acidic polymers grow poorly on a solid alkaline medium (Aono and Ohtani, 1990; Ito, Tabata and Aono, 1994).

These studies suggest that the acidic polymers in cell wall may be important for bacterial growth in an alkaline environment. Aono *et al.* (1995) examined cell wall-defective mutants grown using a pH control device. In their paper, they describe in detail the pH-dependent alteration of the chemical composition of cell walls of the C-125 wild-type and the mutants, the pH-dependent specific growth rates of the strains, and the ultrastructure of cell walls of strains grown at several pH values.

The structural components in cell walls of three mutants (C-125-11, C-125-90 and C-125-F19) of a facultative alkaliphile, *Bacillus lentus* C-125, defective in certain cell wall components, were characterized in detail. The cell wall of the wild-type C-125 was thick and increased in thickness when grown at high pH. Electron microscopy showed that triple layers developed when the bacteria were grown in an alkaline environment. In contrast, cell walls of teichuronopeptide (TUP)-defective mutants consisted of a single layer. For both the wild type and mutants, the cell wall concentrations of the acidic structural polymers teichuronic acid and TUP increased with respect to peptidoglycan as culture pH increased. The results listing the levels of L-glutamic acid, uronic acids and fucosamine relative to DAP in the cell walls are given in Table 6.4, together with the total anionic charge concentrations. The results show that the cell walls of the four alkaliphilic strains grown in high alkaline environments were highly anionic compared to the cell walls from the bacteria grown in neutral environments. In the case of C-125, L-glutamic acid was the predominant acidic compound, and this increased almost threefold with

Table 6.4 Culture–pH dependence of the non-peptidoglycan components of cell walls

Strain	Culture pH	Amounts relative to DAP (μmol·μmol^{-1})			
		L–Glu	UA	FucN	Total anionic compounds
C-125	6·8	1·50	0·51	0·08	2·0
	7·0	2·10	0·69	0·12	2·8
	8·0	2·90	1·70	0·36	4·6
	9·0	3·70	1·90	0·49	5·6
	10·0	4·10	2·10	0·49	6·2
	10·5	4·40	2·40	0·46	6·8
C-125-11	6·8	3·40	0·74	ND	4·1
	7·0	3·40	0·79	ND	4·2
	8·0	4·10	0·90	ND	5·0
	9·0	4·30	0·91	ND	5·2
	10·0	4·50	1·00	ND	5·5
	10·5	4·60	1·10	ND	5·7
C-125-F19	6·8	0·47	1·20	0·19	1·7
	7·0	0·53	1·30	0·07	1·8
	8·0	0·50	2·20	0·19	2·7
	9·0	0·53	2·30	0·15	2·8
	10·0	0·80	2·90	0·23	3·7
C-125-90	6·8	0·46	1·10	ND	1·6
	7·0	0·51	1·30	ND	1·8
	8·0	0·65	1·80	ND	2·4
	9·0	0·92	1·80	ND	2·7
	10·5	1·75	2·10	ND	3·8

The cell walls were prepared from bacteria grown at the constant pH indicated and assayed for constituents of TUA, TUP and peptidoglycan. The L-glutamic acid (L–Glu), uronic acids (UA) and fucosamine (FucN) contents relative to the DAP content are shown. ND, Not detected.
(Reproduced with permission from Aono *et al.*, *Microbiology*, **141**, 2959 (1995*b*))

respect to DAP when the culture pH increased from 6.8 to 10.5. The relative concentrations of uronic acids increased over fourfold. The total acidic constituents in the cell walls increased from 1.0 μmol (μmol DAP)$^{-1}$ in cells grown at pH 6.8 to 6.8 μmol (μmol DAP)$^{-1}$ in cells grown at pH 10.5.

The cell-wall density of the negatively charged compounds (uronic acids plus L-glutamic acid) was calculated to be about 3 and 9 equivalents/(1 cell wall region) for C-125 cells grown at pH 7 and 10, respectively. At high pH, the specific growth rates of the two TUP-defective mutants were much lower than those of the wild type. Recently, they cloned a DNA fragment from the wild-type C-125 that complements the defect in TUP synthesis in mutants C-125-90 and C-125-F19 (Aono, personal communication). It is concluded that increased levels of acidic polymers in the cell walls of alkaliphilic bacteria may be a necessary adaptation for growth at elevated pH.

F. Alkaliphily

Although it is premature to discuss in detail the alkaliphily of *Bacillus* sp. C-125, the author here speculates on the pH homeostasis of *Bacillus* sp.C-125 in alkaline environments. The cells have two barriers to reduce pH values from 10.5 to 8 (Fig. 6.7).

1) Cell walls possess acidic polymer function as negatively charged membrane and may reduce the pH values from 10.5 to probably 9. Cell membrane should be kept below pH 9 because it is quite unstable at alkaline pH values below the pH optimum for growth.

2) Cell membrane may maintain pH homeostasis by using a Na^+/H^+ antiporter system, ATPase, etc. These pH regulating systems are active at pH 9.0, but not detected in higher alkaline environments such as pH 10. It is highly possible that cell membrane reduces the pH values from 9 to 8.

3) The cell walls may keep the intracellular pH values in the range of 7 to 8.5, allowing *Bacillus* sp. C-125 to thrive in alkaline environments.

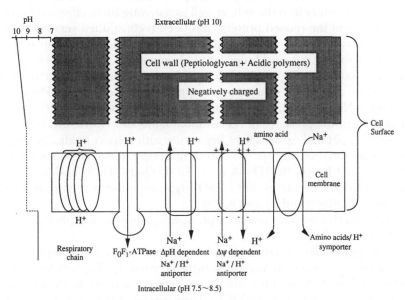

Fig. 6.7   pH Homeostasis of *Bacillus* sp. C-125.

6.2 Alkaliphily of *Bacillus firmus* OF4

Extensive studies on alkaliphily have been reported by Krulwich, focusing on bioenergetics of alkaliphiles.

B. firmus OF4 shows a rapid rate of aerobic growth on malate over the pH range from 7.5 to 10.6, with growth at pH 10.6 being slightly faster (38 min doubling time) than at pH 7.5 (54 min doubling time). When *B. firmus* OF4 and *B. subtilis* are suspended in media of pH 10.5, only the alkaliphile can sustain a large $\Delta\Psi$ across the membrane. As described in section 4.1, Na^+ is required for alkaliphilic *Bacillus* species (Kurono and Horikoshi,

1973; Kitada and Horikoshi, 1977; McLaggan *et al.*, 1984; Ivey *et al.*, 1992) to grow at pH >
9.5; K$^+$ cannot be substituted (Krulwich, Federbush and Guffanti, 1985). The rate of Δp-
dependent Na$^+$ efflux from alkaliphilic *B. firmus* OF4 is much faster than that from *B. subtilis* at pH 8.5–9. Under such conditions both species generate large $\Delta\Psi$ values that can
energize the electrogenic antiporters involved in pH homeostasis (Garcia, Guffanti and
Krulwich, 1983; Ivey *et al.*, 1992). Sturr, Guffanti and Krulwich (1994) also reported that
cells isolated from continuous cultures growing at pH 11.2 include variant strains that have
shorter lag times when re-inoculated into media at pH 11 than the parent strain, and this
correlates with increased Na$^+$/H$^+$ antiport.

One structural gene for an Na$^+$/H$^+$ antiporter, *nhaC*, has been cloned from *B. firmus*
OF4, and at least one other gene encoding a Na$^+$/H$^+$ antiporter exists in this alkaliphile
(*nahJ*) (Ivey *et al.*, 1991). A locus encoding one, or possibly two, Na$^+$/H$^+$ antiporters has
been cloned from alkaliphilic *Bacillus* C-125, and has been characterized (Kudo *et al.*,
1990). The total number of such porters in any one extreme alkaliphile is unknown.

At high pH, Na$^+$/H$^+$ antiporter-dependent pH homeostasis in the alkaliphile requires
a pathway for Na$^+$ re-entry into the cell, as well as a driving force effectively generated by
respiration. None of the related proteins has been reconstituted yet, so the interesting
possibility remains that at high pH, Na$^+$ influx may be enhanced across such porters even
in the absence of solute; this could support pH homeostasis. Alternatively, a pH-regulated Na$^+$ channel has been proposed to account for Na$^+$ re-entry and Na$^+$-dependent pH
homeostasis in the absence of solutes whose uptake is coupled to Na$^+$. Flagellar rotation
in alkaliphiles is Na$^+$-coupled, but flagellar formation of *Bacillus* C-125 was extremely
inhibited at pH 7 in the presence of Na$^+$ (Aono, Ogino and Horikoshi, 1992). Krulwich
also discusses Na$^+$, pH and Δp sensor in her review with respect to alkaliphile pH homeostasis. Synthesis of at least one operon encoding a respiratory chain complex in *B. firmus*
OF4 is up-regulated by low Δp (Quirk *et al.*, 1991; Quirk, Hicks and Krulwich, 1993).
Furthermore, alkaliphiles may exhibit an SOS-type response at the upper edge of their
pH range (Sturr, Guffanti and Krulwich, 1994).

Six possible mechanisms for oxidative phosphorylation in alkaliphiles have been presented by Krulwich (1995).

It was anticipated that oxidative phosphorylation by extreme alkaliphiles would also
be energized using Na$^+$ as a coupling ion to bypass the low Δp (Fig. 6.8(A)). However, the
F$_1$F$_0$-ATP synthesis of *B. firmus* OF4 and *B. alcalophilus* has been shown to be exclusively
proton-translocating, and in *B. firmus* OF4 a single species of the synthase has been detected at both pH 7.5 and pH 10.5 (Ivey *et al.*, 1994).

Intracellular or membrane-associated organelles (Fig. 6.8(B)) would offer a solution,
and have been posited by Skulachev (1991). However, several electron-microscopic studies have failed to indicate the presence of such structures (Khan, Ivey and Krulwich, 1992;
Sturr, Guffanti and Krulwich, 1994).

Another possibility is Fig. 6.8(C). As the Δp falls, the ATP synthase couples with an
increasing stoichiometry of H$^+$ translocated/ATP synthesized. This work is in a totally
chemiosmotic scenario. This possibility leaves the puzzling observation that the
[ATP]/[ADP] ratio was not highest at the pH at which the Δp was highest (Hoffmann and
Dimroth, 1991).

As shown in Fig. 6.8(D), there may be a trapping of protons between the coupling
membrane and the cell wall layer, perhaps dependent upon properties of the cell walls.

A. Use of Na+-motive ATP synthase would bypass the low bulk Δp

B. Sequestration of oxidative phosphorylation, at high pH only?, in intracellular or membrane associated "lens"-like organelles might bypass adverse ΔpH.

C. Use of a very high (up to 13) and exremely variable stoichiometry by the ATP synthase (or multiple synthase species with different stoichiometries) concomitant with pH-dependent changes in H+/O

D. Proton trapping by the cell wall layer(s) produces a lower pH region from which protons enter F_0.

E. Proton movement laterally along membrane surface is rapid enough to support synthesis before protons are lost to the bulk.

F. At high pH (>9) protons are transferred from (a specific?) respiratory chain complex by direct intramembrane transfer to the F_0.

Fig. 6.8 Six possible mechanisms for alkaliphile oxidative phosphorylation. The major features and status of evidence with respect to these possibilities are summarized in the text. Part of this figure was reproduced from an earlier report (Guffanti and Krulwich, 1994)
(Reproduced with permission from Krulwich, *Mol. Microbiol.*, **15**, 407 (1995))

Some alkaliphilic *Bacillus* species such as *Bacillus* C-125 exhibit at least a partially non-alkaliphilic phenotype upon mutational loss of acidic components of the cell wall, and these strains do not form alkali-stable protoplasts (see previous section). On the other hand, equally extreme alkaliphiles such as *B. firmus* OF4 do not have high levels of acidic polysaccharides in their cell walls. Moreover, ADP + Pi-loaded right-side-out membrane vesicles of *B. firmus* OF4 were able to synthesize ATP at pH 10.5 (Guffanti and Krulwich, 1994).

A different sequestration of sorts would involve retention of all, or at least some, of the protons pumped by respiration at the membrane surface, where they may move laterally into the usual entry site for protons in the ATP synthase before they are equilibrated with the bulk. Their retention may depend upon the surface potential or on the surface buffering capacity (Fig. 6.8(E)).

A final proposed mechanism is that the proton transfer occurs not at the surface or in a sequestered space of high proton concentration outside the membrane, but rather as a direct proton transfer within the membrane translocating respiratory-chain complex and the ATP synthase (Fig. 6.8.(F)).

Although the alkaliphily is not yet fully understood, it should be noted that the importance of oxidative phosphorylation in alkaliphilic *Bacillus* strains may be in illustrating the plasticity in energy coupling that can be used to enable an extremophile to meet its special challenges. Other alkaliphiles would presumably have different adaptation mechanisms. It is not surprising, therefore, that some models in Fig. 6.8(A)-(F) may not account for alkaliphiles that adapted to different environments.

Extracellular Enzymes

7.1 Isolation Procedures for Alkaline Enzymes

Enzymes having pH optimum for enzyme action in the alkaline regions are called alkaline enzymes. These include, for example, alkaline proteases (pH 10-12), alkaline amylases (pH 9-11), alkaline cellulases (pH 8-10), and others. Except for alkaline proteases these enzymes had not been isolated before our rediscovery of alkaliphilic microorganisms in 1968. Isolation of alkaline enzymes means the isolation of alkaliphilic microorganisms producing alkaline enzymes. Many screening procedures for these microorganisms have been developed. General procedures are described below with specific methods given in subsequent sections.

7.1.1 Protease Activity

Polypeptone in Horikoshi-I or -II medium is substituted for crude casein such as skim milk (1% w/v) and microorganisms are inoculated on the solidified media. After two or three days incubation at 37°C, peptonization activity is directly observed as colonies hydrolyzing casein surrounded by a white halo. Candidates producing protease are cultured in the liquid media described in Table 7.1, and activity in the culture media are determined by conventional methods. Semi-quantitative assay using pulp disks is also a good method.

7.1.2 Amylase Activity

Isolated microorganisms are grown in solidified Horikoshi-II medium. After two to three days incubation at 37°C, iodine solution is poured onto plates. The colonies surrounded by a white halo are the amylase producers.

7.1.3 Cellulase Activity

Instead of glucose, 1% CMC is added to the Horikoshi-I medium solidified by agar containing 0.03% w/v Congo red, and soil samples spread on the plates. After two to three days incubation at 35-40°C, white halos form around bacterial colonies producing CMCase.

7.1.4 β-1,3-Glucanase

Defatted pachyman powder is used as the β-1,3-glucan. Laminaran is also a good substrate for isolating microorganisms. Congo red (about 0.01-0.03% w/v) is added to the media described above to indicate β-1,3-glucanase activity. The colonies surrounded by white halos are the enzyme producers. Pachyman should be defatted by ether, otherwise very poor growth of microorganisms is observed due to the presence of inhibitory substances in pachyman. Fig. 7.1 shows the media isolating the enzyme-producing microorganisms described above.

Table 7.1   Media for large-scale production

Ingredients (g/l)	Protease			Amylase			Cellulase	β-1,3-Glucanase
Soluble starch	50	40		15	80	80		
CMC							20	
Pachyman								20
Soya meal	20	30	30	20				
Ground barley	50	100	100					
Soya bean extract				10	10			
Polypeptone				10	20		5	5
Fish extract					20			
Na-caseinate	10							
Soya bean oil		5.5						
Polyglycol	0.1	0.1						
Yeast extract							5	5
KH$_2$PO$_4$								1
Na$_2$HPO$_4$·10H$_2$O	9			9				
NaCl						2.5		
CaCl$_2$				1				
CaCO$_3$		5						
MgSO$_4$							0.2	0.2
Na$_2$CO$_3$	10	10		10	10		10	10
NaHCO$_3$				10				

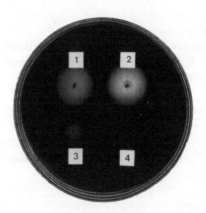

Fig. 7.1   An alkaliphilic *Bacillus* strain producing β-1,3-glucanase on congo red plate.

7.2 Alkaline Proteases

Alkaline proteases are not produced only by alkaliphilic microorganisms; two of the most widely studied enzymes are subtilisins BPN' and Carlsberg produced by neutrophilic strains of *Bacillus subtilis*. They are not inhibited by metal chelators or thiol reagents; the N-terminal amino acid of both enzymes was identified as alanine and both were inhibited by a serine agent such as diisopropylfluorophosphate (DFP). Several alkaliphilic *Streptomyces* strains and *Bacillus* strains produce alkaline proteases capable of hydrolyzing keratinous proteins such as wool, hair, feather and silk under alkaline conditions (in 0.1N NaOH).

7.2.1 Alkaline Protease of Alkaliphilic *Bacillus* Strains

In 1971 Horikoshi (1971*a*) reported the production of an extracellular alkaline serine protease from alkaliphilic *Bacillus* sp. No. 221. This strain, isolated from soil, was found to produce large amounts of alkaline protease which differed from the subtilisin group. The enzyme was purified by DEAE and CM cellulose column chromatography and crystallized from ammonium sulfate solution. The optimum pH of the enzyme was 11.5 with 75% of the activity maintained at pH 13.0 (Fig. 7.2). The enzyme was completely inhibited by DFP or 6 M urea, but not by EDTA or *p*-chloromercuribenzoate. The molecular weight of the enzyme was 30,000, which is slightly higher than those of other reported alkaline proteases. Calcium ions affected both activity and stability of the enzyme. The addition of 5 mM calcium ions reflected a 70% increase in activity at the optimum temperature (60 °C).

Subsequently, two more *Bacillus* species, AB42 and PB12, were also reported to pro-

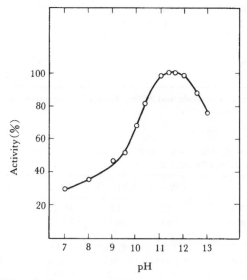

Fig. 7.2   Effect of pH on enzyme activity.
(Reproduced with permission from Horikoshi, *Agric. Biol. Chem.*, **35**, 1410 (1971))

duce alkaline protease (Aunstrup *et al.*, 1972). Molecular weights were about 20,000 and 26,000 and both had an isoelectric point of 11.0. They exhibited a broad pH range of 9.0 to 12.0 and the temperature optimum of the strains was 60 °C for AB42 and 50 °C for PB 12. Both enzymes were inhibited by phenylmethylsulfonyl fluoride, indicating that they are serine enzymes. Further properties are shown in Table 7.2. In 1976 Kitada and Horikoshi reported an alkaline protease produced by a methylacetate utilizing alkaliphilic *Bacillus* sp. No. 8-1, although enzymatic properties were not good enough for industrial applications (Kitada and Horikoshi, 1976*a*, *b*).

In addition to their protease activity, both enzymes No. 221 and AB42 exhibited esterase activity toward amino acid ethylesters. *Bacillus* sp. No. ES 2-2-5 and No. LA 1-1-1 were screened from 300 colonies isolated from soils of Taipei in Taiwan (Nomoto *et al.*, 1984*a*). *Bacillus* sp. No. NKS21 (FERM No. BP-93) was isolated from soil by Tsuchida *et al.* (1986), and *Bacillus* sp. No. B21-2 was isolated from 400 soil samples by Fujiwara and Yamamoto (1987). The four *Bacillus* proteases described above had pH optima, like most other bacterial alkaline proteases, at 10–12. Other enzymatic properties are summarized in Table 7.3. NKS-21 and B21-2 proteases were inhibited by diisopropyl fluorophosphate, indicating that they were serine enzymes, as well as subtilisin. Optical rotation measure-

Table 7.2   Properties of alkaline proteases from alkaliphilic *Bacillus* sp. No. 221, AB42, and PB 12, and from *B. subtilis* strains

Property	Alkaliphilic *Bacillus* strain			*B. subtilis* strain	
	No. 221	AB 42	PB12	Subtilisin BPN'	Subtilisin Carlsberg
$E_{1cm}^{1\%}$ at 280 nm	12.5	—	—	11.7	8.6
S_{20}	3.5	2.8	2.8	2.77	2.85
Molecular weight	30,000	20,000	26,000	27,700	27,600
Isoelectric point	>11	11.0	11.0	7.8	9.8
Optimum pH	11.5	—	—	10.5	10.5
Specific activity, units/mg·protein	18,000			2,200	6,500
N-Terminal	Alanine			Alanine	Alanine

(Reproduced with permission from Horikoshi and Akiba, *Alkalophilic Microorganisms*, p. 94, Springer-Verlag: Japan Scientific Societies Press (1982)).

Table 7.3   Properties of proteinases from *Bacillus* sp. No. ES2-2-5 No. LA1-1-1, 1 No. NKS-21 and No. B21-2

Strain	ES2-2-5	LA1-1-1	NKS-21	B21-2
Optimum pH (40 °C)	12	12	10–11	11.5
Optimum temperature (pH)	60 °C	60 °C	50–55	60
pH stability (40 °C)	7–11	6–11	7–11	
Thermal stability (pH 10)	< 50 °C	< 50 °C	< 45 °C	< 50 °C
Ca^{2+} conc. for stabilization (mM)	~ 2	~ 2	< 10^{-3}	5
Molecular weight	—	—	3.1×10^4	3.0×10^4

ment showed that the α-helix content of No. 221 enzyme was approximately 34% to 37%. No change was observed in the helix content by shifting the pH from the alkaline to the neutral range or by the addition of 6.6 M urea or guanidine hydrochloride. These facts indicate that the structure of the enzyme is quite rigid and stable. In the case of NKS-21 enzyme, the α-helix content was calculated to be 10% from CD spectrum data. This is lower than the α-helix content of No. 221 enzyme and subtilisin BPN (by about 20%). There was a slight difference in the specificity of cleavage sites of oxidized insulin among the serine alkaline proteases. Tsai *et al.* (1983; 1984; 1986) attempted to obtain alkaline elastase from alkaliphilic bacteria. They isolated an alkaline elastase from alkaliphilic *Bacillus* sp. No. Ya-B, which showed a marked preference for elastin over other proteins. The alkaliphilic *Bacillus* sp. No. Ya-B was aerobically cultured in a medium consisting of 2% glucose, 2% soybean casein, 2% glycerin, 0.5% yeast extract, 0.2% KH_2PO_4, 0.02% $MgSO_4$ and 1% Na_2CO_3 at 37 °C for 20 h. The culture broth was adjusted to pH 6.5 with acetic acid; ammonium sulfate was brought to 80% saturation. The precipitate was dissolved in a minimal amount of 20 mM Tris-buffer (pH 8.0) and dialyzed against the same buffer. The dialysate was passed through a DEAE-Sephadex column equilibrated with 20 mM Tris-buffer (pH 8.0) followed by a CM-Sephadex column. The molecular weight was estimated to be 25,000 and the specific activity of the purified enzyme was 12,400 for casein and 2,440 for elastin at pH 11, which is the optimum pH for enzyme action. The isoelectric point was 10.6 and the enzyme was stable in the range of pH 5.0 to 10.0. This elastase, which is a serine enzyme, was strongly inhibited by the alkaline proteinase inhibitor and *Streptomyces* subtilisin inhibitor, but not by metalloproteinase inhibitor or elastatinal. One of the characteristic properties of this elastase is that the enzyme activity was inhibited in the presence of SDS. Sequence homology was exhibited among these four enzymes, as shown in Fig. 7.3. However, this enzyme has several different properties: high elastin binding ability and substrate specificity, as shown in Table 7.4.

Takami, Akiba and Horikoshi (1989) isolated a new alkaline protease, which was extremely thermostable, from alkaliphilic *Bacillus* sp. No. AH-101 (FERM No. P-10531).

	1				5					10					15					20
Subtilisin BPN'	Ala	Gln	Ser	Val	Pro	Tyr	Gly	Val	Ser	Gln	Ile	Lys	Ala	Pro	Ala	Leu	His	Ser	Gln	Gly
Subtilisin Carlsberg	Ala	Gln	Thr	Val	Pro	Tyr	Gly	Ile	Pro	Leu	Ile	Lys	Ala	Asp	Lys	Val	Gln	Ala	Gln	Gly
Alkaline Protease 221	Ala	Gln	Ser	Val	Pro	Trp	Gly	Ile	Ser	Arg	Val	Gln	Ala	Pro	Ala	Ala	His	Asn		
Alkaline Elastase	-	Gln	Thr	Val	Pro	Trp	Gly	Ile	Asn	Arg	Val	Gln	Ala	Pro	Ile	Ala	Gln	Ser	Arg	Gly

Fig. 7.3 N-terminal amino acid sequences of alkaline proteases.

Table 7.4 Hydrolysis of elastin, keratin, and collagen

Enzyme	Relative activity		
	elastin/casein	keratin/casein	collagen/casein
Alkaline elastase	0.58	0.55	0.18
Subtilisin BPN'	0.034	0.17	0.086
Subtilisin Carlsberg	0.097	0.25	0.067
Pronase	0.19	0.26	0.071
Pancreatic elastase	0.54	n.d.	n.d.

n.d. : not determined

The isolate was aerobic, spore-forming, gram-positive, motile, rod-shaped, catalase-positive, oxidase-positive, and citrate-utilizing Koser medium. The GC content of the DNA was 43.2 mol%. Further properties are shown in Table 7.5. The temperature for growth was 20-55 °C and the pH for growth was in the range of pH 7 to 11. The alkaliphilic *Bacillus* sp. No. AH-101 was grown aerobically at 37 °C for two days in a 500-ml flask containing 50 ml of alkaline medium (pH 9.5). The medium consisted of 2% soluble starch, 2.4% corn steep liquor, 1% soybean meal, 0.1% K_2HPO_4, 0.003% $MgSO_4 \cdot 7H_2O$, 0.03% $ZnSO_4$, 0.03% $CaCl_2$ and $NaHCO_3$. A protease from alkaliphilic *Bacillus* sp. No. AH-101 was purified from the culture fluid by only two steps, passing through a DEAE Toyopearl 650M column followed by CM-Toyopearl 650M column chromatography. The enzyme was most active toward casein at pH 12-13 (Fig. 7.4) and stable under 10-min incubation at 60 °C from pH 5-13. Calcium ions were effective in stabilizing the enzyme molecule especially at higher temperatures. The optimum temperature was about 80 °C in the presence of 5 mM calcium ions. The enzyme was stable at 30-70 °C in the presence of 5 mM calcium ions. The enzyme is more stable against both temperature and high alkaline conditions in the presence of detergents than other protease so far reported. Therefore, this enzyme is a good candidate for a detergent additive. The *Bacillus* sp. No. AH-101 alkaline protease showed higher hydrolyzing activity against insoluble fibrous natural proteins such as elastin and keratin compared with subtilisins Proteinase K. The optimum pH of the enzyme toward elastin and keratin was pH 10.5 and pH 11.0-12.0, respectively. The specific activity toward elastin and keratin was 10,600 units/mg protein and 3970 units/mg protein, respectively. The enzymatic activity was not inhibited by *p*-chloromercuribenzoic acid and iodoacetic acid. Carbobenzoxy-glycyl-glycyl-L-phenylalanyl chloroethyl ketone completely inhibited the caseinolytic activity, but 36% elastolytic activity remained. No

Table 7.5   Characteristics of strain No. AH-101

Morphological characteristics	
Form	Rods
Size	0.5×1.5-4.0 μm
Motility	Motile
Gram stain	Positive
Sporangia	Swollen
Spores	0.5-1.0 $\times$ 1.5 μm (oval, terminal)
Cultural and biochemical characteristics	
Nutrient broth (pH 10)	Good growth
Reduction of nitrate	Negative
Voges-Proskauer test	Negative
Indole test	Negative
Hydrolysis of starch	Positive
Hydrolysis of gelatin	Positive
Utilization of citrate	Positive
Oxidase activity	Positive
Urease activity	Negative
Catalase activity	Positive
Temperature for growth	22-55 °C
pH for growth	pH 7-11
Growth on NaCl	Below 7%
GC content of DNA	43.2 mol%

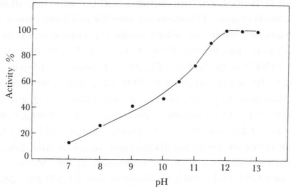

Fig. 7.4 Effect of pH on enzyme activity.
(Reproduced with permission from Takami *et al.*, *Appl. Microbiol. Biotechnol.*, **38**, 103 (1992))

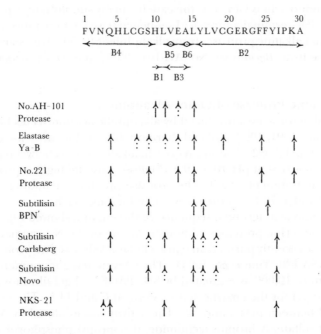

Fig. 7.5 Cleavage patterns of No. AH-101 alkaline protease on oxidized insulin B-chain and its comparison with subtilisins (Novo, BPN', Carlsberg), No. 221 protease, Ya-B elastase, and NKS-21 protease. Thick arrows signify major cleavage sites and dashed arrows signify minor sites.
(Reproduced with permission from Takami *et al.*, *Biosci. Biotechnol. Biochem.*, **56**, 334 (1992))

inhibitory effect on caseinolytic and elastolytic activity was shown by tosyl-L-phenylalanyl-chloromethyl ketone, tosyl-L-lysine chloromomethyl ketone, and elastatinal. Extensive sequence homology was exhibited by these enzymes. The AH101 protease has high elastolytic activity like the Ya-B elastase from *Bacillus*, and did not hydrolyze oxidized insulin A chain but digested the B chain. The cleavage specificity of the enzyme oxidized insulin B chain was much narrower than any other alkaline protease so far tested, as shown in Fig. 7.5 (Takami, Akiba and Horikoshi, 1990; Takami *et al.*, 1992*a*).

Shimogaki *et al.* (1991) reported an alkaline protease (protease BYA) isolated for industrial applications. Its optimum pH was 10.0-12.5, and it was stable towards various surfactive agents and towards members of the subtilisin family, e.g. subtilisin BPN', subtilisin Carlsberg, and No. 221. The enzyme, which is a serine protease, was most active at 70°C. The isoelectric point was about 10.1. Fujiwara, Masui and Imanaka (1993) purified thermostable alkaline protease from an alkaliphilic thermophile *Bacillus* sp. B18 using DEAE- and CM-Toyopearl 650M column chromatography. Molecular weight of the enzyme determined by SDS-PAGE and gel filtration was 30,000 and 28,000, respectively. The optimum pH and temperature toward the hydrolysis of casein were pH 12-13 and 85 °C, both of which are higher than those of a mesophilic alkaline protease from an alkaliphile, *Bacillus* sp. B21-2. Thermostability of the enzyme was enhanced by Ca^{2+}. The enzyme activity was inhibited by DFP, suggesting that the enzyme is a serine protease. The NH_2-terminal amino acid is Gln, as is the case for many subtilisin-type proteases. The 20 residues of the NH_2-terminal amino acid sequence have a comparatively high homology with those of other alkaline proteases from alkaliphiles (40-50%), especially thermostable alkaline protease from *Bacillus* sp. No. AH-101 (95%) and *Thermoactinomyces* sp. HS682 (95%).

7.2.2 Alkaline Protease of Other Alkaliphiles

Several alkaline proteases produced by alkaliphilic actinomycetes have been reported. Tsuchiya *et al.* (1991; 1992) isolated thermostable alkaline protease from alkaliphilic *Thermoactinomyces* sp. HS682. The strain grew in alkaline media between pH 7.5 and 11.5. The growth rate was best at pH 10.3-10.5. Maltose gave the highest productivity of protease at the concentration of 10 g/l. The protease gave the maximum proteolytic activity around pH 11.0 and at 70 °C. In the presence of Ca ions, maximum activity was observed at 80 °C. The amino acid terminal sequence of the enzyme showed high homology with those of microbial serine proteases, although alanine of the N-terminal amino acid was deleted. Yum *et al.* recently purified an extracellular alkaline serine protease produced by *Streptomyces* sp. YSA-130 (Yum *et al.*, 1994). The enzyme was a monomeric protein with a molecular weight of 19,000 as estimated by SDS-PAGE and gel filtration. The optimum temperature and pH for the enzyme activity were 60 °C and 11.5. The enzyme was stable up to 50°C, and between pHs 4 and 12. The activity was inhibited by Ag^+, Hg^{2+}, Co^{2+}, sodium dodecyl sulfate, *N*-bromosuccinimide, diisopropyl phosphorofluoridate (DFP), 2.3-butanedione, 5,5-dithiobis-(2-nitrobenzoic acid) (DTNB), iodoacetate, *N*-ethylmaleimide (NEM), phenylmethanesulfonyl fluoride (PMSF), and phenylglyoxal. The amino acid composition and amino-terminal sequence were similar to those of other bacterial serine proteases, *i.e.*, *Streptomyces griseus* proteases A and B, Lysobacter enzymogenes α-lytic protease and *Nocardiopsis dassonvillei* subsp. *prasina* OPC-210 alkaline serine protease NDP-1.

Kwon *et al.* (1994) isolated alkaline proteases, named VapT and VapK, from gramnegative alkaliphilic *Vibrio metschnikovii* strain RH530, purified and characterized. Both enzymes had optimum pH and temperature of 10.5 and 60°C, respectively. VapT and

VapK retained 40% and 80%, respectively, of their initial activities at pH 12 after 24 h incubation at 25 °C. The half-life of VapT and VapK was 10 min and 24 min, respectively, at pH 8 and 60 °C. Addition of Ca^{2+} extended their half-life more than 20-fold. VapT and VapK retained 30% and 90%, respectively, of their activities in the presence of 5% SDS and 8 M urea. Analysis of amino acid composition showed that VapT contained seven cysteine residues and that VapK had two. The N-terminal amino acid sequences of the proteases were determined and compared with those of *Bacillus licheniformis* subtilisin Carlsberg, *Vibrio alginolyticus* exoprotease A, and *Tritirachium album* proteinase K. Vazquez *et al.* (1995) isolated 840 protease-producing bacterial strains from different sources of the Antarctic ecosystem during the austral summers of 1989/90 and 1991/92 . These were analyzed in skim-milk agar plates. Thirty-four psychrotrophic strains containing alkaliphilic *Bacillus* strains were selected, classified at genus level and tested for proteolytic activity by the azocasein method from the cell-free supernatant of submerged cultures. The results suggest that these psychrotrophic strains may be useful in the isolation of new types of proteases.

7.2.3   Cloning of Alkaline Protease

The first cloning of an alkaline protease (elastase YaB) was done by Kaneko *et al.* in 1989. By the conventional colony hybridization method, they cloned a 2.2 kb *Hin*d III fragment in *E. coli* using pUC 18 plasmid (designated as pED11). Analysis of the insert showed that the C-terminal sequence was not included. A DNA fragment for the C-terminal fraction was isolated from the *Hap* II digest of the chromosomal DNA using the insert in pRDII as a probe. Finally, they isolated a plasmid, pED103, having a whole gene for their alkaline elastase, as shown in Fig. 7.6. This enzyme gene was not well expressed in the *E. coli* system, but well expressed in the *B. subtilis* system using *E. coli/B. subtilis* shuttle vector pH Y300PLK. The expression was assayed on an elastin plate containing 1% elastin. A protease-less mutant *B. subtilis* DB104 did not form halos but *B. subtilis* DB104 carrying the enzyme gene in pH Y300PLK formed clear halos. The mature enzyme (268 amino acids) was preceded by a putative signal sequence and a prosequence (27 and 83 amino acids, respectively). The mature enzyme was 55% homologous to subtilisin BPN'. Almost all the positively charged residues are predicted to differ between elastase YaB and subtilisin BPN'. Takagi *et al.* (1992a; 1992b) studied the effect of amino acid deletion in Subtilisin E by structural comparison with a microbial alkaline elastase for substrate specificity and catalysis. Chang *et al.* (1996) revealed that the alkaline elastase YaB is synthesized as a 378-amino acid preproenzyme and secreted into the culture medium as a 265-amino acid matured protease. From their results, the pro-peptide of subtilisin YaB functions in trans to guide the folding of secreted subtilisin YaB *in vivo*.

Takami *et al.* (1992a; 1992b) discovered two alkaline proteases from *Bacillus* sp. No. AH-101 during shotgun cloning experiments. The gene of a major thermostable alkaline serine protease that has a high optimum pH (pH 12–13) was cloned in *E. coli* and expressed in *B. subtilis*. The cloned protease was identical to the AH-101 protease in optimum pH and thermostability at high alkaline pH. An open reading frame of 1083 bases, identified as the protease gene, was preceded by a putative SD sequence (AAAGGAGG) with a spacing of 11 bases. The deduced amino acid sequence revealed a pre-pro-peptide of 93 residues followed by the mature protease comprising 268 residues. AH-101 protease showed slightly higher homology with alkaline proteases from alkaliphilic bacilli (61.2%

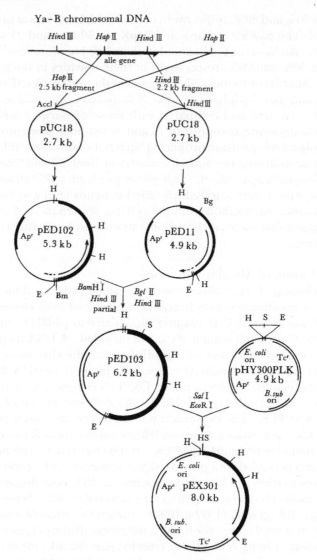

Fig. 7.6   Construction of the plasmid pED103.

and 65.3%) than with those from neutrophilic bacilli (54.9–56.7%). Also AH-101 protease and other proteases from alkaliphilic bacilli shared common amino acid changes and a four amino acid deletion when compared with the proteases from neutrophilic bacilli. AH-101 protease, however, was distinct among the proteases from alkaliphilic bacilli, showing the lowest homology with the others. A minor serine alkaline protease of molecular weight 36,800 was also cloned and expressed in *E. coli.* The plasmid-borne enzyme had an optimum pH for enzyme action of 10.5 and was less thermostable (50 ℃) than that of the major protease.

Another alkaline protease of No. 221 was also cloned by Takami *et al.* (1992*b*). The gene encoding an alkaline serine protease from alkaliphilic *Bacillus* sp. 221 was cloned in *E. coli* and expressed in *Bacillus subtilis*. An open reading frame of 1,140 bases, identified as the protease gene, was preceded by a putative SD sequence (AGGAGG) with a spacing of 7 bases. The deduced amino acid sequence had a pre-pre-peptide of 111 residues followed by the mature protease comprising 269 residues. The alkaline protease from *Bacillus* sp. No. 221 had higher homology with the protease from alkaliphilic bacilli (82.1% and 99.6%) than those from neutrophilic common amino acid changes, and four amino acid deletions that seemed to be related to characteristics of the enzyme of alkaliphilic bacilli when compared to the proteases from neutrophilic bacilli. The alkaliphilic *Bacillus alcalophilus* PB92 gene encoding an alkaline protease was cloned and characterized (Teplyakov *et al.*, 1992; van der Laan *et al.*, 1991). Sequence analysis revealed an open reading frame of 380 amino acids composed of a signal peptide 927 amino acids), a prosequence (84 a. a.), and a mature protein of 269 amino acids. Amino acid comparison with other serine proteases shows good homology with protease YaB. The prosequence of PB92 protease has no significant homology with prosequences of subtilisins. Amino acid sequences of the alkaline proteases listed above are shown in Fig. 7.7. Another alkaline protease, subtilisin ALP1, has been studied extensively by Yamagata *et al.* (1995a ; 1995b ; 1995c). A serine protease (subtilisin ALP I) from alkaliphilic *Bacillus* sp. NKS-21 was cloned and encoded. The molecular weight was 34,624. The protease, subtilisin ALP I, showed about 50% homology with those of serine proteases from *Bacillus subtilis*, *B. polymyxa*, and alkaliphilic *Bacillus* sp. No. 221. The gene contained an open reading frame of 1125 bp, encoding a primary product of 374 amino acids. The mature protease, composed of 272 amino acids, had a putative signal sequence of 37 amino acids and a prosequence of 65 amino acids.

7.2.4   Industrial Applications

A.   Detergent Additives

The main industrial application of alkaliphilic enzymes is in the detergent industry. Detergent enzymes account for approximately 30% of total worldwide enzyme production and these necessary alkaliphilic or alkalitolerant proteins represent a good example of a successful commercial product. As the trend continues toward lower and lower washing temperature, one can expect the penetration of enzyme detergents to increase. Detergents usually have a pH in solution of between 8 and 10.5. For an enzyme to be useful as a detergent additive, it must be active in solution at alkaline pH values and stable in the presence of detergent additives such as bleaching agents, bleach activators, surfactants, perfumes and so on. Furthermore, an enzyme must exhibit long-term stability in the detergent product. Proteolytic enzymes, classified as serine proteases, are widely used in detergent composition. Not all of these are produced by alkaliphilic bacteria.

Several alkaline proteases have been produced by alkaliphilic *Bacillus* strains and are commercially available. After washing, rinsing, and ironing, the effect of the wash was determined as the difference between the reflection of a cloth swatch washed with an enzyme and the reflection without the enzyme. As shown in Fig. 7.8, the addition of alkaline enzymes resulted in high washing efficiency, although each alkaline enzyme showed different values (Aunstrup *et al.*, 1972). The total market for alkaline protease in 1994 was estimated to be 15,000 million yen in Japan.

```
                       1              10                      20
A. AH-101  - Q T V P W G I S F I S Y Q Q H N R G I F G N G A R
B. YaB     - Q T V P W G I N R V Q A P I A Q S R G F T G T G V R
C. PB92    A Q S V P W G I S R V Q A P A H N R G L T G S G V K K
D. BPN'    A Q S V P Y G V S Q I K A P A L H S Q G Y T G S N V K
E. AML     A Q S V P Y G I S Q I K A P A L H S Q G Y T G S N V K
F. SUB     A Q S V P Y G I S Q I K A P A L H S Q G Y T G S N V K
G. CRL     A Q T V P Y G I P L I K A D K V Q A Q G F K G A N V K
```

```
              30                 40                  50
A. V A V L D T G I - A S H P D L R I A G G A S F I S S E - P S Y
B. V A V L D T G I - S N H A D L R I R G G A S F V P G E - P N I
C. V A V L D T G I - S T H P D L N I R G G A S F V P G E - P S T
D. V A V I D S G I D S S H P D L K V A G G A S M V P S E T N P F
E. V A V I D S G I D S S H P D L N V R G G A S F V P S E T N P Y
F. V A V I D S G I D S S H P D L N V R G G A S F V P S E T N P Y
G. V A V L D T G I Q A S H P D L N V V G G A S F V A G E A Y N -
```

```
            60                70                80
A. H D N N G H G T H V A G T I A A L N N S I G V L G V A P S A D
B. S D G N G H G T Q V A G T I A A L N N S I G V L G V A P N V D
C. Q D G N G H G T H V A G T I A A L N N S I G V L G V A P N A E
D. Q D N N S H G T H V A G T V A A L N N S I G V L G V A P S A S
E. Q D G S S H G T H V A G T I A A L N N S I G V L G V S P S A S
F. Q D G S S H G T H V A G T I A A L N N S I G V L G V S P S A S
G. T D G N G H G T H V A G T V A A L D N T T G V L G V A P S V S
```

```
          90            100             110                 120
A. L Y A V K V L D R N G S G S L A S V A Q G I E W A I N N M N M E
B. L Y G V K V L G A S G S G S I S G I A Q G L Q W A A N N G M H
C. L Y A V K V L G A S G S G S V S S I A Q G L E W A G N N G M E
D. L Y A V K V L G A D G S G Q Y S W I I N G I E W A I A N N M D
E. L Y A V K V L D S T G S G Q Y S W I I N G I E W A I S N N M D
F. L Y A V K V L D S T G S G Q Y S W I I N G I E W A I S N N M D
G. L Y A V K V L N S S G S G T Y S G I V S G I E W A T T N G M D
                                   S4
```

```
                130              140                 150
A. I I N M S L G S T S G S S T L E L A V N R A N N A G I L L V G
B. I A N M S L G S S A G S A T M E Q A V N Q A T A S G V L L V V
C. V A N L S L G S P S P S A T L E Q A V N S A T S R G V L L V V
D. V I N M S L G G P S G S A A L K A A V D K A V A S G V V V V V
E. V I N M S L G G P T G S T A L K T V V D K A V S S G I V V A A
F. V I N M S L G G P T G S T A L K T V V D K A V S S G I V V A A
G. V I N M S L G G P S G S T A M K Q A V D N A Y A R G V V V V V
       S1 S2 S3
```

```
                     160                  170
A. AH-101  A A G N T G R Q G - - - - V N Y P A R Y S G V H A V A
B. YaB     A S G N S G A G N - - - - V G F P A R Y A N A M A V G
C. PB92    A S G N S G A G S - - - - I S Y P A R Y A N A M A V G S
D. BPN'    A A G N E G S T G S S S T V G Y P G K Y P S V I A V G G
E. AML     A A G N E G S S G S S S T V G Y P A K Y P S T I A V
F. SUB     A A G N E G S S G S T S T V G Y P A K Y P S T I A V
G. CRL     A A G N S G S S G N T N T I G Y P A K Y D S V I A V G
```

```
          180                190              200
A. A V D Q N G Q R A S F S T Y G P E I E I S A P G V N V Y S R
B. A T D Q N N R A T F S Q Y G A G L D I V A P G V G V Q S T T
C. A T D Q N N N R A S F S Q Y G A G L D I V A P G V N V Q S T
D. A V D S S N Q R A S F S S V G P E L D V M A P G V S I Q S T
E. A V N S S N Q R A S F S S A G S E L D V M A P G V S I Q S T
F. A V N S S N Q R A S F S S A G S E L D V M A P G V S I Q S T
G. A V D S N S N R A S F S S V G A E L E V M A P G A G V Y S
```

```
      210              220                230
A. Y T G N R Y V S L S G T S M A T P H V A G V A A L V K S R Y P
B. V F G N G Y A S F N G T S M A T P H V A G V A A L V K Q K N P P
C. Y P G S T Y A S L N G T S M A T P H V A G A A A L V K Q K N P P
D. L F G N K Y G A Y N G T S M A T P H V A G A A A L I L S K H P
E. L F G G T Y G A Y N G T S M A T P H V A G A A A L I L S K H P
F. L F G G T Y G A Y N G T S M A T P H V A G A A A L I L S K H P
G. Y P T S T Y A T L N G T S M A S P H V A G A A A L I L S K H P
```

Fig. 7.7 (*Continued*)

```
        240                250                260                270
A.  S Y T N N Q I R Q R I N Q T A T Y L G S S N L Y G N G L V H A
B.  S W S N V Q I R N H L K N T A T N L G N T T Q F G S G L V N A
C.  S W S N V Q I R N H L K N T A T S L G S T N L Y G S G L V N A
D.  N W T N T Q V R S S L Q N T T T K L G D S F Y Y G K G L I N V
E.  T W T N A Q V R D R L E S T A T Y L G N S F Y Y G K G L I N V
F.  T W T N A Q V R D R L E S T A T Y L G N S F Y Y G K G L I N V
G.  N L S A S Q V R N R L S S T A T Y L G S S F Y Y G K G L I N V


A.  G R A T Q
B.  E A A T R
C.  E A A T R
D.  Q A A A Q
E.  Q A A A Q
F.  Q A A A Q
G.  E A A A Q
```

Fig. 7.7 Amino acid sequences of AH-101 mature enzyme and other subtilisin-like enzymes. The amino acid sequences enclosed in the *white boxes* (□) and in the *dark boxes* (▨) are common sequences among all of the subtilisin-like enzymes and among the enzymes from alkaliphilic bacilli, respectively. Those enclosed in the *stippled boxes* (▨) are common sequences among subtilisin-like enzymes except the alkaline protease AH-101: AH-101 protease, 221; 221 protease; YaB elastase (YaB) (Kaneko *et al.* 1989); PB92 protease (PB92) (Laan *et al.* 1991); subtilisin BPN' (BPN') (Wells *et al.* 1983). *B. subtilis* var. *amylosaccariticus* (AML) subtilisin (Yoshimoto *et al.* 1988); *B. subtilis* subtilisin (SUB) (Stahl and Ferrari 1984); subtilisin Carlsberg (CRL) (Jacobs *et al.* 1985) (Reproduced with permission from Takami *et al.*, *Appl. Microbiol. Biotechnol.*, **38**, 105 (1992)

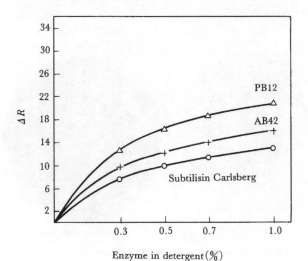

Enzyme in detergent (%)

Fig. 7.8 Relation between enzyme concentration and washing efficiency in phosphate-containing detergent (see text for details) (Aunstrup *et al.*, 1972). PB12, AB42: alkaline proteases.
(Reproduced with permission from Horikoshi and Akiba, *Alkalophilic Microorganisms*, p. 144, Springer Verlag: Japan Scientific Societies Press (1982))

B. Dehairing

Alkaline enzymes have been used in the hide-dehairing process where dehairing is carried out at pH values between 8 and 10. In the old dehairing process, the hides were placed in a bath containing calcium hydroxide and sodium sulfide at a pH of around 12. This process had the disadvantages of being detrimental to the hair and causing swelling of the skin, resulting in difficulties in further processing. These enzymes are now commercially available from several companies.

C. Other Applications

An interesting application of alkaline protease was reported by Fujiwara and Yamamoto (1987). Fujiwara, Yamamoto and Masui (1991) and Ishikawa *et al.* (1993) reported that isolated alkaline protease was used to decompose the gelatinous coating of X-ray films, from which silver was recovered. To improve the processing capacity for the recovery of silver from used X-ray film, a thermostable alkaline protease (Protease B18')

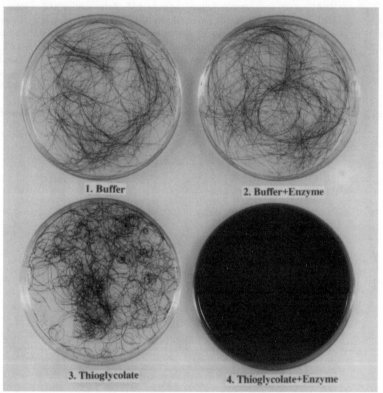

Fig. 7.9 Effects of enzyme on degradation of human hair.

       Human hair (1g) was treated for 2hr at 40 °C under the conditions described in the text. 1, incubation in 50mM glycine-50mM NaCl-NaOH buffer (pH 11.0) (A-buffer); 2, incubation with the alkaline protease AH-101 (3mg) in A-buffer (pH 11.0); 3, incubation in A-buffer containing 1% thioglycolic acid (pH 11.0); 4, incubation with the alkaline protease AH-101 (3mg) in A-buffer containing 1% thioglycolic acid (pH 11.0).

       (Reproduced with permission from Takami *et al.*, *Biosci. Biotechnol. Biochem.*, **56**, 1667 (1992))

was isolated from a thermophilic alkaliphile. It had high optimum pH and temperature, around 13.0 and 85 °C. The enzyme was most active toward gelatin on film at pH 10. As shown in Fig. 7.9, Takami *et al.* (1992*c*) reported degradation of human hair by the alkaline protease from *Bacillus* sp. No. AH-101 that had threefold higher specific activity against keratin than Proteinase K. Human hair was completely hydrolyzed with the enzyme at pH 11.0 in the presence of 1% thioglycolic acid. This enzyme has the potential application to remove clogs in drainpipes, especially in the bathroom since the enzyme degraded human hair at a wide temperature range of 40–90 °C.

7.3   Starch-degrading Enzymes

There are many types of starch-degrading enzymes, e.g. amylase, cyclomaltodextrin glucanotransferase, pullulanase, α-glucosidase, etc. These amylases are widely distributed in living things, where they play an important role in biochemical reactions. Amylase is one of the most familiar enzymes and has been for a long period of time. Its activities have been studied for industrial and pharmaceutical applications and numerous reports published. Until fairly recently all the enzymes reported showed optimum pH for enzyme action in the acid or neutral range. No report concerning an alkaline amylase with optimum activity in the alkaline pH range had been published despite extensive study by many researchers. One noted Japanese enzymologist, Prof. Juichiro Fukumoto of Osaka City University, who investigated amylase, used to say, "There are no alkaline amylases in nature," because he spent over 30 years looking but was not able to discover any.

Horikoshi attempted to isolate alkaliphilic microorganisms producing alkaline amylases, and in 1971 an alkaline amylase was produced in Horikoshi-II medium by cultivating alkaliphilic *Bacillus* sp. No. A-40-2 (Horikoshi, 1971*b*). Several types of alkaline starch-degrading enzymes were subsequently discovered by cultivating alkaliphilic microorganisms (Yamamoto, Tanaka and Horikoshi, 1972; Boyer and Ingle, 1972). No alkaline amylases produced by neutrophilic microorganisms have so far been reported. The above authors classified alkaline amylases into four types according to their pH activity curves,

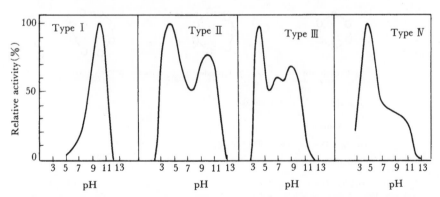

Fig. 7.10   Four types of pH activity curves of alkaline amylases of alkaliphilic *Bacillus* strains.
(Reproduced with permission from Horikoshi and Akiba, *Alkalophilic Microorganisms*, p. 102, Springer-Verlag: Japan Scientific Societies Press (1982))

as shown in Fig. 7.10. The type-I curve has only one peak at pH 10.5; the type-II curve has two peaks at pH 4.0-4.5 and 9.0-10.0; the type-III curve has three peaks at pH 4.5, 7.0 and 9.5-10.0; the type-IV curve has one peak at pH 4.0 with a shoulder at pH 10.0. Two or three peaks in these curves suggested the existence of multiple components in the enzymes, but cloning experiments of the amylase gene revealed that a single enzyme coded in one gene showed a broad peak.

The characteristics of the four types of amylases are summarized in Table 7.6. Type-I amylases are completely adsorbed by DEAE-cellulose at pH 9.0, but the other three types are not. Calcium ions showed protective action against heat inactivation of these amylases, although the extent of the action differed among the four amylases. Of the four amylases, type-I amylases are the most thermolabile. They were completely inactivated by

Table 7.6 Properties of alkaline amylases of alkaliphilic *Bacillus* strains

Type	Strain No.	Optimum pH	Ratio of activity[†1]	Stable pH[†2]	Protection by Ca^{2+}	Adsorption on DEAE cellulose at pH 9.0
I	A-40-2	10.5	1.0	7.0-9.5	+	+
	A-59	10.5	1.0	7.0-9.5	+	+
	27-1	10.5	1.0	7.0-9.0	+	+
	124-1	10.5	1.0	7.0-9.0	+	+
II	135	4.0-4.5 ; 10	0.8	7.0-9.0	+	+
	169	4.0-4.5 ; 10	0.7	7.0-9.0	+	−
III	38-2	4.5 ; 9.0	0.6	5.0-10.5	+	−
IV	13	4.5	0.3	6.5-10	+	−
	17-1	4.5	0.35	6.5-10	+	−

[†1] Ratio of activity at pH 10.0 to maximum activity.
[†2] pH range where 50% of activity remains after incubation at 50°C for 15 min.
(Reproduced with permission from Horikoshi and Akiba, *Alkalophilic Microorganisms,* p.104, Springer-Verlag: Japan Scientific Societies Press (1982))

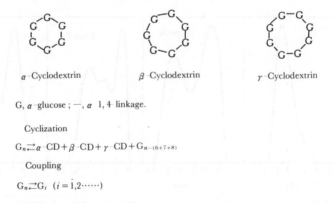

α-Cyclodextrin β-Cyclodextrin γ-Cyclodextrin

G, α-glucose ; —, α 1, 4 linkage.

Cyclization

$G_n \rightleftarrows \alpha\text{-CD} + \beta\text{-CD} + \gamma\text{-CD} + G_{n-(6+7+8)}$

Coupling

$G_n \rightleftarrows G_i \quad (i = 1,2\cdots\cdots)$

Fig. 7.11 Reactions catalyzed by cyclodextrin glycosyltransferase.
CD, cyclodextrin; *n, i,* number of G units.
(Reproduced with permission from Horikoshi and Akiba, *Alkalophilic Microorganisms,* p. 105, Springer-Verlag: Japan Scientific Societies Press (1982))

heating for 15 min at 55 °C in the presence of 10 mM Ca^{2+}. Type-II and type-IV amylases are relatively more stable than type-I amylases. They retained about 70–95% of their original activity after the same heat treatment. Type-III amylases were the most stable. They retained about 50% of their activity at pH 4.5 or pH 10.0 even when heated for 40 min at 65 °C. Type-III amylase (No. 38-2 enzyme) and type-IV amylase (No. 17 and No. 13 enzymes) have high cyclomaltodextrin glucanotransferase activities which convert starch to cyclodextrins. Cyclomaltodextrin glucanotransferase (EC2.4.1.19, CGTase) catalyzes the degradation of starch to form cyclodextrins which are composed of six to eight glucose units linked by an α-1,4-bond. The corresponding cyclodextrins formed are termed α-, β- and γ-CD (Fig. 7.11).

7.3.1 Amylases

A. *Bacillus* sp. No. A-40-2 Amylase

The production of the alkaline amylase was first achieved in an alkaliphilic *Bacillus* species, strain No. A-40-2 (ATCC21592), that was selected from about 300 colonies of bacteria grown in Horikoshi-II medium (Horikoshi, 1971*b*). The isolated strain was an aerobic, sporeforming, gram-positive, motile, rod-shaped bacterium with peritrichous flagella. It is clear that the bacterium should belong to the genus *Bacillus* (Table 7.7). Although the morphological, cultural and biochemical characteristics of the strain resembled those of *Bacillus subtilis*, the special feature of the bacteria was that growth was very good in alkaline media, and the optimal pH for growth was about 10. No growth was detected in neutral media. Furthermore, the GC content of the strain was 39.8 mol% and menaquinone 6 and 7 were detected in the cells. These results are different from those obtained for *Bacillus subtilis*.

Bacillus sp. No. A-40-2 was grown aerobically at 37 °C in Horikoshi-II medium. After 3 days cultivation, the culture fluid was centrifuged at 6,000 × *g* for 10 min. The supernatant fluid was passed through a DEAE-cellulose column equilibrated with 0.1 M Na_2CO_3 (pH 9.5 adjusted with HCl). The alkaline amylase was eluted with 0.1 M Na_2CO_3 containing 0.2 M NaCl. The enzyme was treated by a hydroxyl apatite column, followed by gel filtration on a Sephadex G-100 and purified 80-fold. The molecular weight was estimated to be about 70,000 by gel filtration method. The enzyme is most active at pH 10.0–10.5 and retains 50% of its activity between pH 9.0 and 11.5. The enzyme is not inhibited by 10 mM EDTA at 30 °C, and completely inactivated by 8 M urea. However, about 95% of the activity is recovered upon removal of urea by dialysis. The enzyme can hydrolyze 70% of starch to yield glucose, maltose and maltotriose. Therefore, the enzyme is a type of saccharifying α-amylase.

During cultivation experiments to increase productivity, some amino acids, β-alanine, DL-norvaline and D-methionine, were effective for the production of alkaline amylase. The addition of 0.5% DL-norvaline and 0.5% D-methionine to the culture medium increased amylase production 1.7-fold, while they repressed microbial growth (Ikura and Horikoshi, 1987*a*).

B. *Bacillus* Strain NRRL B-3881 Amylase

Boyer and Ingle (1972) and Boyer *et al.* (1973) reported alkaline amylase in the strain NRRL B-3881. This was the second report of an alkaline amylase. The B-3881 amylase

Table 7.7 Morphological, cultural and biochemical characteristics of strain No. A-40-2

1. Morphological characteristics

Form	Rods
Size	0.5-0.6 μm × 2-2.5 μm
Motility	Motile
Gram strain	Positive
Sporangia	Slightly swollen
Spores	0.9-1.0 μm × 1.2-1.5 μm ; oval ; central

2. Cultural characteristics

	Growth at pH 7	Growth at pH 10.3
Nutrient broth	−	+
Nutrient agar slant	−	+
Glucose-nutrient broth	−	++
Glucose-nutrient agar slant	−	++
Potato	−	++
Horikoshi-I medium	−	++
Horikoshi-II medium	−	++
Glucose-nitrate agar slant	−	−
Glucose-asparagine agar slant	−	−
Anaerobic growth in glucose broth	−	−
Anaerobic production of gas from nitrate	−	−
Horikoshi-I medium containing 5% NaCl	−	++

3. Biochemical characteristics

Hydrolysis of gelatin and casein	Positive
Hydrolysis of starch	Positive
Utilization of citrate	Utilized
Utilization of ammonium salts	Utilized
Reduction of nitrate to nitrite	Reduced
Voges–Proskauer test	Positive

4. pH and temperature

pH for growth in Horikoshi-I medium[†]	pH 7.5 to pH 11
Temperature for growth in Horikoshi-I medium	up to 45 °C at pH 10.3

[†] pH in Horikoshi-I medium was adjusted by adding HCl or NaOH.
− indicates no growth ; +, normal growth ; ++, abundant growth.
(Reproduced with permission from Horikoshi, *Agric. Biol. Chem.*, **35**, 1785 (1971))

had its optimum pH for enzyme action at 9.2. No. A-40-2 amylase retains 50% of its activity between pH 9.0 and 11.5, and B-3881 enzyme retains the same activity between pH 7.0 and 10.5. Both amylases are relatively more stable against EDTA than either *Bacillus amyloliquefacienc* or *B. subtilis* amylase. The enzyme yields maltose, maltotriose and small amounts of glucose and maltotetraose, all of which have a β configuration. The properties of these amylases are given in Table 7.8.

Table 7.8 Properties of alkaline amylases from alkaliphilic *Bacillus* sp. strain No. A-40-2, No. A-59 and NRRL B-3881

Properties	Amylases		
	A-40-2	A-59	B-3881
Optimum pH	10.5	10.5	9.2
Optimum temperature (°C)	55	50	50
Molecular weight	70,000	50,000	
Hydrolysis product from starch	Glc > Mal > M_3	Glc > Mal > M_3	Mal > M_3 > Glc
Inhibition by EDTA	not inhibited	not inhibited	partially inhibited
Type	Saccharifying α-amylase	Saccharifying α-amylase	Saccharifying α-amylase

Abbreviations : Glc, glucose ; Mal, maltose ; M_3, maltotriose.
[Boyer and Ingle (1972)]

C. *Bacillus* sp. No. A-59 Amylase

Alkaliphilic *Bacillus* sp. No. A-59 (ATCC21591) isolated from soil also produced a type-I alkaline amylase in Horikoshi-II medium (Horikoshi, unpublished data). The properties of the strain are almost the same as those of *Bacillus subtilis*. They can grow either in neutral media or in alkaline media. This differs from *Bacillus* sp. No. A-40-2. The alkaline amylase was purified by passing it through a DEAE-cellulose column in the presence of 0.8% NaCl then chromatographed on a DEAE-Toyopearl 650 M column. The enzyme preparation was further separated into two fractions, a major alkaline amylase having molecular weight of 50,000 and a minor alkaline amylase of molecular weight 70,000. The ratio between them was about 10 to 1 and the other properties were not significantly different from the alkaline amylase of alkaliphilic *Bacillus* sp. No. A-40-2 (Table 7.8).

With our *Bacillus* sp. No. A-59 deposited as ATCC21591, Kelly, O'Reilly and Fogarty (1983; McTigue *et al.*, 1994) studied starch-degrading enzymes. They found that alkaliphilic *Bacillus* sp. No. A-59 (ATCC21591) produces three enzymes associated with the degradation of starch: α-amylase, pullulanase and α-glucosidase. The organism was grown in a medium containing 10 g of soluble starch, 6 g of peptone, 3 g of yeast extract, 1 g of K_2HPO_4, 0.04 g of $MnSO_4 \cdot 4H_2O$, 0.20 g of $MgCl_2 \cdot 7H_2O$, 3 g of $Na_2CO_3H_2O$ and 1000 ml of water (final pH, 9.7). Analysis of the growth curve of the organism relative to α-glucosidase production revealed that the biomass rose rapidly and reached a peak after 18 h. They pH fell sharply from 9.7 to 6.6 after 15 h, and after 18 h rapid cell lysis took place. The level of α-glucosidase reached maximum after 24 h. Optimum enzyme production was reached at initial pH 9.7 for α-glucosidase and pullulanase after 24 h growth. However, the pH of the culture at (initial) pH 10.2 dropped to 9.0 after 48 h cultivation, but no trace of α-glucosidase activity could be detected.

Interestingly, α-amylase and pullulanase activities were detected primarily in media having a pH between 9.7 and 10.4. It is of further interest that the alkaliphilic *Bacillus* sp. No. A-59 produces an α-glucosidase with a pH optimum at 7.0 and that the enzyme had only 15% of optimum activity when assayed at pH 9.5 and was inactive at pH 10.0. The enzyme was substrate-specific for *p*-nitrophenyl-α-D-glucoside, maltose and maltotriose in

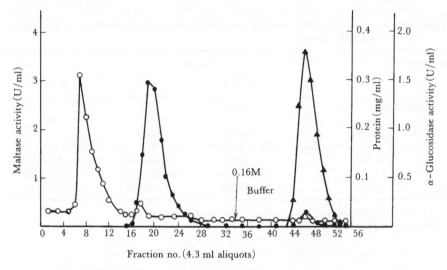

Fig. 7.12 Separation of starch-degrading enzymes from *Bacillus* sp.
Protein ○—○; α-Glucosidase activity ●—●; Maltase activity ▲—▲

that order. Almost all of the activity was located in the cell-free supernatant. Transferase activity was detected using maltose (4%, w/v) as substrate at pH 7.0 at 40 °C for 1 h; the major product was isomaltose.

Another alkaliphilic *Bacillus* sp., NCIB 11203, which was isolated from soil, was found to produce several enzymes in culture broth; these included alkaline amylase, alkaline protease and alkaline phosphatase (Kelly, Brennon and Fogarty, 1987; McTigue *et al.*, 1994). Two extracellular α-glucosidases were detected in the culture broth. The organism was grown in 250-ml Erlenmeyer flasks containing 50 ml of Horikoshi-II medium at 30 °C for 18 h. Cells were removed from the culture medium by centrifugation and the cell-free supernatant was used for purification of the extracellular enzymes. The activity of the crude extracellular α-glucosidase system was examined during cultivation. Essentially, in its extracellular nature and pattern of production, the system is similar to that in the previous report on alkaliphilic *Bacillus* sp. No. A-59 (Kelly O'Reilly and Fogarty, 1983).

The enzyme was purified by fractionation with ammonium sulfate and chromatography on DEAE-Biogel A. As shown in Fig. 7.12, two types of α-glucosidase were isolated. The first of these hydrolyzed *p*-NPG preferentially and had minor activity on isomaltose and isomaltotriose. The second enzyme strongly hydrolyzed maltose and maltotriose and had some activity on *p*-NPG. Both enzymes had pH optimum at 7.0, which is distinct from the enzymes of most other alkaliphilic Bacilli but similar to that of *Bacillus* sp. No. A-59.

Ikura and Horikoshi (1992) studied pH optima of intracellular, membrane-bound, and extracellular amylases from alkaliphilic *Bacillus* No. A-40-2, No. A-59, and neutrophilic *Bacillus licheniformis*. Intracellular, membrane-bound and extracellular amylases of alkaliphilic bacilli cultured in alkaline media exhibited pH optima for enzyme action at 10–10.5, but those amylases from neutrophilic *B. licheniformis* had pH optima for enzyme action at pH 6–6.5. However, the membrane-bound amylase of *Bacillus* No. A-40-2 cultured in the presence of 0.8% NaCl at pH 8.4 had optimum pH around neutrality, indicating that the membrane presumably contained an amylase different from the extracellular one.

D. Amylases of Alkalipsychrotrophic Bacteria

Kimura and Horikoshi (1989) isolated starch-degrading microorganisms from nature. One of them was *Micrococcus* sp. 207 (see p. 28). The main hydrolysis product from amylose, with a crude enzyme preparation, was maltotetraose. The optimum temperature for activity of the amylase was 60 °C and that for pullulanase 55 °C. The activities at 0 to 30 °C exhibited similar activation energy values. In an optimized production medium at pH 8.7 the highest yields of these enzymes were obtained after cell growth at 18 °C for 4 days. At pH 8.5 the yields of amylase and pullulanase became maximum after 3 days cultivation. With more prolonged cultivation the yield of amylase but not that of pullulanase activity decreased.

These enzymes were not produced at temperatures above 30 °C. Sucrose was not effective as an inducer but it stimulated cell growth and enhanced the enzyme productivities with soluble starch. Two starch-degrading enzymes of an alkalipsychrotrophic *Micrococcus* were purified by chromatographies of DEAE-Toyopearl, Butyl-Toyopearl and Shodex WS-2003 (Kimura and Horikoshi, 1990*a*, *d*). Molecular weights and p*I* values of the purified enzymes, I and II, were 185,000 and 125,000 by SDS-PAGE and 4.8 and 4.3 by isoelectric focusing, respectively. Enzyme I had not only amylase but also pullulanase activity. In the presence of Ca^{2+} ions, other properties of the two enzymes were very similar: optimum temperature 55-60 °C (Fig. 7.13), optimum pH 7.5-8.0, K_m value for maltopentaose 0.09 mM. Both amylases were completely inactivated after incubation with EDTA at 30 °C and thereafter could be reactivated by the addition of $CaCl_2$. In the presence of calcium ions, amylase I became thermoresistant, while the thermostability of amylase II decreased. Amylase activity of neither enzyme I nor enzyme II was inhibited by pullulan. Enzyme II hydrolyzed starch and converted to maltotetraose with a yield of 60% in the early stage of

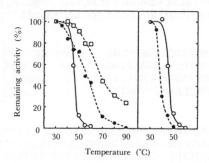

Fig. 7.13   Effect of temperature on the stability of amylase I (left) and II (right) from *Micrococcus* sp. 207.
The enzyme solutions (0.03 U/ml) were incubated for 30 min at various temperatures in 20 mM Tris-acetate (pH 7.5). The remaining activity was assayed in the reaction mixture containing 5 mM $CaCl_2$. —○— Contained 5 mM $CaCl_2$; ⋯●⋯ dialyzed against the buffer without $CaCl_2$; ⋯□⋯ EDTA-treated amylase I described in the text.(Reproduced with permission from Kimura and Horikoshi, *Starch/ Stärke*, **42**, 406 (1990))

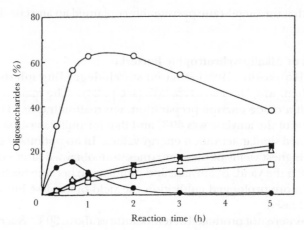

Fig. 7.14 Course of hydrolysis of amylose by enzyme II from *Micrococcus* sp. 207. One ml of 1% amylose in 20mM HEPES buffer (pH 7.5) at 30 °C with 0.2U of purified enzyme II. Symbols; △, glucose; □, maltose; ■, maltotriose; ○, maltotetraose; ●, maltopentaose.
(Reproduced with permission from Kimura and Horikoshi, *J. Ferment. Bioeng.*, **70**, 134 (1990))

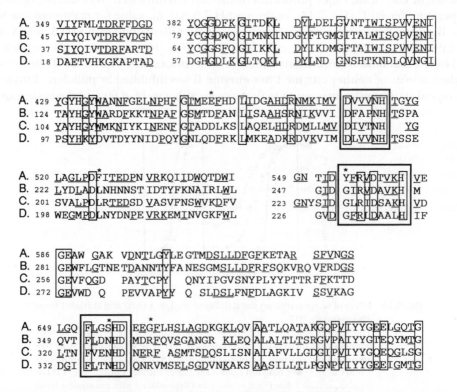

```
A. 349 VIYFMLTDRFFDGD    382 YQGGDFK GITDKL   DYLDELGVNTIWISPVVENI
B.  45 VIYQIVTDRFVDGN     79 YCGGDWQ GIMNKINDGYFTGMGITALWISQPVENI
C.  37 SIYQIVTDRFARTD     64 YCGGSFQ GIIKKL   DYIKDMGFTAIWISPVVENI
D.  18 DAETVHKGKAPTAD     57 DGHGDLK GLTQKL   DYLND GNSHTKNDLQVNGI


A. 429 YGYHGYWANNFGELNPHF GTMEEFHD LIDGAHDRNMKIMV DVVVNH TGYG
B. 124 TAYHGYWARDFKKTNPAF GSMTDFAN LISAAHSRNIKVVI DFAPNH TSPA
C. 104 YAYHGYWMKNIYKINENF GTADDLKS LAQELHDRDMLLMV DIVTNH    YG
D.  97 PSYHKYDVTDYYNIDPQY GNLQDFRK LMKEADKRDVKVIM DLVVNH TSSE


A. 520 LAGLPDFITEDPN VRKQIIDWQTDWI   549 GN TID YFRMDTVKH VE
B. 222 LYDLADLNHNNST IDTYFKNAIRLWL   247    GID GIRMDAVKH M
C. 201 SVALPDLRTEDSD VASVFNSWVKDFV   223 GNYSID GLRIIDSAKH VD
D. 198 WEGMPDLNYDNPE VRKEMINVGKFWL   226    GVD GFRIDAALH IF


A. 586 GEAW GAK VDNTLGYLE GTMDSLLDFGFKETAR  SFVNGS
B. 281 GEWFLGTNE TDANNTYFA NESGMSLLDFRFSQKVRQ VFRDGS
C. 256 GEVFQGD  PAYTCPY   QNYIPGVSNYPLYYPTTR FFKTTD
D. 272 GEVWD Q  PEVVAPYY  Q SLDSLFNFDLAGKIV  SSVKAG


A. 649 LGQ FLGSHD EEGFLHSLAGDKGKLQV AATLQATAKGQPVIYYGEELGQTG
B. 349 QVT FLDNHD MDRFQVSGANGR KLEQ ALALTLTSRGVPAIYYGTEQYMTG
C. 320 LTN FVENHD NERF ASMTSDQSLISN AIAFVLLGDGIPVIYYGQEQGLSG
D. 332 DGI FLTNHD QNRVMSELSGDVNKAKS AASILLTLPGNPYIYYGEEIGMTG
```

Fig. 7.15 (see figure legend p. 169)

the reaction at 30 °C (Fig. 7.14). This activity was significantly depressed under cold and alkaline reaction conditions (Kimura and Horikoshi, 1990*b*).

The enzyme II gene was cloned into *E. coli* JM101 using the vector pHSG399 (Kimura and Horikoshi, 1990*c*). The constructed recombinant plasmid pYK63 contained a 4.8 kb chromosomal DNA fragment derived from strain 207 DNA. The cloned amylase isolated from *E. coli* JM101 (pYK63) produced mainly maltotetraose from starch, and exhibited temperature and pH activity profiles closely similar to those of the enzyme from the original strain. Nucleotide sequence analysis of the cloned DNA fragment revealed one open reading frame containing a gene which consisted of 3312 bp (1104 amino acids). When compared with several other α-amylases, three consensus sequences were identified in the region of the active site. About 300 amino acid residues were present both upstream and downstream of the active site region (Fig. 7.15).

E.   Amylase of Haloalkaliphiles

A haloalkaliphilic archaebacterium, *Natronococcus* sp. strain Ah-36, produced extracellularly a maltotriose-forming amylase (Kobayashi *et al.*, 1992). The amylase was purified to homogeneity by ethanol precipitation, hydroxylapatite chromatography, hydrophobic chromatography, and gel filtration. The molecular weight of the enzyme was estimated to be 74,000 by SDS-PAGE. The amylase exhibited maximal activity at pH 8.7 and 55 °C in the presence of 2.5 M NaCl. The activity was irreversibly lost at low ionic strength. KCl, RbCl, and CsCl could partially substitute for NaCl at higher concentrations. The enzyme was stable in the range of pH 6.0 to 8.6 and up to 50 °C in the presence of 2.5 M NaCl (Fig. 7.16). The enzyme activity was inhibited by the addition of 1 mM $ZnCl_2$ of 1 mM *N*-bromosuccinimide.

The amylase hydrolyzed soluble starch, amylose, amylopectin, and more slowly, glycogen to produce maltotriose with small amounts of maltose and glucose of α-configuration. Maltotetraose to maltoheptaose tested were also hydrolyzed. Transferase activity was detected. However, maltotriose and maltose were not hydrolyzed. The enzyme hydrolyzed γ-cyclodextrin. α- and β-Cyclodextrins were not hydrolyzed. These compounds acted as competitive inhibitors to the amylase activity. Amino acid analysis showed that the amylase was characteristically enriched in glutamic acid or glutamine and in glycine. This was the first report on a purified amylase of an archaebacterium.

Kobayashi *et al.* (1994) cloned the α-amylase expressed in *Haloferax volcanii*. The α-amylase gene (1,512 bp) of the *Natronococcus* sp. contained a signal peptide of 43 amino acids. *Haloferax volcanii* expressed the gene and cleaved the signal peptide accurately. The

---

Fig. 7.15   Comparison of the amino acid sequences of the homologous regions in α-amylases of *Micrococcus* sp. 207 (A), *Bacillus circulans* (B), *Saccharomyces fibuligera* (C) and *Bacillus megaterium* (D). Homologous sequences are underlined and especially boxed when amino acid residues of all proteins are conserved. Asterisks indicate the amino acid residues only differing in the Micrococcal amylase. Thick boxes indicate the three homologous regions of the amylases. (p.168)
(Reproduced with permission from Kimura and Horikoshi, *FEMS Microbiol. Lett.*, **71**, 40 (1990))

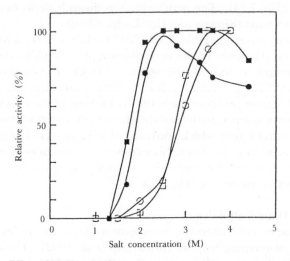

Fig. 7.16 Effect of NaCl and KCl on the amylase stability. Amylase activity was measured after incubation at 50 °C for 30 min at various concentrations of NaCl (closed circles and squares) or KCl (open circles and squares) in the absence (circles) or presence (squares) of 0.5% soluble starch.
(Reproduced with permission from Kobayashi *et al.*, *J. Bacteriol.*, **174**, 3442 (1992))

signal peptide shared an extremely high amino acid sequence identity with that of a protease from the halophilic archaeon 172Pl.

F. Amylases of Other Alkaliphiles

Kim *et al.* (1995) isolated a maltotetraose-forming alkaline α-amylase from an alkaliphilic *Bacillus* strain, GM8901. An alkaliphilic bacterium, *Bacillus* sp. strain GM8901, grown at pH 10.5 and 50 °C, produced five alkaline amylases in culture broth. At an early stage of the bacterial growth, amylase I (Amyl I) was produced initially and then, as cultivation progressed, four alkaline amylases, Amyl II, Amyl III, Amyl IV, and Amyl V, were produced from proteolytic degradation of Amyl I. A serine protease present in the culture medium was believed to be involved in Amyl I degradation. The Amyl I from the culture supernatant was purified by ammonium sulfate precipitation, heparin-Sepharose CL-6B column chromatography, phenyl-Toyopearl column chromatography, and Mono Q HR5/5 high-performance liquid chromatography. The molecular weight of Amyl I was estimated to be sulfate-polyacrylamide gel electrophoresis. Amyl I had an extremely high optimal pH of 11.0 to 12.0 and was stable in a broad pH range of 6.0 to 13.0. Amyl I had an optimal temperature of 60 °C and was stable to 50 °C. Thermostability was increased in the presence of Ca^{2+} and soluble starch. The enzyme required metal ions such as Ca^{2+}, Mg^{2+}, Cu^{2+}, Co^{2+}, Ag^+, Zn^{2+}, and Fe^{2+} for its enzyme activity and was inhibited by 1 mM EDTA and 1 mM phenylmethylsulfonyl fluoride. According to the mode of action of Amyl I on starch, Amyl I was classified as an α- and exo-amylase. Amyl I produced maltotetraose predominantly from starch *via* intermediates such as maltohexaose and maltopentaose.

An alkaline amylase was also isolated from alkaliphilic *Pseudomonas* sp. KFCC 10818 (Na *et al.*, 1996 ; Kim *et al.*; 1996). A gene coding for a new amylolytic enzyme from

Pseudomonas sp. KFCC10818 was cloned and its nucleotide sequence was determined. The coding region for the amylase precursor contained 1,692 nucleotides. The presumed Shine-Dalgarno sequence, AAGG, was located at 8 nucleotides upstream from the ATG initiation codon. The precursor protein had a putative signal peptide of 25 amino acid residues at its amino terminus. There was an open reading frame composed of 1,398 bases in the sequence. A deduced amino acid sequence contained four highly conserved regions of α-amylases. Cloned amylase was purified from *Escherichia coli.* NH_2-terminal sequencing of the enzyme showed the presence of a signal peptide composed of 23 amino acids. Maltose and maltotriose were major products from starch by the enzyme action. pH and temperature optima of the α-amylase were pH 8 and 45 °C, respectively.

7.3.2 CGTases

A. Purification and Properties of No. 38-2 and No. 17-1 Enzymes

Two strains, No. 38-2 and No. 17-1, were selected from approximately 1,000 strains as the best enzyme producers. These bacteria were aerobic, spore-forming, motile and rod-shaped. All the strains belong to the genus *Bacillus.* Further properties are listed in section 2.3 on taxonomy. The organisms were aerobically grown for three days in Horikoshi-II medium at 37 °C. The cells were removed by centrifugation. Three volumes of cold ethanol (-20 °C) were added to one volume of the supernatant fluid. The mixture was stored in a cold room overnight. The precipitates formed were collected by centrifugation, washed with cold ethanol and dried *in vacuo.* The crude enzymes thus prepared were used for the following experiments.

The enzymes of *Bacillus* sp. No. 38-2 and *Bacillus* sp. No. 17-1 were similarly purified (Nakamura, Watanabe and Horikoshi, 1975; Nakamura and Horikoshi, 1976*a, b, c, d*).

(1) Starch adsorption: The crude enzyme powder was dissolved in tap water and dialyzed against tap water for two days. About 3% (w/v) corn starch and 20% ammonium sulfate (w/v) were added to the dialysate and stirred for 60 min at 4 °C. The corn starch which adsorbed the enzyme was collected by filtration and washed thoroughly with 10 mM phosphate buffer (pH 7.0) containing 20% ammonium sulfate (w/v) and NaCl (1.0 N). The CGTase was eluted from the starch with the same buffer containing 3 N NaCl and 0.1 M maltose then dialyzed against 10 mM glycine-NaOH-NaCl buffer (pH 8.5).

(2) DEAE-cellulose chromatography: The dialysate was loaded onto a DEAE-cellulose column equilibrated with the same buffer. The enzyme was eluted with a linear gradient of NaCl (0-0.3 N). Active fractions were collected and concentrated by Ficoll 400 followed by gel filtration on a Sephadex G-100 column. The enzyme fraction was then applied on a preoperative gel electrophoresis apparatus.

The crude enzyme of *Bacillus* sp. No. 38-2 was a mixture of three enzymes: acid CGTase having optimum pH for enzyme action at 4.6; neutral CGTase, pH 7.0 and alkaline CGTase, pH 8.5, as shown in Table 7.9. Southern hybridization experiment showed that only one band hybridized with *Bacillus* sp. No.38-2 CGTase gene. Mäkela *et al.* (1988) purified the CGTase of alkaliphilic *Bacillus* sp. No. 38-2 by a two-step procedure involving affinity chromatography followed by HPLC anion exchange. The HPLC anion exchange chromatogram exhibited at least six active fractions from the sample purified by affinity chromatography, but except for the isoelectric point ($pI = 4.9$-4.6) their enzymatic properties were identical. The main fraction comprising 70% of the total activity had a mole-

Table 7.9   Properties of cyclodextrin glycosyltransferases from alkaliphilic *Bacillus* sp. No. 38-2

Property	Acid CGT	Neutral CGT	Alkaline CGT
Optimum pH	4.5–4.7	7.0	8.0–9.0
Optimum temperature, °C	45	50	
Molecular weight	88,000	85,000–88,000	85,000–88,000
Isoelectric point	5.4		5.4
Stable pH[†1]	6.0–10.0	6.0–9.0	
Stable temperature[†2], °C	Up to 65	Up to 60	
Predominant product	β-Cyclodextrin	β-Cyclodextrin	β-Cyclodextrin

[†1] For 30 min at 60 °C.
[†2] Treated for 30 min.
(Reproduced with permission from Horikoshi and Akiba, *Alkalophilic Microorganisms*, p.107, Springer-Verlag: Japan Scientific Societies Press (1982))

Table 7.10   Yields of cyclodextrins from potato starch

	Yield (%)		
Enzymes	α	β	γ
Bacillus No. 38-2	4	54	16
Bacillus No. 17-1	0	45	15
Bacillus No. 13	1	44	13
Bacillus macerans	12	11	4

cular weight of 70,500 (pI = 4.9). The N-terminal amino acid sequence of 32 amino acids showed that their sequence and ours were identical except for the 30th amino acid, which they reported to be arginine instead of proline. This difference may be caused by the mutation of the second letter which corresponds to either Arg or Pro. Why were many CGTases detected? No definitive experiment has been conducted yet, but the following hypothesis is highly possible. (1) During cultivation, the CGTase must be processed by protease/s and processed protein molecules form. (2) These protein molecules must have different thermal stabilities under specified pH values. (3) Therefore, these CGTases show different pH optima for enzyme action.

The crude enzyme of *Bacillus* sp. No. 17-1 was a mixture of two enzymes: acid CGTase, optimum pH 4.5, and alkaline CGTase, optimum pH 9.5 (Horikoshi, unpublished data). Soluble starch solution (14% w/v) was incubated with the CGTases at pH 8.5 or 4.7 at 40 °C for 10 h and the CDs formed were analyzed by the HPLC method. Table 7.10 summarizes the formation of CDs at pH 8.5. In the case of the No. 38-2 enzyme, the yield of CD ranged 85–90% from amylose, 70–80% from potato starch, 65–70% from amylopectin, and 10–15% from maltose, which is the minimum substrate for CD formation. Further properties such as thermal stability, effect of substrate concentration, etc. are discussed in section 7.2.4 on industrial applications.

Since Nakamura and Horikoshi discovered and isolated bacterium alkaliphilic *Bacillus* sp. No. 38-2 for the first time, many alkaliphilic microorganisms producing CGTases have been reported. Nomoto *et al.* (1984*b*; 1986) found a CGTase produced by alkaliphilic *Bacillus* sp. No. HA3-3-2 isolated from soil from Taipei, Taiwan. The enzyme showed maximum CD-forming activity in the pH range of 6.5 to 8.0 and was stable between pH 6 to 11. Abelyan *et al.*(1994*a*, *b*) developed an isolation method for CGTase using cyclodextrin polymers and their derivatives. CGTases were directly purified from culture broth of mesophilic, thermophilic, and alkaliphilic bacilli by affinity chromatography on a β-CD polymer. The enzymes were further purified by gel filtration and chromatography on diethylaminoethyl (DEAE) derivatives of poly(β-CD) or poly(cellulose-β-CD). This method may open the way to the isolation and identification of CGTase isoenzymes and active fractions.

The source of microorganisms has been expanded to include the deep sea from the

Table 7.11   Characteristics of isolates

Shape	No. 1-7 Rod	No. 3-22 Rod
Size (μm)	0.5–0.7 × 2.5–3.0	0.7–0.8 × 3.0–5.0
Spores	+	+
Ellipsoid	+	+
Round	−	−
Swelling of the sporangium	−	−
Gram reaction	+	+
Catalase	+	+
Anaerobic growth	−	+
VP reaction	−	−
Maximum temperature		
growth positive at (°C)	37	40
growth negative at (°C)	40	45
Growth in		
medium pH 5.7	−	−
NaCl 5%	+	−
7%	+	−
10%	+	−
Acid from		
glucose	−	+
L-arabinose	−	+
xylose	−	+
mannitol	−	+
Gas from glucose	−	−
Lecithinase	+	−
Hydrolysis of		
starch	+	+
gelatin	+	+
casein	−	n.d.
Use of		
citrate	+	−
propionate	−	−
Decomposition of tyrosine	−	−
Nitrate to Nitrite	+	−
Indol	−	−
Phenylalanine deaminase	−	+
Arginine dihydrolase	−	−

n.d. : not done

surface of the earth. Georganta *et al.* (1993) isolated CGTase-producing psychrophilic alkaliphilic bacteria from deep-sea bottom mud samples. Isolates No. 3-22 and 1-7 were identified as an alkaliphilic *Bacillus* sp. Isolate No. 3-22 grew at 4 °C, whereas No. 1-7 did not. The crude enzymes of the both strains showed activity in both broad temperature and broad pH (5–9) ranges. No. 3-22 CGTase produced predominantly β-CD, and minor products were α- and γ-CDs. However, No. 1-7 enzyme produced β-CD as the main product and a small amount of α-CD. No α-CD was detected in the hydrolyzate (Table 7.11).

Padkovyrov and Zeikus found cyclomaltodextrinase from an aerobic microorganism, *Clostridium thermohydrosulfuricm*-39E. A CGTase gene was cloned and expressed in *E. coli* (Podkovyrov and Zeikus, 1992).

B.   Cloning and Nucleotide Sequencing of No. 38-2 CGTase Gene

Hamamoto *et al.* (1987) and Kaneko, Hamamoto and Horikoshi (1988) have characterized a CGTase of alkaliphilic *Bacillus* sp. No. 38-2. This is a unique enzyme, especially in having wide pH optimum, pH or thermal stability and high productivity, compared with other microbial CGTases. Conventional purification methods showed that the crude enzyme preparation contained three enzymes: acid-, neutral- and alkaline-CGTase. Therefore, it is very interesting to investigate the genetic information of the enzyme(s) using gene cloning methods.

The alkaliphilic *Bacillus* sp. No. 38-2 was aerobically grown in Horikoshi-II medium overnight at 37 °C and the chromosomal DNA extracted. The chromosomal DNA was partially hydrolyzed with the *Sau*3AI restriction enzyme. The DNA fragments were inserted into the *Bam*HI site and ligated with T$_4$ ligase. Chimeric DNAs were used for transformation experiments. The transformants were plated on LB-starch agar medium containing 50 µg/mg Ampicillin. About 104 Ap^r Tc^s transformants per µg of DNA were obtained. One transformant containing the gene of a starch hydrolyzing enzyme was isolated after exposure to iodine. This transformant carrying a plasmid pSC8, which contains a 5.3 kb DNA fragment, was found to possess CGTase activity by analyzing the digest. The Southern hybridization experiment showed that the insert was hybridized with the chromosomal DNA of *Bacillus* sp. No. 38-2. However, no complementary sequences were detected in *E. coli* DNA fragments. Since CGTase activity was not observed when this plasmid was hydrolyzed with *Hinc* II, the nucleotide sequence around the *Hinc* II site was studied. The DNA sequence and the deduced amino acid sequence are shown in Fig. 7.17. There was a single open reading frame of 2,136 bp, which encoded a polypeptide of 712 amino acids. A putative ribosome binding site, a GAGGAGG sequence that highly complemented the 3'end of 16S ribosomal RNA, was observed upstream of the open reading frame. The nucleotide sequence and the amino acid sequence of this CGTase has strong homology with those of the CGTase of *B. macerans*. At the amino acid level 448(68%) of the aligned amino acids were identical and the overall homology of the aligned nucleotide sequences was about 64%.

E. coli carrying pCS8 was aerobically grown in LB-broth for 24 h at 37 °C. The CGTase activity localized in the extracellular fraction, periplasmic fraction and cellular fraction was 5.3%, 58.8% and 35.9%, respectively. The synthesis of the enzyme was constitutive and no effect of supplements such as glucose and starch was observed.

The CGTase produced by the *E. coli* gave one line of precipitation which fused completely with that for the *Bacillus* sp. No. 38-2 CGTase by the Ouchterlony double diffusion

```
            GGCAACGCACGCCCAATCTTCCTTCCAGCTCTGTAATCTCTCGACTGCTATTCGGTGCACACCGGTCCACGTCAGGAT

AACGGGCATGAATTGGCGGATAAAATCTTTAACATTCATGGCGTCGATCCCCCTATAAGGTAGTCTTCCTGATCCGTCTCGACTCCTTAA
                                                          EcoRI
TCCCACTCCCTCGATCATACTATATATCTGAGAATATTGTTATATATTGACATTTGAATTCGCTTTCATATAAAATGAACAAGAACACAT

CACTATACTTACATACAAGCTAAGGGCTATGCATTCCTTACCTTACCCCGGTATGGAACAACCCCGGTATCTCTATTAGAGACGCGGGGT

TTTTTATGTAGCCGAGATGAAGGAGGTGATCCCCAAAGCGACGGACAGGCCTGTTATCCCCAAGCACTGTATACGATGAGGAGGTATAGT
                                                                              (SD)
        10        20        30        40        50        60        70        80 |      90
ATGAAAAGATTTATGAAACTAACAGCCGTATGGACACTCTGGTTATCCCTCACGCTGGGCCTCTTGAGCCCGGTCCACGCAGCCCCGGAT
MetLysArgPheMetLysLeuThrAlaValTrpThrLeuTrpLeuSerLeuThrLeuGlyLeuLeuSerProValHisAlaAlaProAsp
(-27)                                                                              (1)
        100       110       120       130       140       150       160       170 |     180
ACCTCGGTATCCAACAAGCAGAATTTCAGCACGGATGTCATATATCAGATCTTCACCGACCGGTTCTCGGACGGCAATCCGGCCAACAAT
ThrSerValSerAsnLysGlnAsnPheSerThrAspValIleTyrGlnIlePheThrAspArgPheSerAspGlyAsnProAlaAsnAsn

        190       200       210       220       230       240       250       260       270
CCGACCGGCGCGGCATTTGACGGATCATGTACGAATCTTCGCTTATACTGCGGCGGCGACTGGCAAGGCATCATCAACAAAATCAACGAC
ProThrGlyAlaAlaPheAspGlySerCysThrAsnLeuArgLeuTyrCysGlyGlyAspTrpGlnGlyIleIleAsnLysIleAsnAsp

        280       290       300       310       320       330       340 BclI 350       360
GGTTATTTGACCGGCATGGGCATTACGGCCCATCTGGATTTCACAGCCTGTCGAGAATATCTACAGCGTGATCAACTACTCCGGCGTCCAT
GlyTyrLeuThrGlyMetGlyIleThrAlaHisLeuAspPheThrAlaCysArgGluTyrLeuGlnArgValIleAsnTyrSerGlyValHis

        370       380  .    390       400       410       420       430       440       450
AATACGGCTTATCACGGCTACTGGGCGCGGGACTTCAAGAAGACCAATCCGGCCTACGGAACGATGCAGGACTTCAAAAACCTGATCGAC
AsnThrAlaTyrHisGlyTyrTrpAlaArgAspPheLysLysThrAsnProAlaTyrGlyThrMetGlnAspPheLysAsnLeuIleAsp

        460       470       480       490       500       510       520       530       540
ACCGCGCATGCGCATAACATAAAAGTCATCATCGACTTTGCACCGAACCATACATCTCCGGCTTCTTCGGATGATCCTTTCCTTTGCAGAG
ThrAlaHisAlaHisAsnIleLysValIleIleAspPheAlaProAsnHisThrSerProAlaSerSerAspAspProSerPheAlaGlu

        550       560       570       580       590       600       610       620       630
AACGGCCGCTTGTACGATAACGGCCAACCTGCTCGGCGGATACACCAACGATACCCAAAATCTGTTCCACCATTATGGCGGCACGGATTTC
AsnGlyArgLeuTyrAspAsnGlyAsnLeuLeuGlyGlyTyrThrAsnAspThrGlnAsnLeuPheHisHisTyrGlyGlyThrAspPhe

        640       650       660       670       680       690       700       710       720
TCCACCATTGAGAACGGCATTTATAAAAACCTGTACGATCTGGCTGACCTGAATCATAACAACAGCAGCGTCGATGTGTATCTGAAGGAT
SerThrIleGluAsnGlyIleTyrLysAsnLeuTyrAspLeuAlaAspLeuAsnHisAsnAsnSerSerValAspValTyrLeuLysAsp

        730       740       750 HincII760       770       780       790       800       810
GCCATCAAAATGTGGCTCGACCTCGGGGTTGACGGCATTCGCGTGGACGCGGTCAAGCATATGCCATTCGGCTGGCAGAAGAGCTTTATG
AlaIleLysMetTrpLeuAspLeuGlyValAspGlyIleArgValAspAlaValLysHisMetProPheGlyTrpGlnLysSerPheMet

        820       830       840       850       860       870       880       890 ,     900
TCCACCATTAACAACTACAAGCCGGTCTTCAACTTCGGGCGAATGGTTCCTTGGCGTCAATGAGATTAGTCCGGAATACCATCAATTCGCT
SerThrIleAsnAsnTyrLysProValPheAsnPheGlyGluTrpPheLeuGlyValAsnGluIleSerProGluTyrHisGlnPheAla

        910   .   920       930       940       950       960       970       980       990
AACGAGTCCGGGATGAGCCTGCTCGATTTCCCGTTTGCCCAGAAGGCCCGGCAAGTGTTCAGGGACAACACCGACAATATGTACGGCCTG
AsnGluSerGlyMetSerLeuLeuAspPheProPheAlaGlnLysAlaArgGlnValPheArgAspAsnThrAspAsnMetTyrGlyLeu

        1000      1010      1020      1030      1040      1050      1060      1070      1080
AAAGCCGATGCTGGAGGGCTCTGAAGTAGACTATGCCCAGGTGAATGACCAGGTGACCTTCATCGACAATCATGACATGGAGCGTTTCCAC
LysAlaMetLeuGluGlyGlySerGluValAspTyrAlaGlnValAsnAspGlnValThrPheIleAspAsnHisAspMetGluArgPheHis

        1090      1100      1110      1120      1130      1140      1150      1160      1170
ACCAGCAATGGCGACAGACGGAAGCTGGAGCAGGCGCTGGCCTTTACCCTGACTTCACGCGGTGTGCCTGCCATCTATTACGGCAGCGAG
ThrSerAsnGlyAspArgArgLysLeuGluGlnAlaLeuAlaPheThrLeuThrSerArgGlyValProAlaIleTyrTyrGlySerGlu

        1180      1190      1200      1210      1220      1230      1240      1250      1260
CAGTATATGTCTGGCGGGAATGATCCGGACAACCGTGCTCGGATTCCTTCCTTCTCCACGACGACGACGCCATATCAAGTCATCCAAAAG
GlnTyrMetSerGlyGlyAsnAspProAspAsnArgAlaArgIleProSerPheSerThrThrThrAlaTyrGlnValIleGlnLys

        1270      1280      1290      1300      1310      1320      1330 BclI 1340      1350
CTCGCTCCGCTCCGCAAATCCAACCCGGCCATCGCTTACGGTTCCACACAGGAGCGCTGGATCAACAACGATGTGATCATCTATGAACGC
LeuAlaProLeuArgLysSerAsnProAlaIleAlaTyrGlySerThrGlnGluArgTrpIleAsnAsnAspValIleIleTyrGluArg

        1360      1370      1380      1390      1400      1410      1420      1430      1440
AAATTCGGCAATAACGTGGCCGTTGTTGCCATTAACCGCAATGAACACACCGGCTTCGATTACCGGCCTTGTCACTTCCCTCCCGCAG
LysPheGlyAsnAsnValAlaValValAlaIleAsnArgAsnMetAsnThrProAlaSerIleThrGlyLeuValThrSerLeuProGln

        1450      1460      1470 EcoRI 1480      1490      1500      1510      1520      1530
GGCAGCTATAACGATGTGCTCGGCGGAATTCGTGAACGGCAATACGCTAACCGTGGGTGCTGGCGGTGCAGCTTCCAACTTTACTTTGGCT
GlySerTyrAsnAspValLeuGlyGlyIleLeuAsnGlyAsnThrLeuThrValGlyAlaGlyGlyAlaAlaSerAsnPheThrLeuAla

        1540      1550      1560      1570      1580      1590      1600      1610      1620
CCTGGCGGCACTGCTGTATGGCAGTACACACAACCGATGCCACAGCTCCGATCAACGGCAATGTCGGCCCGATGATGGCCAAGGCAGGGGTC
ProGlyGlyThrAlaValTrpGlnTyrThrThrAspAlaThrAlaProIleAsnGlyAsnValGlyProMetMetAlaLysAlaGlyVal

        1630      1640      1650      1660      1670      1680      1690      1700      1710
ACGATTACGATTGACGGCCGCGCTTCGGCTCGGCAAGGAACGGTTTACTTCGGTCAACGGCAGTCACTGGCGCGGACATCGTAGCTTGG
ThrIleThrIleAspGlyArgAlaSerAlaArgGlnGlyThrValTyrPheGlyThrThrAlaValThrGlyAlaAspIleValAlaTrp
```

Fig. 7.17   (Continued)

Fig. 7.17   The nucleotide sequence of the CGTase gene from alkaliphilic *Bacillus* sp.
No. 38-2.

The DNA sequence of the coding strand is given from 5' to 3', numbered
from nucleotide 1 at the putative initiation site. The proposed ribosomal
binding site (SD) is underlined with a dotted line. The predicted amino
acid sequence is given below the DNA sequence. The position of processing of the signal peptide is indicated by an arrowhead. The boxed amino
acids have been determined by automated Edman sequencing of the purified CGTase. The sequence containing an inverted repeat structure downstream from the termination codon TAA is designated by ⟶ ⟶ .

(Reproduced with permission from Kaneko *et al.*, *J. Gen. Microbiol.*, **134**, 100
(1988))

test. A protein with a molecular weight of 75,159 could be translated from the open reading frame of 2,136 bp; the molecular weight was slightly higher than that estimated by the
SDS-PAGE method (68,000). The NH_2-terminal sequence of the purified CGTase of
Bacillus sp. No. 38-2 was determined to be as follows by an automatic sequencer up to the
17th residue: Ala-Pro-Asp-Thr-Ser-Val-Ser-Asn-Lys-Gln-Asn-Phe-Ser-Thr-Asp-Val-Ile-. This
amino acid sequence was identical to that deduced from the DNA sequence. Therefore,
27 amino acid residues (residues–27 to –1) may be a signal peptide which is removed during the secretion process. No significant difference was observed between the CGTase of
E. coli carrying pCS8 and the crude CGTase of alkaliphilic *Bacillus* sp. No. 38-2, and both
CGTases could produce CD from starch at pH 10.0.

The products were analyzed by HPLC and the major product was β-CD (α-CD : β-CD
: γ-CD = 10 : 70 : 20). Why were we able to isolate three CGTases from the culture fluid?
There are three possibilities: (1) artifacts during the purification process, (2) three genes
in the chromosomal DNA, but not found, and (3) one gene makes several gene products
in *Bacillus* sp. No. 38-2, although in *E. coli* the HB101 system exhibited one protein band.
Akino, Kato and Horikoshi (1989*b*) reported that one gene of β-mannanase of *Bacillus* sp.
No. AM-001 produced two enzymes having different C-terminals in *E. coli* (see p. 252).
We have tried to clone acid-, neutral-, or alkaline-CGTase gene(s) from *Bacillus* sp. No. 38-
2, but have not yet succeeded. Since we isolated three cellulase genes from *Bacillus* sp. No.
N-4, it is highly possible that *Bacillus* sp. No. 38-2 has other CGTase genes besides the
enzyme we isolated.

A CGTase gene of *Bacillus* sp. No. 17-1 has been cloned and sequenced. As shown in

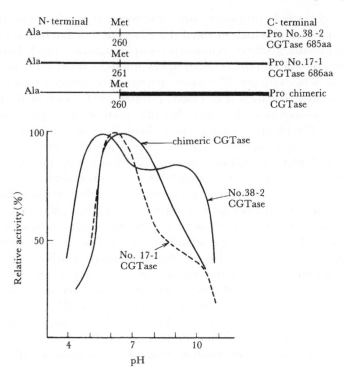

Fig. 7.18 pH activity curves No. 38-2, No. 17-1 and chimeric CGTases.

Fig. 7.18, chimeric experiments between No. 17-1 CGTase and No. 38-2 CGTase revealed that the pH-activity profile was affected by the C-terminal region of the protein molecule (Kaneko *et al.*, 1989). The CGTase of *E. coli* carrying pSC8 could be changed to type-IV CGTase having optimum pH for enzyme action at pH 6, when the N-terminal fraction of pSC8 was substituted by the N-terminal fraction of pUCP1 having the CGTase gene of *Bacillus* sp. No. 17-1 (Kaneko, Kudo and Horikoshi, 1990; Kaneko *et al.*, 1990). Using 12 chimeric CGTases constructed from two genes, the CGTase gene from the alkaliphilic *Bacillus* sp. strain No. 38-2 and the CGTase gene from the alkaliphilic *Bacillus* sp. No. 17-1, we compared the effect of those chimeric enzymes on cyclodextrin (CD) production, especially on the composition of the CDs produced. It was found that the N-terminal and the C-terminal segments were important for CD production. Chimeric enzymes that contained the N-terminal and the C-terminal segments derived from CGTase 38-2 produced large amounts of CD and, in particular, a higher proportion of α-CD than other chimeric enzymes.

Although no actual industrial production has been reported, Hellman and Maentsaelae (1992) over expressed CGTase 38-2 in *E. coli*. A novel export-affinity fusion vector having the gene encoding CGTase from *Bacillus* No. 38-2 (ATCC21783) was constructed. The CGTase fusion protein vector was constructed by deleting the translational stop codons from the gene encoding CGTase by *in vitro* mutagenesis. *E. coli* alkaline phosphatase (APase; phoA) and *Bacillus stearothermophilus* (ATCC12980) α-amylase were fused to the CGTase gene. Overexpression of a wild-type CGTase and the hybrid proteins under the control of the *lac* promoter caused a leaky phenotype in *E. coli*; the outer membrane became permeable which enabled the adsorption of the fusion protein directly from the culture medium onto α-cyclodextrin coupled agarose. The hybrid proteins were eluted from the column with α-CD under mild conditions at pH 7.5. The CGTase-APase fusion had good *in vivo* stability. The fused enzymes retained their biological activities.

As described above, Mäkela *et al.* (1988) purified the CGTase of alkaliphilic *Bacillus* sp. No. 38-2 by a two-step procedure involving affinity chromatography followed by HPLC anion exchange. The HPLC anion exchange chromatogram exhibited at least six active fractions from the sample purified by affinity chromatography. We found three fractions (acid, neutral and alkaline CGTases) having different pH optima after conventional purification. We have tried to isolate other DNA fragments besides pSC8 but have not yet succeeded. Southern hybridization experiment showed that only one band hybridized with pSC8. No other band was found in chromosomal DNA. Why were many CGTases detected? We have not yet conducted any definitive experiment but the following hypothesis is highly possible. (1) During cultivation, the CGTase must be processed by protease(s) and processed protein molecules formed as shown in Mäkela's paper. (2) These protein molecules must have different thermal stabilities under specified pH values. (3) Therefore, these CGTases show different pH optima for enzyme action. Actually, the CGTase produced by *E. coli* carrying pSC8 exhibited different pH activity curves under different incubation temperatures, as shown in Fig. 7.19 (Kaneko, unpublished data). These results strongly indicate that alkaliphilic *Bacillus* sp. No. 38-2 chromosomal DNA has one CGTase gene which produces one protein molecule. This protein molecule is probably processed during the cultivation or secretion process.

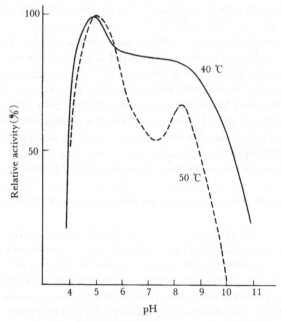

Fig. 7.19 pH activity curves at 40 °C and 50 °C.

C. Cloning of CGTase of *Bacillus* sp. No. 1011

Kimura, Takano and Yamane (1987) cloned the gene for β-CGTase from an alkaliphilic bacterium, *Bacillus* sp. No. 1011, in an *Escherichia coli* phage λ D69. This alkaliphilic bacterium produces extracellular β-CGTase. *Bacillus* sp. No. 1011 was aerobically grown in Horikoshi-II medium and the chromosomal DNA of the bacteria was prepared. λPhages containing the β-CGTase gene were subcloned in *E. coli* using pBR322 and *B. subtilis* by pUB110. An *E. coli* recombinant plasmid, pTUE202, and a *B. subtilis* plasmid, pTUB703, were selected from ten plasmids, because the transformants by each of the two plasmids produced the highest amounts of extracellular β-CGTase in each strain. The major product of hydrolysis from starch by the β-CGTase of *E. coli* carrying pTUB202 and *B. subtilis* carrying pTUB703 was β-CD, as in the case of the enzyme of the parent *Bacillus* sp. No. 1011. A common DNA fragment of approximately 2.5 kb was detected in all plasmids. The enzymatic activity was lost when part of the common region was deleted so the nucleotide sequence of this fragment in the inserted DNA was determined (Kimura *et al.*, 1987). There was an open reading frame of 2,154 bp beginning with the ATG initiation codon and ending with two termination codons, TAA and TGA. The structural gene for the β-CGTase consisted of 2,139 bp (713 amino acids with a total molecular weight of 78,339). The extracellular β-CGTase of *Bacillus* sp. No. 1011 had Ala-Pro-Asp at the N-terminal. These amino acids corresponded to nucleotides 82 to 90. Therefore, the first 27 amino acids, from the initiator methionine to alanine, constitute a signal peptide involved in the secretion of protein. The CGTase was composed of 686 amino acids (molecular weight 75,225), and the molecular weight estimated by SDS-PAGE was 66,000.

Upstream of the ATG initiation codon, the sequence for the ribosome-binding site was detected at nucleotide positions 12 to 6. The deduced amino acid sequence at the N-terminal side of the enzyme showed a high homology with the sequences of α-amylase in the three regions which constitute the active centers of α-amylases.

In contrast, the COOH-terminal region of the β-CGTase was completely different from those of the α-amylases. Many α-amylases are composed of approximately 500 amino acids. Therefore, the COOH-terminal region of β-CGTase would contain an extra 200 to 250 amino acids in addition to the polypeptide exhibiting the amylase activity. Although such a hypothesis may be premature, it is suggested that β-CGTase may have two protein domains: the one on the N-terminal side cleaves the α-1,4-glycosidic bond in starch, and the other on the COOH-terminal side catalyzes other activities, such as cyclization, etc.

Nakamura *et al.* (1992) extensively studied the functional relationships between cyclodextrin glucanotransferase from an alkaliphilic *Bacillus* and α-amylases by site-directed mutagenesis. Two Asp and one Glu residues, which are considered to be the catalytic residues in α-amylases, were also involved in CGTases. The three mutant CGTases, in which Asp^{229}, Glu^{257} and Asp^{328} were individually replaced by Asn or Gln, completely lost both their starch-degrading and β-CD-forming activities. These three inactive enzymes retained the ability to be bound to starch (Nakamura, Haga and Yamane, 1993). Furthermore, they revealed that three histidine residues in the active center are important in the stabilization. On the basis of the three-dimensional structures of CGTases, three histidine residues, which are conserved between CGTases and α-amylases, are located at the active center and are proposed to constitute the substrate binding sites. The three histidine residues (His-140, His-233, and His-327) of CGTase from alkaliphilic *Bacillus* sp. 1011 were individually replaced by site-directed mutagenesis to probe their roles in catalysis. Asparagine-replaced CGTases (H140N-, H233N-, and H327-CGTase) retained cyclization activity but had altered production ratios of α-, β-, and γ-cyclodextrin. Replacement of histidine by asparagine residues strongly affected the k(cat) for β-cyclodextrin-forming, coupling, and hydrolyzing activities, whereas it barely affected the K(m) values. The activation energies for α-cyclodextrin hydrolysis were increased more than 12 kJ/mol by the replacement. Furthermore, the K(i) values of acarbose, which is thought to be a transition-state analog of glycosidase catalysis, were 2–3 orders of magnitude larger in asparagine-replaced CGTases than that in wild-type CGTase. Therefore, the three histidine residues participate in the stabilization of the transition state, whereas they participate little in ground-state substrate binding. H327N-CGTase showed decreased activity over an alkaline pH range indicating that His-327 is important for catalysis over an alkaline pH range.

Furthermore, Nakamura, Haga and Yamane (1993) found that four aromatic residues, which are highly conserved among CGTases but not found in α-amylases, are located in the active center. To analyze the roles of these aromatic residues, Phe-183, Tyr-195, Phe-259, and Phe-283 of *Bacillus* sp. 1011 CGTase were replaced by site-directed mutagenesis and the effects of this procedure were examined. Y195L-CGTase, in which Tyr-195 was replaced by a leucine residue, underwent a dramatic change in its cyclization characteristics: it produced considerably more γ-cyclodextrin than the wild-type enzyme and virtually no α-cyclodextrin. Y195L-CGTase demonstrated increased K_m values for cyclodextrins, whereas the values for a linear maltooligosaccharide donor were insignificantly changed. Taken together with the structural information of CGTase crystals soaked with substrates, the authors of the report proposed that Tyr-195 plays an important role in the spiral binding of substrate. Replacing either Phe-183 or Phe-259 with leucine induced increased K_m values for acceptors.

Furthermore, the double mutant F183L/F259L-CGTase showed considerably decreased cyclization efficiency, but the intermolecular transglycosylation activity remained normal. These results indicate that Phe-183 and Phe-259 are cooperatively involved in acceptor binding, and that they play a critical role in cyclization when the nonreducing end of amylose binds to the active center of CGTase. Replacing Phe-283 with a leucine residue induced a decrease in k(cat) and an affinity for acarbose, suggesting that Phe-283 is involved in transition-state stabilization.

D. γ-CD-forming CGTase

Although the strain isolated was not alkaliphile, Kato and Horikoshi demonstrated that one strain of *Bacillus subtilis* produced a γ-CD-forming CGTase in culture broth (Kato and Horikoshi, 1986*a*). The isolate, *Bacillus subtilis* No. 313, was grown aerobically for five days at 37 °C in a cultivation medium containing 1% potato starch, 1% polypeptone, 0.1% yeast extract, 0.3% KH_2PO_4, 1% $MgSO_4$ and 0.02% $CaCO_3$. Maximum cell growth was observed after 30 h cultivation and γ-CD-forming activity reached maximum after five days. The culture broth was centrifuged at 6,000 × *g* for 10 min to remove cells, and the supernatant fluid was used as a crude sample. The γ-CD-forming activity in the crude sample was assayed by the BGC method (Kato and Horikoshi, 1984*a*). The crude enzyme preparation exhibited a relatively broad pH activity curve for CD formation with a pH opti-

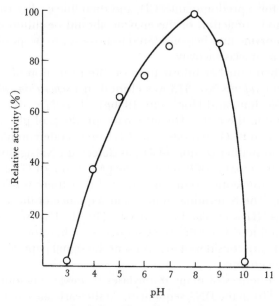

Fig. 7.20   Effect of pH on activity.

mum at 8.0, as shown in Fig. 7.20. Analysis of the enzymatic digest by HPLC showed the product to be only γ-CD. The crude enzyme solution (3,000 ml) was brought to 30% saturation with ammonium sulfate. The precipitate formed was removed by centrifugation and the enzyme was precipitated from the supernatant fluid by adding ammonium sulfate (70% saturation). The precipitated enzyme was dialyzed against 10 mM acetate buffer (pH 6.0) for 24 h. The dialysate was loaded onto a DEAE-Sephadex A-50 column. The γ-CGTase was passed through the column and concentrated *in vacuo* followed by dialysis against 25 mM Tris-acetic acid buffer (pH 8.3). The sample was loaded on a Polybuffer exchanger PBE 94 column and eluted by applying pH gradient (pH 8.3-6.0). Enzyme activity was detected in the fractions of pH 7.0-7.1. After gel filtration by Sephacryl S-200 column, the enzyme solution was dialyzed against 10 mM acetate buffer (pH 5.5) and charged on a CM-Toyopearl column equilibrated with the same buffer described above. The γ-CGTase was eluted at a concentration of 0.05 M NaCl. About 1 mg of the purified enzyme was obtained from 3,000 ml of culture broth.

The isoelectric point of the enzyme was 7.1 and its molecular weight was estimated to be 64,000 by the SDS-PAGE method. The NH_2-terminal sequence of the purified enzyme was determined by automated Edman degradation up to the 13th residue to be Ser-Val-Lys-Asn-Gly-Thr-Ile-Leu-His-Ala-Trp-Asn-Trp-. The enzyme converted starch to γ-CD with an optimum pH 8.0 and was stable over the range pH 5.5-8.5 at 50 °C (30-min incubation). Enzyme activity was stable up to 50 °C, and was inactivated at 70 °C in the presence or absence of calcium ions.

Although this enzyme produced only γ-CD from starch the yield was about 5%, which was lower than that of other CD-forming enzymes. Such a low yield may be caused by a different mode of action to starch. During the course of CD production, formation of both CD and oligosaccharide was observed simultaneously. *B. macerans* and *Bacillus* sp.

No. 38-2 enzymes form predominantly CD, and then linear oligosaccharides are formed. Therefore, the mode of action of the enzyme should be different from that of other enzymes and the enzyme may be considered to possess two properties: CD-forming activity and liquefying α-amylase activity.

In order to obtain further information on the formation of γ-CD from starch, the CGTase of *Bacillus subtilis* No. 313 was cloned in *Escherichia coli* using pBR322 and expressed in *E. coli* (Kato and Horikoshi, 1986*b*). *E. coli* carrying pMT2 produced 140 μg/ml of γ-CGTase in the cells. The enzymatic products of pMT2-borne CGTase were analyzed by HPLC and BCG methods. The CGTase encoded on pMT2 produced only γ-CD from starch and neither α- nor β-CD was detected. No significant differences were observed between the enzyme of *E. coli* carrying pMT2 and γ-CGTase of *B. subtilis* No. 313.

Analysis of the nucleotide sequence showed that there was a single open reading frame of 1,632 bp. The N-terminal amino acid sequence of the γ-CGTase from *Bacillus subtilis* No. 313 was NH$_2$-Ser-Val-Lys-Asn-Gly-Thr-Ile-His-Ala-Trp-Asn-Trp, which is identical to that deduced from the DNA sequence. Therefore, 45 amino acid residues (residues -45 to -1) are considered to represent the signal peptide which is removed during secretion of the enzyme.

Amino acid sequences of many α-amylases from various sources have been determined or predicted from the DNA sequences. At the early stage of hydrolysis of starch the γ-CGTase produced γ-CD and acyclic dextrin at the same time. On the other hand, other CGTases producing predominantly CD and acyclic dextrin could be detected after long incubation. These results strongly suggest that the γ-CGTase may be an intermediate type of α-amylase and true CGTases (Kato, 1989). The DNA sequence of the enzyme gene supports this point because analysis demonstrated that the DNA sequence has strong homology with the gene of liquefying α-amylase of *Bacillus subtilis*.

E. Industrial Production of Cyclodextrins

The cyclodextrins, Schardinger dextrins, are a group of homologous oligosaccharides obtained from starch by the action of cyclomaltodextrin glucanotransferase (CGTase). These compounds are fascinating to carbohydrate chemists because they have properties markedly different from other dextrins. Some of these unique properties are as follows: (1) Cyclodextrins (CD) are homogeneous cyclic molecules composed of six or more glucose units linked α-1,4 as in amylose; (2) as a consequence of cyclic arrangement they have neither a reducing end-group nor a nonreducing end-group and are not decomposed by hot aqueous alkali; (3) they are rather resistant to acid hydrolysis and the common starch-splitting amylases; (4) they crystallize very well from water and from aqueous alcohols; (5) they form an abundance of crystalline complexes called inclusion compounds with organic substances; and (6) they form a variety of inorganic complexes with neutral salts, halogens and bases. Properties of cyclodextrins are shown in Table 7.12. Mass production of these unique compounds on an industrial scale has been attempted several times in the past. In 1969, Corn Products International Co., U.S.A., produced β-CD using *B. macerans* CGTase. Teijin Ltd. of Japan also produced β-CD using the macerans enzyme in a pilot plant. However, there were serious problems in both production processes. (1) CGTase from *B. macerans* is not suitable for industrial use because the enzyme is not thermostable enough. (2) Yield of CD from starch is not high, usually 20% to 30% on an industrial scale. (3) Toxic organic solvents such as trichloroethylene, bromobenzene, toluene, etc. were

Table 7.12 Properties of cyclodextrins

	α-CD	β-CD	γ-CD
Number of glucose units	6	7	8
Molecular weight	973	1,135	1,297
Cavity diameter	5–6Å	7–8Å	9–10Å
Cavity depth	7–8Å	7–8Å	7–8Å
Crystal from (from H_2O)	Needle	Prism	Prism
$[\alpha]^{25}{}_D(H_2O)$	+150.5°	+162.5°	+177.4°
Solubility (g/100 ml·H_2O, 25 °C)	14.5	1.85	23.2

Table 7.13 Some properties of various CGTases

	Optimum pH	Optimum temp.	Molecular weight	pI	CDs composition from starch	Yield (%)
Bacillus macerans	5.0–5.7	55 °C	65,000	4.6	α-CD > β-CD > γ-CD	50
Klebsiella pneumoniae	6–7	35–40	–	–	α-CD > β-CD > γ-CD	50–60
Bacillus stearothermophilus	6.0	70	68,000	4.5	β-CD $\geqq$ α-CD > γ-CD	50
Bacillus megaterium	5.2–6.2	55	66,000	6.1, 6.8	β-CD > α-CD > γ-CD	62
Bacillus circulans	5.2–7.2	55	–	5.8, 6.6	β-CD > α-CD > γ-CD	–
Alkaliphilic *Bacillus*						
No. 38-2	5–9	65–70	75,160	5.4	β-CD > γ-CD > α-CD	75–85
No. 17-1	5–9	60	74,140	–	β-CD > γ-CD > γ-CD	70

used to precipitate CD due to the low conversion rate (Table 7.13). The use of such harmful organic solvents is now prohibited in various fields, especially food processing. Therefore, the development of large-volume use is quite limited. We isolated several CGTases from alkaliphilic *Bacillus* strains as described in the following section. One of them, a CGTase produced by alkaliphilic *Bacillus* sp. No. 38-2, overcame all these weak points and we succeeded in mass-producing crystalline α-, β-, γ-CD and CD-mixture at low cost without using any organic solvents.

1. Analysis of CDs

To determine CGTase activity, three methods had been reported prior to our investigation: the blue-value method, Tilden-Hudson method and glucoamylase method. These were not accurate enough to determine cyclodextrin content in the hydrolysate. Sato *et al.* (1985) reported direct measurement of cyclodextrin using high performance liquid chromatography (HPLC).

Kaneko *et al.* (1987) developed a new colorimetric determination method for CDs. Reduction in the color intensity to phenolphthalein after complexation with α-, β- or γ-CD is involved. The color intensities of inclusion complexes of β- and γ-CDs decreased linearly by increasing amounts of the CDs up to 0.5 mg β-CD and 2.0 mg γ-CD. However, no absorbance change was observed in the case of α-CD. Neither soluble starch nor maltooligosaccharides affected the color intensity of phenolphthalein. The cyclization activity was determined as follows. A reaction mixture containing 40 mg of each substrate in 1.0 ml of 0.1 M phosphate buffer (pH 6.0) and 0.1 ml of CGTase solution suitably diluted was incubated at 60 °C for 20 min. The reaction was stopped by adding 3.5 ml of 30 mM

Table 7.14 Effect of various carbohydrates on the BCG method

γ-CD	α-CD	β-CD	Glucose	Soluble starch	Amount found, µg	% Error[†]
440	0	0	0	0	440	
440	200	0	0	0	430	− 2.3
440	400	0	0	0	470	+ 6.8
440	4000	0	0	0	460	+ 4.5
440	0	200	0	0	440	0
440	0	400	0	0	460	+ 4.5
440	0	4000	0	0	450	+ 2.3
440	0	0	200	0	450	+ 2.3
440	0	0	400	0	440	0
440	0	0	4000	0	450	+ 2.3
440	0	0	0	200	440	0
440	0	0	0	400	450	+ 2.3
440	0	0	0	4000	460	+ 4.5
440	200	200	0	0	440	0
440	400	400	0	0	440	0
440	4000	4000	0	0	410	− 6.8
440	400	400	400	0	460	+ 4.5
440	200	200	200	200	430	− 2.3
440	400	400	400	400	460	+ 4.5
440	4000	4000	2000	2000	440	0

† Error (%) = (found/(γ-CD − 1)) × 100.

NaOH solution and 0.5 ml of 0.02% (w/v) phenolphthalein in 5 mM Na_2CO_3 solution. As a reference 0.5 mg of β-CD in 0.1 ml of water was used. One unit of enzyme activity was defined as the amount of enzyme that formed 1 mg of β-CD per minute under the conditions described above.

A simple colorimetric determination method for γ-CD was developed by Kato and Horikoshi (1984*a*). One milliliter of sample solution (up to 700 µg as γ-CD) is mixed with 0.1 ml of Bromocresol green (BCG) (5 mM) and 2 ml of 0.2 M citrate buffer (pH 4.2), and the absorbance is measured at 630 nm. Addition of various carbohydates, such as a α-, β-CDs, glucose, or soluble starch, did not cause significant error, as shown in Table 7.14.

2. Industrial Production of β-CD

The optimum concentration represents a compromise of several factors. Theoretically, the best yields are expected from the most diluted solutions. Since starch is a low-cost material, the use of high concentrations of starch has significant economic advantages for production on an industrial scale. As a result, the optimum substrate concentration in our plant was set at about 15%. One ton of potato starch suspension (15%) containing 5 mM $CaCl_2$ was liquefied by the No. 38-2 enzyme at 82 °C at pH 8.5 then cooled to 65 °C. The liquefaction was readjusted to pH 8.5 with $Ca(OH)_2$ and 1,500 g of the enzyme was added. CD formation was maintained at 65 °C with continuous stirring for a further 30 h. After the reaction, the enzyme was inactivated by heating at 100 to 120 °C. Then the pH of the reaction mixture was brought to 5.5 to 5.7, and an appreciable amount of bacterial α-amylase was added to hydrolyze the saccharides which were not converted to CD. The digest was decolorized with active charcoal and filtered off. The filtrate was passed through ion-exchange resin to remove ions. The hydrolysate was concentrated to about 50% (w/v) under reduced pressure and crystallized by the addition of a small amount of mother crystalline β-CD lowering the temperature gradually. Crystalline material was separated by a basket type centrifuge and washed with a small amount of water. The cured β-CD was recrystallized by the conventional method with hot water, if necessary. The filtrate contained α-, β-, γ-CDs, glucose, maltose, and oligosaccharides. The filtrate thus obtained is able to form inclusion complexes with many organic substances because about 20% of the filtrate is a mixture of CDs (Fig. 7.21).

The first attempt to immobilize *Bacillus* sp. No. 38-2 CGTase was made using an ion-exchange resin of vinylpyridine copolymer (Nakamura and Horikoshi, 1977), and more recently using a synthetic absorption resin Diaion HP-20 purchased from Mitsubishi Chemical Industrial Co. (Kato and Horikoshi, 1984*b*). In the first attempt the crude enzyme preparation was succinylated in a preliminary step and vinylpyridine copolymer resin was activated by 0.1 M phosphate (KH_2PO_4–Na_2HPO_4) buffer (pH 8.0). This process resulted in immobilization of the enzyme with an activity yield of about 210 units. In the next experiment the immobilization of CGTase was achieved by passing 10 ml of the crude enzyme preparation through a Diaion HP-20 column equilibrated with 5 mM Tris-HCl buffer (pH 8.5) containing 10 mM $CaCl_2$. The yield of the enzyme in the immobilization was about 70% based on the enzyme protein.

Several reports on improved production processes to reduce production costs have been published. Kim, Lee and Kim (1993) developed an ultrafiltration membrane bioreactor to increase CD production. Cyclodextrin glycosyltransferase (CGTase) was found to be severely inhibited by cyclodextrins. In order to increase the conversion yield by reducing product inhibition and reuse the CGTase in the production of cyclodextrins from milled corn starch, an ultrafiltration membrane bioreactor system was employed. In a batch operation with ultrafiltration the conversion yield was increased 57% compared with that without ultrafiltration. Moreover, Okada, Ito and Hibino (1994) reported immo-

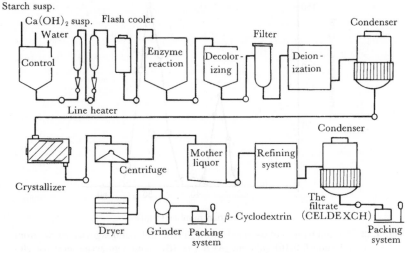

Fig. 7.21   Flow chart of production of β-cyclodextrin.

bilization of CGTase on capillary membrane. The percentage of cyclodextrin (CD) to total sugar obtained in the permeate was slightly more than 60% under most operating conditions and a reaction temperature of 50 °C.

In practical industrial scale plants, however, only batch-wise production has been conducted as described above, and no immobilized enzyme system has yet been used.

Abraham (1996) reported a continuous production of cyclodextrins using immobilized CGTase. Continuous operation at 55 °C of a cylindrical reactor of 141 ml capacity using the immobilized enzyme (80 g of high-silica fabric containing 114 mg of purified enzyme) gave a maximum productivity of 10.2 g of cyclodextrins/l/h at a dilution rate of 0.32/h and a substrate concentration of 20 g/l. The half life of the biocatalyst was found to be 22 days, which could be further improved by using a lower operating temperature.

3. Production of α- and γ-CDs from the filtrate

The filtrate was treated with bacterial amylase (saccharifying) and glucoamylase to hydrolyze β-, γ-CDs, and maltooligosaccharides, decolorized, concentrated, and passed through a column of Diaion FRK 101 Na (Mitsubishi Chemicals Industrial Co., Tokyo) at 60–80 °C to separate the α-CD and glucose. Fig. 7.22 shows an elution profile of α-CD, glucose and maltooligosaccharides on a laboratory scale. The α-CD solution was concentrated to up to 45–50% and α-CD crystallized. γ-CD was also crystallized from the filtrate as follows. The filtrate was incubated with glucoamylase at 50 °C for 24 h to hydrolyze acyclic dextrins. The CD mixture and glucose were separated with a column of FRK 101, and γ-CD was isolated by passing it through a Toyopearl HW-40 column at 60–70 °C. Then γ-CD fractions were collected and concentrated to about 60% (w/v) under reduced pressure. γ-CD in the concentrate was crystallized directly.

Specifications: Crystalline CDs (α-, β- and γ-), and two types of filtrate are now com-

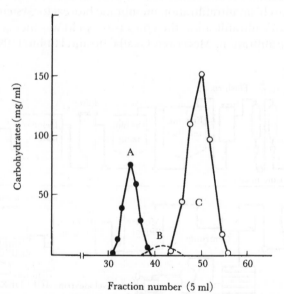

Fig. 7.22 Elution pattern of α-cyclodextrin, glucose and maltooligosaccharides from a Diaion FRK101 column: (A) α-CD; (B) maltooligosaccharides; (C) glucose.

Table 7.15 Specification of CDs and the filtrates.

	α-CD	β-CD N	β-CD P	γ-CD
Appearance		White crystals or crystalline powder		
Transparency		Colorless and transparent		
	at 10% (w/v)	at 1% (w/v)		at 15% (w/v)
$[\alpha]_D^{20}$ (H_2O, 1%, w/v)	$+147.0 - +152.0°$	$+160.0 - +164.0°$		$+172.0 - +177.7°$
Loss on drying (%)	<10	<12	<10	<10
CD content (%)	>98.5	>98.0	>99.0	>98.5
Other CDs (%)	None	None		None
Dextrose equivalent (D.E.)	<0.5	<1.0	<0.5	<0.5
Ash (%)	<0.10	<0.10	<0.10	
Heavy metals (ppm)	<4	<4		<4
Arsenic (ppm)	<1	<1		<1
Viable counts Bacteria	$<300/g$	$<300/g$		$<300/g$
Yeasts and Molds	$<50/g$	$<50/g$		$<50/g$
Coli form bacteria	$(-)$	$(-)$		$(-)$

	CH-20	CH-30	CH-30H
Appearance		Viscous	
Transparency		Colorless and transparent[†]	
	—	—	
$[\alpha]_D^{20}$ (H_2O, 1%, w/v)	—	—	—
Loss on drying (%)	<25	<25	<25
CD concent. (%)	18–22	18–22	>15
Dextrose equivalent (D.E.)	18–22	28–32	<0.5
Ash (%)	<0.05	<0.05	<0.05
Heavy metals (ppm)	None	None	None
Arsenic (ppm)	None	None	None
Viable counts Bacteria	$<300/g$	$<300/g$	$<300/g$
Yeasts and Molds	$<50/g$	$<50/g$	$<50/g$
Coli form bacteria	$(-)$	$(-)$	$(-)$

† White precipitates (CD) are formed if stored at low temperatures.

mercially available. Specifications of the products are summarized in Table 7.15. Solubility: Crystalline β-CD has low solubility but the filtrates (CH) are freely soluble in water. Toxicological studies: Data were obtained using Sprague-Dawley strain albino rats with body weight ranging between 150 and 200 g (Table 7.16). Comparative studies of metabolism have been made with radioactive starch and β-CD. β-CD was metabolized at a slower initial rate than starch, but after 24 h the total amounts metabolized were about the same. The distribution of radioactivity in the animals was similar to that obtained with starch. The simple methods described above can reduce the cost of β-CD from ¥100,000 to ¥2,000/kg, and that of α-CD to within ¥30,000/kg. Also, the filtrate, CH, has come to be marketed at approximately ¥300/kg. This has paved the way for its use in large quantities in foodstuffs, chemicals and pharmaceuticals.

Table 7.16 Toxicological and metabolic studies of α-, β- and γ-CDs

1. Toxicological studies

(a) α-CD (cyclohexaamylose)

≪ Acute ≫

Route of	LD$_{50}$ (g/kg)			
	Rats		Mice	
Administration	♀	♂	♀	♂
Oral	> 10	> 10	9.24	8.03
Intravenous	1.070	0.842	1.015	0.921

≪ Sub-acute ≫

No toxicological symptoms were observed during 30 days oral and hypodermic administration to mice at daily doses of 2.5 g and 0.42 g/kg, respectively.

(b) β-CD (Cycloheptaamylose)

≪ Acute ≫

Route of	LD$_{50}$ (g/kg)			
	Rats		Mice	
Administration	♀	♂	♀	♂
Oral	> 10	> 10	> 10	> 10
Intravenous	0.453	0.500	0.299	0.216

≪ Chronic ≫

No toxicological symptoms were observed during 6 months oral administration to rats at daily doses of 0.1, 0.4, and 1.6 g/kg.

(c) γ-CD (Cyclooctaamylose)

≪ Acute ≫

Route of	LD$_{50}$ (g/kg)			
	Rats		Mice	
Administration	♀	♂	♀	♂
Oral	> 8	> 8	> 16	> 16
Subcutaneous	> 2.4	> 2.4	> 4	> 4
Intravenous	> 2.4	> 2.4	> 4	> 4

2. Metabolic studies

Comparative metabolic studies have been done with radioactive starch, glucose and cyclodextrins. Cyclodextrins were metabolized at a slower initial rate than starch and glucose, but after 24 h the total amounts of metabolized materials were about the same.

The distribution of radioactivity within the animals was similar to that obtained with starch.

4. Uses of CD and CH

The following uses are suggested. (1) Stabilization of volatile materials (a) Conversion to dry form from toxic liquid chemicals (b) Stabilization of flavors and spices (c) Deodorization of medicines and foods (2) Protection against oxidation and UV-degradation during storage or processing (3) Modification of physical and chemical properties (a) Increase solubility of water-insoluble medicines (b) Masking of shifting of colors and fluorescences (c) Masking of bitterness in foods and medicines (d) Stabilization of deliquescent chemicals (e) Promotion of hydrolysis of some esters (4) Emulsification of steroids, hydrocarbons, oils, fats and fatty acids (5) Solidification of hydrocarbons, oils, fats and fatty acids. Using the above properties, many industrial applications of CD have

Table 7.17 Industrial application of cyclodextrins

Functions	Guests	End Products
Foods		
(1) Emulsification	Oils and fats	Margarine, Cake, Whipping cream, French dressing
(2) Stabilization	Flavors, Spices, Colors and pigments	Horseradish paste, Mustard paste. Cakes and cookies, Pickled vegetables, Dried vegetables
(3) Masking of taste and odor		Juices, Soy milk, Bone powder, Boiled rice
(4) Improvement of quality		Hard candy, Cheese, Soy sauce, Canned citrus fruits and juices
(5) Reduce volatility	Ethanol	Food preservatives
(6) Others		Breath mints
Cosmetics and toiletries		
(1) Emulsification	Oils and fats	Face cream, Face lotion, Toothpaste
(2) Stabilization	Flavors and fragrances	Bath refresher crystals
Agrochemicals		
(1) Stabilization	Pyrolnitrin	Fungicide
	Pyrethroids	Insecticide
(2) Reduce volatility	Organic phosphates (DDVP)	Insecticide
	Thiocarbamic acid	Herbicide
(3) Reduce toxicity	2-Amino 4-methyl-phosphynobutyric acid	Fungicide

Functions	Guests and End Products
Pharmaceuticals	
(1) Improve solubility	Prostaglandins, Steroids, Cardiac glycosids, Non-steroidal antiinflammatory agents, Barbiturates, Phenytoin, Sulfornamides, Sulfonylureas, Benzodiazepines
(2) Chemical stabilization	
(A) Hydrolysis	Prostacylin, Cardiac glycosides, Aspirin, Atropine, Procaine
(B) Oxidation	Aldehydes, Epinephrine, Phenothiazines
(C) Photolysis	Phenothiazines, Ubiquinones, Vitamins
(D) Dehydration	Prostaglandin E_1, ONO-802
(3) Improve bioavailability	Aspirin, Phenytoin, Digoxine, Acetohexamide, Barbiturates, Nonsteroidal antiinflammatories
(4) Powdering	ONO-802, Clofibrate, Benzaldehyde, Nitroglycerin, Vitamin K_1, K_2, Methylsalicylate
(5) Reduce volatility	Iodine, Naphthalene, d-Camphor, l-Menthol, Methylcinnamate
(6) Improve taste, smell	Prostaglandins, Alkylparabens
(7) Reduce irritation to stomach	Nonsteroidal antiinflammatory agents
(8) Reduce hemolysis	Phenothiazines, Flufenamic acid, Benzylalcohol, Antibiotics

been reported and some of them are now commercially available. Since we first succeeded in the industrial production of CD in 1978, commercial applications of these unique compounds have been investigated by numerous companies, and many of them are now in use in various fields. Table 7.17 summarizes the industrial application of CDs.

F.   Novel Applications of CGTase

A novel application of CGTases as transglucosidase was reported by Kometani *et al.* (1994*a, b*). A CGTase from an alkaliphilic *Bacillus* species was purified by starch adsorption and Q-Sepaharose chromatography. The purified enzyme had cyclizing activity, transglycosylating (coupling) activity, and starch-hydrolyzing activity, and their pH-activity curves had a single peak (pH 5.5 as the optimum pH) with a broad shoulder at alkaline pHs (Type IV enzyme). Transglycosylation to various saccharides and flavonoids at alkaline pH was more effective than that at neutral pH. Among flavonoids, those containing rutinose (diosmin and hesperidin) were transglycosylated more effectively than those containing neohesperidose (naringin and neohesperidin).

Furthermore, the CGTase produced hesperidin monoglucoside and a series of its oligoglucosides by the transglycosylation reaction with hesperidin as an acceptor and soluble starch as a donor. The formation of the glycosides was more effective at alkaline pHs than at neutral or acidic pHs because of the higher solubility of the acceptor. The structure of the purified monoglucoside was identified as 4(G)-α-D-glucopyranosyl hesperidin by FAB-MS, α-, β-glucosidase and glucoamylase treatments, and methylation analysis. The solubility of both hesperidin mono- and diglucoside in water was about 300 times higher than that of hesperidin and they were found to have a stablizing effect on the yellow pigment crocin, from fruit of *Gardenia jasminoides*, against ultraviolet radiation. Therefore, CGTase is used not only for production of CDs but also has potential application as transglucosidase.

Kometani and his colleagues (1996*a, b*) synthesized neohesperidin glycosides and naringin glycosides by cyclodextrin glucanotransferase from an alkaliphilic *Bacillus* sp. An alkaline pH was very effective for solubilizing neohesperidin, the amount of glycosides formed was increased. As a result, its amount with β-CD at pH 10 was about seven times greater than that with soluble starch at pH 5. The enzyme from an alkaliphilic *Bacillus* sp. had a wider acceptor specificity than that from *B. macerans*.

7.3.3   G6-forming Enzyme

From the begining of the 1970s, many bacterial strains which produce amylases catalyzing the degradation of starch to malto-oligosaccharides (Gn-amylase) have been isolated and the enzymes characterized.

A.   *Bacillus* sp. No. 707

In order to obtain hyperproducers of enzymes for industrial applications, an alkaliphilic bacterium, *Bacillus* sp. No. 707, a producer of Gn-amylase, was isolated from soil and the gene for maltohexaose-producing amylase (G6-amylase) cloned (Kimura *et al.*, 1988). An alkaliphilic bacterium, *Bacillus* sp. No. 707, produces at least five Gn-amylase components in Horikoshi-II medium.

The gene for G6-amylase of the chromosomal DNA of *Bacillus* sp. No. 707 was cloned in an *E. coli* phage, λ D69, and recloned in *E. coli* plasmid pBR322 and *B. subtilis* plasmid pUB110 using the same cloning method as that utilized for the genes for CGTase. Plasmid pTUE306 for *E. coli* and plasmid pTUB812 for *B. subtilis* were isolated. A common DNA region of approximately 2.5 kb was defined among the inserted DNA's. The enzymatic activity was lost when a part of the common region was deleted. The plasmids were stably maintained and the gene was well expressed in *B. subtilis*, which produced more than 70 times higher activity in the culture medium than did *Bacillus* sp. No. 707. The major hydrolysis product from soluble starch by the enzyme from *Bacillus* sp. No. 707 was G4. In contrast, the product of the enzyme from *B. subtilis* (pTUB812) and *E. coli* (pTUE306) was G6. The content of G6 in the hydrolyzate was approx. 50-60%. Three other major bands (110, 75 and 60 kDa) of *Bacillus* sp. No. 707 in SDS polyacrylamide gels were extracted and incubated with soluble starch. The major hydrolysis products by 110, 75 and 60 kDa bands were G4, G5 and G6, respectively.

The nucleotide sequence of the gene for the maltohexaose-producing enzyme was determined to pinpoint the location of the enzyme gene (Tsukamoto *et al.*, 1988). Several deletion plasmids of pTUE306 were constructed and their enzyme production assayed by measuring conventional amylase activity. The limit of the DNA region for the expression of the activity was approximately 2.5 kb. There was an open reading frame of 1,554 bp beginning with the ATG codon at the nucleotide positions $+1$ to $+3$, ending with a termination codon TAA at nucleotide position 1,555 to 1,557 (515 amino acids with a molecular weight of 59,007). The N-terminal analysis of the enzyme of *B. subtilis* 207-25 carrying pTUB812 indicated that the first 33 amino acids from Met to Ala were a peptide involved in the secretion of the exported proteins. Upstream of the ATG initiation codon, there are AGGAGG (-11 to -5) for the ribosome-binding sequence and TTG-CCAATTGATATTTAAGTCGAGTGAAAT (-128 to -99), which seemed to be the most probable promoter. The deduced amino acid sequence of the extracellular mature enzyme was more than 60% homologous to those of the liquefying type α-amylases but not to those of the saccharifying type α-amylases. This suggests that the enzyme gene is derived from a common ancestor gene of other liquefying α-amylases. However, the signal peptide and N-terminal, which are essentially not responsible for the enzyme activity, are completely different from other α-amylases. It is of interest that the G6 enzyme may be one kind of liquefying type α-amylase as was found in the γ-CGTase described above.

B.   *Bacillus* sp. No. H-167 Enzyme

Kimura *et al.* (1988) reported a maltohexaose-forming enzyme from alkaliphilic *Bacillus* sp. No. 707 and cloned its enzyme gene. However, no report on microbiological properties has been presented. Independently, we have isolated alkaliphilic *Bacillus* strains producing maltohexaose-forming enzymes in their culture broths (Hayashi, Akiba and Horikoshi, 1988*a*, *b*).

One of the strains isolated, alkaliphilic *Bacillus* sp. No. H-167 (Table 7.18), produced three α-amylases which yielded maltohexaose as the main product from starch. The optimum culture conditions for enzyme production were: initial medium, pH 9.4; culture temperature, 37°C; and 50-60 h cultivation under aerobic conditions. The enzymes (H-I-1, H-I-2, and H-II) were separated completely and purified to homogeneity on disc electrophoresis using chromatographies on DEAE-Toyopearl 650 M and DEAE-Sepharose CL-6B, followed by gel filtration on Sephadex G-100 or Ultrogel AcA-34. The optimum and

Table 7.18 Morphological and biochemical properties of the isolated bacterium, strain H-167

Cells	Rods (2–3 μm × 0.4–0.7 μm)
Motility	Motile
Spores	Central
Sporangia	Swollen
Gram stain	Variable
pH for growth	pH 7–12
Temperature for growth	15–52 °C
Catalase test	+
Oxidase test	+
Reduction of nitrate	+
V–P test	−
Utilization of citrate	−
Urease test	−
Production of indole	−
Growth in NaCl	Up to 12%
GC content	46.3%

Acid formation from carbohydrates

	Arabinose	Glucose	Xylose	Mannitol
Aerobic	+	+	+	−
Anaerobic	+	+	+	+

(Reproduced with permission from Hayashi *et al.*, *Agric. Biol. Chem.*, **52**, 445 (1988))

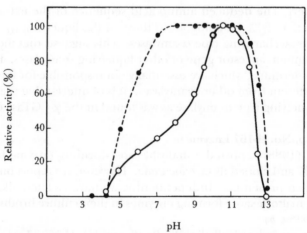

Fig. 7.23 Effect of pH on activity and stability of the amylase H-I-1. Each pH was obtained with the following buffer system: acetate buffer, pH 3.5–6.0; Tris-HCl buffer, pH 7.0–8.0; glycine-NaOH buffer, pH 9.0–11.5; KCl-NaOH buffer, pH 12.0–13.0. —○— pH activity curve; ·—●— pH stability curve. (Reproduced with permission from Hayashi *et al.*, *Appl. Microbiol. Biotechnol.*, **28**, 283 (1988))

Table 7.19 Properties of three amylases (H-I-1, H-I-2, and H-II) from alkaliphilic
Bacillus H-167

Property	H-I-1	H-II-2	H-II
Molecular weight	73,000	59,000	80,000
Isoelectric point	4.1	3.5	4.3
Optimum pH	10.5	10.5	10.5
Optimum temperature	60 °C	60 °C	60 °C
pH stability	7–12	7–12	7–12
Heat stability	55 °C	55 °C	55 °C
K_m for amylose (DP 17)	0.35 mM	0.43 mM	0.40 mM
Inhibitors	Hg^{2+}, Zn^{2+}, Pb^{2+}, Co^{2+}, Ni^{2+}		

(Reproduced with permission from Hayashi *et al.*, *Appl. Microbiol. Biotechnol.*, **28**, 284
(1988))

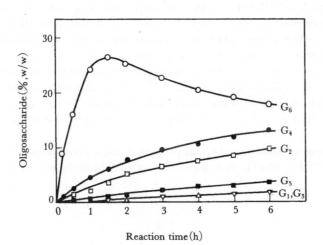

Reaction time (h)

Fig. 7.24 Time course of hydrolysis of starch by enzyme H-I-1.
One ml of 5% soluble starch in 20 mM glycine-NaOH buffer (pH 10.5) was
incubated at 50 °C with 1.0 U of purified enzyme H-I-1. Symbols: —▽—, glucose (G_1); —□—, maltose (G_2);—△—, maltotriose (G_3); —●—, maltotetraose (G_4); —■—, maltopentaose (G_5); —○—, maltohexaose (G_6).
(Reproduced with permission from Hayashi *et al.*, *Agric. Biol. Chem.*, **52**, 447
(1988))

stable pH values of amylase H-I-1 are shown in Fig. 7.23. No distinct differences were
found either in the pH activity curves or pH stability curves among H-I-1, H-I-2, and H-II.
All enzymes were most active at pH 10.5 and stable in the range pH 7.5–12.0 on standing
at 50 °C. Other properties are shown in Table 7.19, and all enzymes produced maltohexaose in the early stage of hydrolysis, as shown in Fig. 7.24. The maximum yield was
about 25% to 30%.

C. G6-amylase Gene of *Bacillus* sp. No. H-167
In order to obtain genetic information on multi-enzyme systems and to elucidate the

regulation of their formation, we cloned the gene of G6-amylases of *Bacillus* sp. No. H-167 and expressed it in the *E. coli* system (Shirokizawa, Akiba and Horikoshi, 1989). Chromosomal DNA of *Bacillus* sp. No. H-167 was partially digested with Sau3AI. Then DNA fragments longer than 2 kb were separated by sucrose density gradient centrifugation. The cloning vector, pBR329, was cleaved with *Bam*H I then treated with alkaline phosphatase. The plasmid and the chromosomal DNA were ligated with T₄ DNA ligase overnight at 16 °C and introduced into *E. coli* HB101. Amylase activity of the transformants was detected as the formation of large halos around the colonies on LB agar plates supplemented with 0.2% starch azure and 20 μg chloramphenicol (LB-SA agar).

One colony formed a large clear halo on an LB-SA agar plate. A plasmid (pSB404) having an insert of approximately 3.0 kb DNA was isolated from the transformant. The restriction map of pSB404 is shown in Fig. 7.25. Southern blot analysis showed that the insert was derived from the chromosomal DNA of *Bacillus* sp. No. H-167. This plasmid was able to transform *E. coli* HB101 into an amylase-producing strain. An overnight culture of *E. coli* HB101 (pSB404) was centrifuged at 8,000 rpm for 10 min, the cells washed once with 50 mM Tris-HCl buffer (pH 8.0), resuspended in the same buffer and finally disrupted by ultrasonic treatment. Most of the enzyme was found in the intracellular fraction on growth in LB medium or LB medium containing glycerol, whereas in LB medium containing glucose, maltose or starch (0.2-0.5%), the enzyme was detected not only in the intracellular fraction but also in the extracellular fraction (Table 7.20). About 60% of the enzyme was found in the culture broth supplemented with 0.5% maltose. The mechanism is not yet known.

HPLC analysis showed that the component of the hydrolysis products with the amylase of *E. coli* HB101 (pSB404) maltohexaose was formed as the main product from starch,

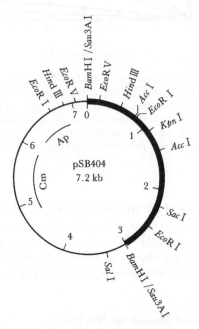

Fig. 7.25 Restriction map of pSB404.

Table 7.20   Effect of carbohydrate on production of G6-amylase from *E. coli* HB101 (pSB404)

Medium		Growth (A_{660}) and Activity (units/ml)				
		4 h	8 h	16 h	24 h	36 h
LB	Growth	0.86	2.49	2.59	2.49	2.63
	Extra.	0.00	0.02	0.03	0.02	0.11
	Intra.	0.12	0.52	0.78	0.68	0.81
	Total	0.12	0.54	0.81	0.70	0.92
LB + glucose	Growth	1.21	2.58	2.73	2.67	2.87
	Extra.	0.00	0.68	0.48	0.60	0.62
	Intra.	0.17	0.48	0.49	0.51	0.63
	Total	0.17	1.16	0.97	1.11	1.25
LB + maltose	Growth	0.90	2.58	2.76	2.72	2.88
	Extra.	0.02	0.83	0.68	0.85	0.96
	Intra.	0.12	0.49	0.51	0.50	0.66
	Total	0.14	1.32	1.19	1.35	1.62
LB + starch	Growth	0.88	2.67	2.76	2.79	2.83
	Extra.	0.02	0.47	0.37	0.49	0.53
	Intra.	0.14	0.75	0.78	0.78	0.90
	Total	0.16	1.22	1.15	1.27	1.43
LB + glycerol	Growth	0.78	2.63	2.80	2.76	2.91
	Extra.	0.00	0.00	0.00	0.05	0.19
	Intra.	0.06	0.84	1.01	0.97	1.17
	Total	0.06	0.84	1.01	1.02	1.36

Growth was determined by measuring absorbance at 660 nm.
Extra., extracellular ; Intra., intracellular

and minor peaks of maltotetraose and maltose were also observed. The pH activity profile of the plasmid-borne amylase was essentially the same as that of the amylases from *Bacillus* sp. No. H-167 (Fig. 7.26). After SDS-PAGE, three bands having amylase activity were detected for the supernatant of *E. coli* HB101 (pSB404). The molecular weights were about 90,000, 73,000 and 60,000. Obviously, the inserted 3.0 kb DNA fragment on pSB404 is too short to encode the three amylase genes. Multiple enzymes are usually generated through the following mechanisms: (i) multiple genes, (ii) proteolytic processing of enzymes inside the cell or (iii) multiple transcription of a gene. The cultivation of *Bacillus* sp. No. H-167 in the presence of inhibitor (antipain, 1 mg/ml) indicated that the 73-kDa and 60-kDa amylases were the products of proteolytic degradation of the 90-kDa amylase. Thus mechanism (ii) is the most likely one for the formation of multiform G6-amylases, although we have not conducted definitive experiments showing proteolysis in the *E. coli* system. Furthermore, Shirokizawa, Akiba and Horikoshi (1990) sequenced the 3.0-kb DNA fragment which is responsible for the G6-amylase activity. An open reading frame of 2,865 bp was detected in the fragment and no other open reading frame was observed. Furthermore, no homology was exhibited between the DNA sequence of maltohexaose-forming enzyme of *Bacillus* sp. No. 707 and ours except for the A, B, and C domains found

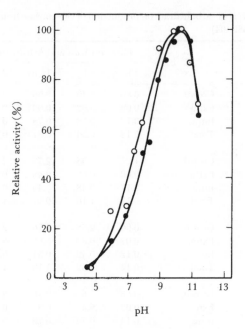

Fig. 7.26 pH activity of the amylases.
○: *Bacillus* ; ●: plasmid-borne

in amylases. These results indicate that our G6-forming amylase is different from that previously reported (Kimura *et al.*, 1988).

7.3.4 Pullulanases

Pullulan, a polysaccharide originally produced by the yeast *Pullularia pullulans*, is composed of linear chains of maltotriose units attached by an α-1,6-linkage. Pullulanase [EC3.2.1.41], which causes cleavage of the α-1,6-links of pullulan, is distributed in several bacteria. Two other new enzymes, "isopullulanase" [EC3.2.1.57] and "new-type α-amylase" have been discovered in Japan; these can hydrolyze pullulan to yield isopanose and panose, respectively. The actions of these enzymes on pullulan are summarized in Fig. 7.27.

A. Pullulanase of *Bacillus* sp. No. 202-1

In 1975, while screening alkaline amylases from alkaliphilic *Bacillus* strains, Nakamura *et al.* discovered that an alkaliphilic bacterium, *Bacillus* sp. No. 202-1, produced an extracellular pullulanase in an alkaline medium of pH 10 (Nakamura,Watanabe and Horikoshi, 1975). Strain No. 202-1 did not grow at neutral pH. The pullulanase from strain No. 202-1 was purified 290-fold by DEAE cellulose adsorption at pH 8.0, precipitation with 75% saturated acetone and with 80% saturated $(NH_4)_2SO_4$, column chromatography on DEAE cellulose, and gel filtration on a Sephadex G-200. The purity of the enzyme was confirmed by disc gel electrophoresis. The molecular weight of the alkaline pullulanase was 92,000, and the isoelectric point below 2.5. The enzyme has an optimum pH at 8.5-9.0 and is stable for 24 h at pH 6.5-11.0 at 4 °C. It is most active at 55 °C, and is

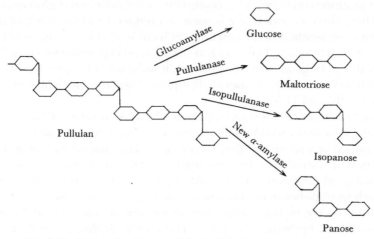

Fig. 7.27   Modes of action of pullulan-hydrolyzing enzymes.
(Reproduced with permission from Horikoshi and Akiba, *Alkalophilic Microorganisms*, p. 134, Springer-Verlag: Japan Scientific Societies Press (1982))

Table 7.21   Combined action of pullulanase and β-amylase on branching polysaccharides

Enzyme	Degree of hydrolysis (%)	
	Amylopectin	Glycogen
β-Amylase	52	35
No. 202-1 pullulanase + β-amylase	102	97
A. aerogenes pullulanase + β-amylase	103	49

(Reproduced with permission from Horikoshi and Akiba, *Alkalophilic Microorganisms*, p.135, Springer-Verlag: Japan Scientific Societies Press (1982))

stable up to 50 °C for 15 min in the absence of substrate. The enzyme is inhibited by Hg^{2+} and Zn^{2+}, but not by sulfhydryl reagents or by chelating agents such as EDTA. These results indicate that no sulfhydryl or serine is involved in the catalytic site of the enzyme.

The hydrolysis of amylopectin (from rice) and glycogen (from oysters) by alkaline pullulanase followed by β-amylase resulted in complete degradation of these substrates, while hydrolysis by β-amylase alone resulted in partial degradation. On the other hand, *Aerobacter aerogenes* pullulanase and the subsequent action of β-amylase caused complete degradation of amylopectin but incomplete degradation of glycogen, as summarized in Table 7.21. The debranching activity of alkaline pullulanase towards glycogen was further confirmed by a high increase (70%) in iodine coloration at 470 nm. The increase in iodine color by the action of *A. aerogenes* pullulanase glycogen was observed to be low (13%). These results indicate that alkaline pullulanase, unlike *A. aerogenes* pullulanase, can hydrolyze almost all the α-1,6-linkages involved in the compact glycogen molecule.

B.   Pullulanase of Other Alkaliphiles
Kelly, O'Reilly and Fogarty (1983) found that alkaliphilic *Bacillus* sp. No. A-59

(ATCC21591) produced three enzymes, α-amylase, pullulanase and α-glucosidase, in culture broth. These three enzymes were separately produced and the levels of α-glucosidase and pullulanase reached maximum after 24 h cultivation at the initial pH 9.7. But the highest biomass was attained after 48 h at pH 10.4; no α-glucosidase could be detected. Although this pullulanase was not purified, it indicated pH optimum at 7.0. This enzyme thus differs from other enzymes of alkaliphilic *Bacillus* strains which indicate optimum between pH 9.5 and 11.5.

In screening alkaline cellulases for detergent additives, Ara *et al.* (1992) and Igarashi *et al.* (1992) isolated a novel alkaline pullulanase from alkaliphilic *Bacillus* sp. KSM-1876, which was identified as a relative of *Bacillus circulans*. The enzyme had an optimum pH for enzyme action of around 10.0–10.5, the highest pH for optimum pullulanase activity. This enzyme is a good candidate for use as detergent additive in dish washers, especially to remove starch from dishes in the presence of detergents. The gene was isolated from KSM-1876. However, the plasmid-borne enzyme was not thermostable and not alkaline pullulanase, having an optimum pH of 7.5. The enzyme hydrolyzed pullulan 3.0-fold faster than soluble starch, but hydrolyzed amylose and amylopectin less efficiently. Therefore, the gene cloned was not the major alkaline pullulanase gene, but a minor one.

Ara *et al.* isolated new alkaliphilic *Bacillus* sp. KSM-1378 and reported two independent active sites for the α-1,4 and α-1,6 hydrolytic reactions (Ara *et al.*, 1995 ; Ara *et al.*, 1996 ; Hatada *et al.*, 1996). Alkaliphilic *Bacillus* sp. KSM-1378 produces an alkaline amylopullulanase that hydrolyzes both α-1,4 linkages in amylose, amylopectin and glycogen, and α-1,6 linkages in pullulan. The hydrolytic activities against amylose and pullulan were specifically inhibited by maltotriose ($K-i = 0.5$ mM), isomaltitol ($K-i = 5.2$ mM), and methyl α-D-galactoside ($K-i = 40$ mM) and by β-cyclodextrin ($K-i = 0.9$ mM), α-cyclodextrin ($K-i = 11$ mM), and raffinose ($K-i = 31$ mM), respectively, in a competitive manner in each case. Inhibition by *N*-bromosuccinimide of the α-amylase activity was prevented by amylose but not by pullulan, while inhibition by *N*-bromosuccinimide of the pullulanase activity was prevented by pullulan but not by amylose. Kinetics of reactions in the simultaneous presence of amylose and pullulan indicated that the observed rates of formation of products closely matched those predicted by a kinetic model in which the α-1,4 and α-1,6 hydrolytic reactions were catalyzed at two independent active sites. Incubation of the enzyme at 40 °C and pH 9.0 caused complete inactivation of the amylase activity within 4 days, but the pullulanase activity remained at the original level under the same conditions. Then they treated the pullulanase with papain and separated two functional domains. The intact, pure 210 kDa pullulanase was partially digested with papain for a short time, yielding simultaneously two smaller non-overlapping active fragments, designated an amylose-hydrolyzing fragment (AHF114, 114 kDa) and a pullulan-hydrolyzing fragment (PHF102, 102 kDa). The partial amino-terminal sequences of the intact pullulanase and AHF114 were both Glu–Thr–Gly–Asp–Lys–Arg–Ile–Glu–Phe–Ser–Tyr–Glu–Arg–Pro and that of PHF102 was Thr–Val–Pro–Leu–Ala–Leu–Val–Ser–Gly–Glu–Val–Leu–Ser–Asp–Lys–Leu. This alkaline amylopullulanase can, therefore, be considered to be a bi-headed enzyme molecule. Furthermore, they reported the molecular cloning and sequencing of the gene for and the structure of this enzyme and show that its dual hydrolytic activities are associated with two independent active sites. The structural gene contained a single, long open reading frame of 5,814 base pairs, corresponding to 1,938 amino acids that included a signal peptide of 32 amino acids. The molecular mass of the extracellular mature enzyme (Glu(33) through Leu(1938)) was calculated to be 211,450 Da, a value close to the 210 kDa determined for the amylopullulanase produced by *Bacillus* sp. KSM-1378. The amylase and the pullulanase domains were located in the amino-terminal half and in the carboxyl-terminal half of the enzyme, respectively, being separated by a tandem repeat of a sequence of 35 amino acids. Four regions, designated I, II, III, and IV, were highly conserved in each catalytic domain, and they included a putative catalytic triad Asp(550)-Glu(579)-Asp(645) for the amylase activity and Asp(1464)-Glu (1493)-Asp(1581) for the pullulanase activity. The purified enzyme was rotary shadowed at a low angle and observed by transmission electron microscopy; it appeared to be a "castanet-like" or "bent dumbell-like" molecule with a diameter of approximately 25 nm.

From the industrial point of view two highly alkaliphilic bacteria and potent producers of alkaline pullulanase were isolated from Korean soil (Kim, Choi and Lee, 1993*b*). The two isolates, identified as *Bacillus* sp. S-1 and *Micrococcus* sp. Y-1, grow on starch under alkaline conditions and effectively secrete extracellular pullulanases. The two isolates were extremely alkaliphilic since bacterial growth and enzyme production occurred at pH values ranging from pH 6.0 to 12.0 for *Micrococcus* sp. Y-1 and pH 6.0 to 10.0 for *Bacillus* sp. S-1. Both strains secrete enzymes that possess amylolytic and pullulanolytic activities. Extracellular crude enzymes of both isolates gave maltotriose as the major product formed from soluble starch and pullulan hydrolysis. Compared with other alkaliphilic microbes such as *Micrococcus* sp. (0.57 units/ml), *Bacillus* sp. KSM-1876 (0.56 units/ml) and *Bacillus* No. 202-1 (1.89 units/ml), these isolates secreted extremely high concentrations (7.0 units/ml for *Bacillus* sp. S-1 and 7.6 units/ml for *Micrococcus* sp. Y-1) of pullulanase in batch culture. The extracellular enzymes of both bacteria were alkaliphilic and moderately thermoactive; optimal activity was detected at pH 8.0–10.0 and between 50 and 60 °C. Even at pH 12.0, 65% of original Y-1 pullulanase activity and 10% of S-1 pullulanase activity was retained.

Furthermore, the same authors purified the S-1 pullulanase and reported the properties (Kim, Choi and Lee 1993*a*). The purified enzyme had a molecular mass of about 140 kDa under denatured and natural conditions. The p*I* was 5.5. The pullulanase, when resolved by SDS-PAGE, was negative for Schiff staining, suggesting that the enzyme is not a glycoprotein. The N-terminal amino acid sequence of the enzyme was Phe-Leu-Asn-Met-Ser- (Trp-Phe). The enzyme displayed a temperature optimum of around 60 °C and a pH optimum of around 9.0. The presence of of pullulan protected the enzyme from heat inactivation, the extent depending upon the substrate concentration. The activity of the enzyme was stimulated by Mn^{2+} ions. Ca^{2+} ions and EDTA did not inhibit the enzyme activity. The enzyme hydrolyzed the α-1,6-linkages of amylopectin, glycogens, α-,β-limited dextrin, and pullulan. The enzyme caused the complete hydrolysis of pullulan to maltotriose. The activity was not inhibited by α, β, or γ-cyclodextrins. These results strongly indicate that these new bacterial isolates have potential for use as producers of pullulanases in the starch industry.

Kim and Kim (1995) then analyzed substrate specificity of a bifunctional pullulanase of alkaliphilic *Bacillus circulans* F-2. *Bacillus circulans* F-2 amylase-pullulanase enzyme (APE) displayed dual activity with respect to glycosidic bond cleavage. The enzyme was active on α-1,6 bonds in pullulan, amylopectin, and glycogen, while it showed α-1,4 activity against malto-oligosaccharides, amylose, amylopectin, and soluble starch, but not pullulan. Kinetic analysis of the purified enzyme in a system which contained both pullulan and amylose as two competing substrates was used to distinguish the dual specificity of the enzyme from the single-substrate specificity known for pullulanases and α-amylases. Enzyme activities were inhibited by some metal ions, and by metal-chelating agents with a different mode. The enzyme-inhibitory results of amylase and pullulanase with Hg^{2+} and Co^{2+} ions were different, indicating that the activation mechanisms of both enzyme activities are different. Cyclomaltoheptaose inhibited both α-amylase and pullulanase activities. The N-terminal sequence of the enzyme was Ala–Asp–Ala–Lys–Lys–Thr–Pro–Gln–Gln–Gln–Phe–Asp–Ala–Leu–Trp–Ala–Ala–Gly–ILe–Val–Thr–Gly–Thr–Pro–Asp–Gl–Phe. When both amylose and pullulan were simultaneously present, the observed rate of product formation closely fitted a kinetic model in which the two substrates are hydrolyzed at different active sites. These results suggest that amylopullulanases, which possess both α-1,6 and α-1,4 cleavage activities at the same active site, should be distinguished from APEs, which contain both activities at different active sites on the same polypeptide. Also, it is proposed that the Enzyme Commission use the term "amylase-pullulanase enzyme" to refer to enzymes which act on starch and cleave both α-1,6-bonds in pullulan and α-1,4 bonds in amylose at different active sites.

C. Pullulanases of Thermophilic Alkaliphiles

Lin, Tsau and Chu (1994) isolated an alkaliphilic strain of *Bacillus* sp., designated TS-23, from a soil sample collected at a hot spring area in Tainan, Taiwan. During growth in a medium containing 1% soluble starch as the sole source of carbon the fermentation broth exhibited both pullulanase and amylase activity. Pullulanase and amylase activities were maximal at 65 °C. The pH optima were 8.8 to 9.6 for pullulanase and 7.5 to 9.4 for amylase. Under optimal conditions a crude preparation hydrolyzed pullulan, generating maltotriose as the major product. Native corn starch was partially digested by adsorbed amylases during the course of 12 h at 50 °C, with initiation of granular pitting. Further incubation of the reaction mixture resulted in considerable morphological changes in corn starch granules, and the main soluble products were maltose, maltotriose and higher oligosaccharides.

Lee *et al.* isolated a thermophilic and alkaliphilic *Bacillus* sp. strain, XAL601, from soil (1994). It produces a thermostable and alkaline-stable enzyme with both α-amylase and pullulanase activities. The α-amylase-pullulanase gene (*aapT*) from this *Bacillus* strain was cloned, and its nucleotide sequence determined. A very large open reading frame composed of 6,096 bases, which encodes 2,032 amino acid residues with an M(r) of 224,992, was found. The deduced amino acid sequence revealed that the four highly conserved regions that are common among amylolytic enzymes were well conserved. These include an active center and common substrate-binding sites of various amylases. In the C-terminal region, a six-amino-acid sequence (Gly–Ser–Gly–Thr–Thr–Pro) is repeated 12 times. The *aapT* gene was then subcloned in *Escherichia coli* and over expressed under the control of the *lac* promoter. Purification of AapT from this recombinant *E. coli* was performed, and it was shown that the *aapT* gene product exhibits both α-amylase and pullulanase activities with one active site. The optimum temperature and pH for enzyme activity were found to be 70 °C and pH 9, respectively. Furthermore, AapT was found to strongly adsorb to crystalline cellulose (Avicel) and raw corn starch. Final hydrolyzed products from soluble starch ranged from maltose (G2) to maltotetraose (G4). Only maltotriose (G3) was produced from pullulan. The enzyme also hydrolyzes raw starch under a broad range of conditions (60 to 70 °C and pH 8 to 9).

D. Pullulanase of an Alkalipsychrotrophic Strain

In some food industries there is demand for enzymes showing activity at lower temperatures for food processing. Psychrotrophic bacteria are thought to be potential producers of these enzymes. Kimura and Horikoshi (1989; 1990*b*) reported that an alkalipsychrotrophic strain, *Micrococcus* sp. 207 (Table 7.22), inducibly and extracellularly produced amylase and pullulanase. A pullulanase-hydrolyzing enzyme of *Micrococcus* sp. 207 was purified to an electrophoretically homogeneous state by chromatography on DEAE-Toyopearl, α-cyclodextrin-Sepharose and Asahipak GS-520P. The purified enzyme was free of α-amylase activity. The molecular weight of the enzyme as estimated by SDS-PAGE was 120,000 and the p*I* value as determined by isoelectric focusing was 4.9. The enzymes showed pH optimum for enzyme action at 7.5–8.0 and were relatively thermostable (stable up to 45 °C). The enzyme could hydrolyze the α-1,6-linkages of amylopectins, glycogens and pullulan, and the K_m value for pullulan was about 0.018%.

Table 7.22 Properties of two pullulan hydrolyzing enzymes (PHEs) from *Micrococcus* sp. 207

Property	PHE	Enzyme 1
Molecular weight	120,000	185,000
Isoelectric point	4.9	4.8
Optimum pH	8.0	8.0
Optimum temperature	50 °C	50 °C
K_m for pullulan	0.018%	0.018%
pH stability	5.0–10.0	5.0–10.0
Temperature stability		
40 °C ($+ Ca^{2+}$)	100%	88%
70 °C ($- Ca^{2+}$)	55%	59%
Activator	Ca^{2+}	Ca^{2+}
Inhibition of	Co^{2+}, Mn^{2+}, Cu^{2+}, Cd^{2+}, Hg^{2+},	
metal ion	Pb^{2+}, Sn^{2+}, Ni^{2+}, Zn^{2+}	
SDS (0.1%, $+ Ca^{2+}$)	55.4%	26.1%
EDTA (5 mM)	30.3%	94.9%
Pullulan (0.1%)	28.7%	35.7%
Cyclodextrins	Competitive	Competitive

7.4 Cellulases of Alkaliphilic *Bacillus* Strains

Commercially available cellulases display optimum activity over a pH range from 4 to 6. No enzyme with an alkaline optimum pH for activity (pH 10 or higher) has been reported. We have found newly isolated bacteria (*Bacillus* sp. No. N-4 and No. 1139) producing extracellular carboxymethylcellulases (CMCases) in alkaline growth media. One of these, alkaliphilic *Bacillus* sp. No. N-4, produced multi CMCases which were active over a broad pH range (pH 5 to 10). These CMCases have been partially purified and characterized.

Another bacterium, *Bacillus* sp. No. 1139, produced one CMCase which was entirely purified and shown to have optimum pH for activity at pH 9.0. However, all alkaline cellulases so far reported can not or only barely hydrolyze cellulose fibers. An *Aeromonas* sp. No. 212 (ATCC No. 31085) enzyme that we found had the optimum pH for enzyme action at about 8 and is very stable in the presence of metal ions and hydrogen sulfate. The enzyme strongly hydrolyzed native cellulose.

7.4.1 Cellulases of *Bacillus* sp. No. N-4

A. Isolation, Characterization and Cultivation of Alkaliphilic *Bacillus* sp. No. N-4

An alkaliphile, *Bacillus* sp. No. N-4 (ATCC21833), isolated from soil is a facultative aerobic, spore-forming (spherical, terminal, 1.4 μm), gram-positive, motile, and rod-shaped bacterium (0.3-0.4 μm × 2.0-3.0 μm). It is clear that the bacterium belongs to the genus *Bacillus* group 3. The GC content of the DNA of the isolate was 36.5%. Its characteristic feature was that the growth rate of the isolate was best at pH 10.0 to 10.5 in the presence or absence of 5% NaCl. The growth rate was slight or zero at pH 7.0. Other properties are shown in Table 7.23. The isolate, therefore, is quite similar to *Bacillus pasteurii*, although its optimum pH for growth is higher (Horikoshi *et al.*, 1984).

Bacillus sp. No. N-4 was cultivated with continuous shaking on a rotary shaker at 37 °C for 50 to 70 h in a modified Horikoshi-II medium (2% CMC w/v instead of starch). The culture broth was centrifuged at $10,000 \times g$ for 30 min at 4 °C and the supernatant fluid collected. The CMCase was precipitated by the addition of two volumes of cold ethanol, collected by centrifugation, and dried *in vacuo*. This preparation was used as the crude enzyme. Very weak activity toward Avicel was detected. However, very strong activity was exhibited on CMC at pH 6.7 and 10.0, as shown in Table 7.24. The crude enzyme preparation had a broad pH activity curve (pH 5 to 10), as shown in Fig. 7.28. The enzyme was very stable at pH values ranging from 6 to 10. As shown in Fig. 7.29, thermal stability at various temperatures suggests that the crude enzyme preparation is a mixture of several CMCase enzymes (thermostable and thermounstable).

B. Partial Purification of the Enzymes

The CMCases were separated by passing through a Sephadex G-150 column (Fig. 7.30) followed by hydroxyl apatite column chromatography. Two alkaline CMCases (enzymes El and E2) having an optimum pH for enzyme action at pH 10.0 were partially purified from the crude enzyme preparation. Their molecular weight estimated by the Sephadex gel filtration method was approximately 50,000. The enzyme E2 was stable up to 80 °C and El up to 60 °C. No difference between the two enzymes has been observed so far in the type of products formed. Cellobiose was not hydrolyzed at all. Cellotriose was hydrolyzed, and glucose and cellobiose were detected. Cellotetraose was also hydrolyzed: almost all the samples were converted to cellobiose, and about 5% to glucose and cellotriose. In comparison with other cellulases, these two cellulases, namely El and E2 having high pH optima, are believed to be the first finding of alkaline CMCases. Such multiplicity of CMCase components has also been reported by many investigators. However, no crucial experiment on the multiplicity of cellulase molecules has been conducted. To remove the effect of proteolytic degradation of cellulase molecule, Sashihara *et al.* cloned the cellulase genes of *Bacillus* sp. No. N-4 in *Escherichia coli* HB101 with pBR322. Several

Table 7.23 Properties of alkaliphilic *Bacillus* sp. No. N-4

1. Morphological characteristics

Form	Rods
Size, μm	0.1–0.2 × 4.0–6.0
Motility	Motile
Gram stain	Positive
Sporangia	Swollen
Spores	1.5–2.0 μm ; spherical ; terminal

2. Cultural characteristics

	Growth at[†]	
	pH 7	pH 10.3
Nutrient broth	−	−
Nutrient agar slant	−	−
Glucose-nutrient broth	−	++
Glucose-nutrient agar slant	−	++
Gelatin broth	−	−
Peptone water	−	−
Potato	−	++
Horikoshi-I medium	−	++
Horikoshi-II medium	−	++
Anaerobic growth in glucose broth	−	+
Anaerobic production of gas from nitrate	−	−
Horikoshi-I medium containing 5% NaCl	−	+

3. Biochemical characteristics

Hydrolysis of gelatin and casein	Weak
Hydrolysis of starch	Positive
Utilization of ammonium	Very poor
Utilization of citrate	Utilized
Reduction of nitrate to nitrite	Positive
Voges-Proskauer test	Negative

4. pH and temperature

pH for growth in Horikoshi-I medium	pH 8-10.7
Temperature for growth in Horikoshi-I medium	Up to 42 °C at pH 10.3

[†] −, no growth ; +, normal growth ; ++, abundant growth.
(Repoduced with permission from Horikoshi and Akiba, *Alkalophilic Microorganims*, p.127, Springer-Verlag: Japan Scientific Societies Press (1982))

Table 7.24 Enzyme activities on various substrates

	pH of reaction mixture	
Substrate	6.7	10.0
CMC	0.81	0.60
Avicel SF	0.003	0.002

Note: The values are expressed as units per milliliter of culture broth.
(Reproduced with permission from Horikoshi *et al.*, *Can. J. Microbiol.*, **30**, 775 (1984))

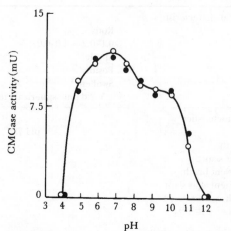

Fig. 7.28 Effect of pH on unpurified CMCase activity. ○, Culture fluid; ●, Ethanol-dried enzyme.

The following buffer systems were used: 0.05 M acetate buffer (pH 4.0-6.7); 0.05 M Tris-HCl buffer (pH 7.0-9.0); 0.05 M glycine-NaOH-NaCl buffer (pH 9.0-12.0). Other conditions were the same as those of the standard assay method.

(Reproduced with permission from Horikoshi *et al.*, *Can. J. Microbiol.*, **30**, 775 (1984))

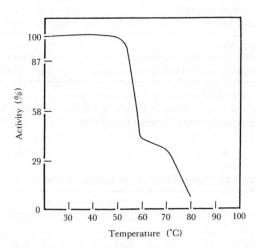

Fig. 7.29 Effect of temperature on stability of cellulase.

(Reproduced with permission from Horikoshi and Akiba, *Alkalophilic Microorganisms*, Springer-Verlag: Japan Scientific Societies Press (1982))

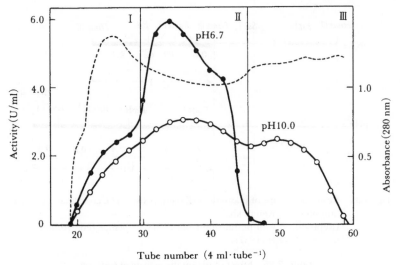

Fig. 7.30 Elution profile of cellulose.○, CMCase activity assayed at pH 10.0; ●, CMCase activity assayed at pH 6.7; broken line, absorbance at 280 nm. (Reproduced with permission from Horikoshi *et al.*, *Can. J. Microbiol.*, **30**, 777 (1984))

DNA fragments which have different DNA sequences were obtained (Sashihara, Kudo and Horikoshi, 1984).

C.   Molecular Cloning of Cellulase Genes of Alkaliphilic *Bacillus* sp. No. N-4

Alkaliphilic *Bacillus* sp. No. N-4 was aerobically grown to the early stationary phase at 37 °C in Horikoshi-I medium. Bacterial chromosomal DNA was digested with *Hin*d III at 37 °C for 1 h (plasmid DNA) or for 14 h (chromosomal DNA). After the digestion, 1 μg of plasmid and 3 μg of bacterial chromosomal DNA were mixed and ligated with T_4 DNA ligase overnight at room temperature. This ligation mixture was used for transformation, and about 2×10^4 Ap^{r}Tc^s transformants per mg of DNA were obtained. The CMCase activity could be tested directly on the plates because a shallow crater formed around a colony producing CMCase on an LB agar plate containing CMC. Several plasmids were isolated (Sashihara, Kudo and Horikoshi, 1984; Fukumori *et al.*, 1989): plasmid pNK1 containing a 2.0 kb *Hin*d III fragment, plasmid pNK2 having 2.8 kb *Hin*d III fragment and plasmid pNK3. Restriction maps of these plasmids are shown in Fig. 7.31. Genomic hybridization experiments showed partial homology among these fragments and these results were confirmed by sequence analysis of these CMCase genes.

E. coli HB101 carrying pNK1 and pNK2 plasmids were aerobically grown in LB broth for 24 h at 37 °C. A significant amount of the enzymatic activity was found in the periplasmic space in *E. coli* carrying the plasmids and the synthesis of CMCase in *E. coli* was constitutive. Ouchterlony double-diffusion analysis showed that plasmid-borne CMCases gave lines of precipitation that fused with that for CMCase from the alkaliphilic *Bacillus* sp. No. N-4. As shown in Table 7.25, the CMCase thus produced had broad pH activity curves (pH of 5.0 to 10.9) and was stable up to 75 °C. It is well known that some cellulases pro-

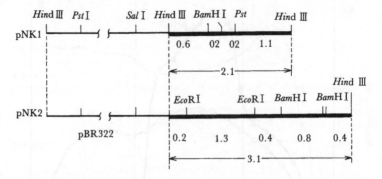

Fig. 7.31   Restriction maps of plasmids pNK1 and pNK2. The coordinates of restriction sites are given in kb.
(Reproduced with permission from *FEMS Symposium* (eds. Aubert, J.-P., *et al.*), No. 43, 204 (1987))

Table 7.25   Effect of pH on the activity of CMCase

| | CMCase activity (%) | | |
pH	pNK1	pNK2	Crude enzyme from N-4
4.4	18.4	30.3	27.8
5.0	81.0	82.4	98.1
7.5	96.7	79.7	86.4
8.0	106.8	89.1	103.3
9.2	102.3	87.9	92.3
10.1	100.0	100.0	100.0
10.9	107.7	100.0	119.9
11.7	36.8	62.1	54.7
12.8	7.4	10.6	14.7

(Reproduced with permission from Sashihara *et al.*, *J. Bacteriol.*, **158**, 505 (1984))

duced by microorganisms have multi-enzyme systems. Two possibilities for the multi-enzyme formation should be considered: (i) processing or modification during production; (ii) the DNA contains genes for multi-enzyme formation. It is quite possible, given the extensive homology of these CMCase genes, that the cellulase gene itself is duplication or triplication present on the chromosomal DNA. This point will be discussed in the following section. Another interesting result is that plasmid-encoded cellulases have very broad pH activity curves, as was observed for the enzymes of the parent strain.

D.   Nucleotide Sequences of Two Cellulase Genes, *celA* and *celB*

Two plasmids, pNK1 carrying *celB* and pNK2 carrying *celA*, have been sequenced (Fig. 7.32). pNK1 contained an open reading frame of 1,464 base pairs (bp) which encoded 488 amino acids (*celB*). pNK2 contained an open reading frame of 1,227 bp, which encoded 409 amino acids (*celA*). The calculated molecular weights of the pNK1- and pNK2-encoded cellulase were 54,267 (*celB*) and 45,695 (*celA*) respectively, and were similar to those estimated using Sephacryl S-200 gel filtration (Fukumori *et al.*, 1986). The predicted amino acid sequences analyzed on a dot plot matrix showed that the two proteins are almost the same, except that the *celB* had a direct repeat sequence of a stretch of 60(59) amino acids which was absent in the *celA*. Two direct repeat peptides exhibited 94% amino acid homology. Also, short direct repeat sequences (Pro–Asp–Pro–Gly–Glu) are observed near the N-terminal of the long direct repeat units in the pNK1 and pNK2 cellulase genes. It has been reported that a stretch of 23 amino acids close to the C-terminal appeared to be repeated with a high degree of homology in a cellulase gene of *Clostridium thermocellum*. In plasmid pNK1 there are two *Nco*I restriction sites near the 3' end of each direct repeat sequence. Therefore, one of the direct repeat units could be removed by the treatment of restriction enzyme *Nco*I. The protein encoded by this derivative of pNK1 from which the DNA fragment of 261 bp had been deleted retained similar cellulase activity. These results suggest that the direct repeat is not important for enzyme activity. When these cellulases were compared for amino acid sequences, 362 (89%) of aligned amino acids were identical, although the signal peptide sequence regions showed less homology. Since the amino acids which differed in the signal peptides had similar chemical properties, these changes probably do not affect the function of the signal peptide.

E. Evolution of the *celA* and *celB*

Two cellulase genes seem to form by gene duplication because they show very strong homology. The method developed for estimating the evolutionary rate of synonymous substitutions from homologous nucleotide sequences was applied to the *celA* and *celB* genes. The phylogenic relationships, K_s^s, were obtained. These K_s^s values did not appear to be significantly different. It was suggested that the duplication of an ancestrail cellulase gene (if it exists) and the formation of the direct repeat sequence near the C-terminal end of the *celB* gene occurred at almost the same time.

Because the duplicated genes (*celA* and *celB* genes) were expected to be located close to each other on the chromosomal DNA, Fukumori *et al.* tried to clone a DNA fragment containing the *celA* and *celB* genes (Fukumori *et al.*, 1987). A plasmid pNK21 having a 5.7 kb *Eco*R I fragment and pBR322 was obtained from cellulase-positive transformants. Sequencing experiments revealed that two cellulase genes, *celA* and *celB*, were located tandemly. Two *Hin*d III fragments of pNK1 and pNK2 were connected by a 0.1 kb *Hin*d III fragment, as shown in Fig. 7.33. To obtain further information, the upstream region of the *celA* gene was sequenced, and non-coding regions of the *celA* and *celB* genes were compared with each other. Homologous repeat sequences were observed at three positions: upstream of the *celA* gene, between the *celA* and *celB* genes, and downstream of the *celB* gene. This result strongly indicates that such a tandem location was caused by gene duplication due to homologous recombination between the direct repeat sequences located upstream and downstream of an ancestral cellulase gene. Furthermore, a space between the first repeat sequence and the *celA* gene, a second space between the second sequence and the *celB* gene, and a third space located downstream of the *celB* gene had strong homology. Therefore, it is highly possible that the two cellulase genes of *Bacillus* sp. No. N-4 located tandemly on the chromosome were formed by gene duplication.

ATTGATTAGGTATATTCACTGATTGAATAAAGAGAGTTTGTTTTATAAAATTAACAACGAAAAAACCATTTTCGTTCGGGAGGAGTTGAT

AGCGCTTACAATTCGCTAGTTTATAAGTGTGTTTAGAAACTAAAACGATTTCGTCTCTAAAGTCTGTAAAAAAATTTT<u>AGGAGG</u>AAAAG
SD

1
ATGAAAAAACTAACGACTATTTTTATTGTATTTACGTTAGCATTATTATTTGTTGGAAACTCTACTTCTGCAAATAATGGTTCCGTTGTT
MetLysLysLeuThrThrIlePheIleValPheThrLeuAlaLeuLeuPheValGlyAsnSerThrSerAlaAsnAsnGlySerValVal
(1)

100
GAGCAAAATGGACAATTAAGTATTCAAAACGGTCAATTAGTGAATGAACATGGAGACCCGGTACAGCTAAAGGGAATGAGTTCACATGGA
GluGlnAsnGlyGlnLeuSerIleGlnAsnGlyGlnLeuValAsnGluHisGlyAspProValGlnLeuLysGlyMetSerSerHisGly

200
TTACAATGGTATGGACAATTTGTAAACTACGATAGTATAAAGTGGTTAAGAGACGATTGGGGAATAACCGTATTCCGAGCAGCGATGTAT
LeuGlnTrpTyrGlyGlnPheValAsnTyrAspSerIleLysTrpLeuArgAspAspTrpGlyIleThrValPheArgAlaAlaMetTyr

300
ACATCTTCAGGAGGGTATATTGAAGATCCTTCAGTAAAAGAAAAAGTAAAAGAGGCTGTTGAAGCTGCGATTGATCTCGGAATATATGTC
ThrSerSerGlyGlyTyrIleGluAspProSerValLysGluLysValLysGluAlaValGluAlaAlaIleAspLeuGlyIleTyrVal
(100)

400
ATTATTGACTGGCATATTCTTTCTGATAATGACCCGAACATTTATAAGGAAGAAGCAAAAGAGTTTTTTGATGAAATGTCTGCACTTTAT
IleIleAspTrpHisIleLeuSerAspAsnAspProAsnIleTyrLysGluGluAlaLysGluPhePheAspGluMetSerAlaLeuTyr

500
GGGGATTATCCGAATGTCATTTATGAGATTGCGAATGAACCAAACGGACATAATGTAAGATGGGATAGCCATATTAAACCATATGCAGAG
GlyAspTyrProAsnValIleTyrGluIleAlaAsnGluProAsnGlyHisAsnValArgTrpAspSerHisIleLysProTyrAlaGlu

600   PvuII
GAAGTAATCCCAGTAATCCGTGCAAACGATCCTAATAATATTGTCATTGTTGGGACAGCAACTTGGAGTCAGGATGTACACGAAGCAGCT
GluValIleProValIleArgAlaAsnAspProAsnAsnIleValIleValGlyThrAlaThrTrpSerGlnAspValHisGluAlaAla
(200)

700   HincII
GATAACCAACTAGATGACCCGAATGTAATGTATGCGTTCCACTTCTACGCTGGAACACATGGTCAGCAATTAAGAAATCAAGTTGACTAT
AspAsnGlnLeuAspAspProAsnValMetTyrAlaPheHisPheTyrAlaGlyThrHisGlyGlnGlnLeuArgAsnGlnValAspTyr

800
GCTTTAAGTCGAGGAGCAGCGATATTCGTTAGTGAGTGGGGAACGAGTGCGGCTACGGGTGATGGTGGCGTATTTTTAGATGAAGCCCAA
AlaLeuSerArgGlyAlaAlaIlePheValSerGluTrpGlyThrSerAlaAlaThrGlyAspGlyGlyValPheLeuAspGluAlaGln

Pst I   900
GTGTGGATTGACTTTATGGATGAGAGAAATTTAAGCTGGGCAAACTGGTCTTTAACACATAAAGATGAGTCATCTGCAGCGTTAATGCCT
ValTrpIleAspPheMetAspGluArgAsnLeuSerTrpAlaAsnTrpSerLeuThrHisLysAspGluSerSerAlaAlaLeuMetPro
(300)

GGTGCAAACCCAACTGGTGGGTGGACAGCGGCTGAATTGTCTCCATCTGGTGCATTTGTGAGGGAAAAAATAAGAGAATCAGCGTCTATT
GlyAlaAsnProThrGlyGlyTrpThrAlaAlaGluLeuSerProSerGlyAlaPheValArgGluLysIleArgGluSerAlaSerIle

1000
CCGCCAAGCGATCCAACACCTCCATCTGAT|CCAGATCCAGGTGAGCCGGACCCAACGCCACCAAGTGATCCAGGGGAGTATCCAGCATGG
ProProSerAspProThrProProSerAsp|ProAspProGlyGluProAspProThrProProSerAspProGlyGluTyrProAlaTrp

1100
GATCCAAATCAAATTTACACACAAATGAAATTGTGTATCATAACGGTCAGTTATGGCAAGCGAAATGGTGGACACAAAATCAAGAGCCGGGA
AspProAsnGlnIleTyrThrAsnGluIleValTyrHisAsnGlyGlnLeuTrpGlnAlaLysTrpTrpThrGlnAsnGlnGluProGly

NcoI   1200
GCTAATCAATATGGGCCATGGGAGCCTTTA|GGAGATGCTCCTCCAAGTGAACCGAGCGATCCACCACCACCATCAGAACCAGAGCCGGAC
AlaAsnGlnTyrGlyProTrpGluProLeu|GlyAspAlaProProSerGluProSerAspProProProSerGluProGluProAsp
(400)

1300 BamHI
CCAGGAGAACCAGATCCAGGTGAC|CCAGACCCAGGAGAACCGGATCCAACGCCACCAAGTGATCCAGGAGAGTATCCAGCATGGGACCCA
ProGlyGluProAspProGlyGlu|ProAspProGlyGluProAspProThrProProSerAspProGlyGluTyrProAlaTrpAspPro

1400
ACGCAAATTTACACGAATGAAATTGTTTACCATAACGGACAATTGTGGCAAGCAAATGGTGGACACAAAATCAAGAACCCGGATATCCT
ThrGlnIleTyrThrAsnGluIleValTyrHisAsnGlyGlnLeuTrpGlnAlaLysTrpTrpThrGlnAsnGlnGluProGlyTyrPro

NcoI   1464
TATGGACCATGGGAGCCGTTA|AACTAACATATTAATCATAAAT<u>GAAAAGGTGCTC</u>AGCAAAGAG<u>GAGCACC</u>TTTTCTAACAAAACTTTAG
TyrGlyProTrpGluProLeu|Asn***
(488)

TCCTTCGCTCTAACAAGAGGCATAGTACAACATCTAAAAGTACCACCGGACTTAATAATTTCTGTAATATCGACTTCTATCACTTCATAA

CCACGAATACGAAGCTGCTGATTGACTTCTCTATTTATAGGTAAGCTAACAATTCTTTTTTTTACCAATAGAAAGTACGTTTGTTCCAAGT

GTGAACTGCTCTTCTTCATTTACTTCAATTAAATCAAAATGCTTTGCTAATATCTTTTCTTCCTTTTTGTCAAAAGCTCGTGGGAAATAT

AACGCTTCATTCTCTGAAATAATATGAACACACAATCTAAGTGTAAATATTTTTCAGTAAAAGGTACGGTTATTACTTCGTACTCGTGTA

HindIII
AAAGCTT

Fig. 7.32   (Continued)

(b)

EcoRI
GAATTCCGCAACACATTTTTTTTGGAAGAGGGTGATAG

CGTTTACATATACTCCTTAAAAAGTGCAAGCGCAGACTAAAACGATTTCGTTTCAGTATGAAAAGCTAAAACATTACCGAGGAGGAAAAT
SD

1
ATGAAAAAGATAACTACTATTTTTGTCGTATTGCTCATGACACTGGCATTGTTCATTATAGGAAACACGACTGCTGCTGATGATTATTCA
MetLysLysIleThrThrIlePheValValLeuLeuMetThrLeuAlaLeuPheIleIleGlyAsnThrThrAlaAlaAspAspTyrSer
(1)

100
GTTGTAGAGGAGCATGGGCAATTAAGTATTAGTAACGGAGAATTAGTCAATGATCGAGGCGAACCAGTTCAGTTAAAAGGGATGAGTTCC
ValValGluGluHisGlyGlnLeuSerIleSerAsnGlyGluLeuValAsnAspArgGlyGluProValGlnLeuLysGlyMetSerSer

200
CATGGTTTACAATGGTACGGTCAATTTGTAAACTATGAAAGCATGAAATGGCTAAGAGATGATTGGGGTATAACTGTATTCCGAGCAGCG
HisGlyLeuGlnTrpTyrGlyGlnPheValAsnTyrGluSerMetLysTrpLeuArgAspAspTrpGlyIleThrValPheArgAlaAla

AccI   300 BamHI
ATGTATACATCTTCGGGAGGATATATTGAGGATCCTTCCGTAAAGGAAAAAGTAAAAGAGGCTGTTGAGGCTGCGATAGACCTTGGTATA
MetTyrThrSerSerGlyGlyTyrIleGluAspProSerValLysGluLysValLysGluAlaValGluAlaAlaIleAspLeuGlyIle
(100)

400
TATGTCATAATTGATTGGCACATCCTTTCAGACAATGACCCGAATATATATAAAGAAGAAGCAAAGGATTTCTTTGATGAAATGTCTGAG
TyrValIleIleAspTrpHisIleLeuSerAspAsnAspProAsnIleTyrLysGluGluAlaLysAspPhePheAspGluMetSerGlu

500
CTGTATGGAGATTACCCGAATGTGATATACGAAATTGCAAATGAACCGAATGGTAGTGATGTTACGTGGGACAATCAAATAAAACCGTAT
LeuTyrGlyAspTyrProAsnValIleTyrGluIleAlaAsnGluProAsnGlySerAspValThrTrpAspAsnGlnIleLysProTyr

600
GCAGAGGAAGTAATTCCGGTTATCCGTAACAATGATCCTAATAACATTATTATTGTAGGTACAGGTACATGGAGTCAGGATGTTCATCAT
AlaGluGluValIleProValIleArgAsnAsnAspProAsnAsnIleIleIleIleValGlyThrGlyThrTrpSerGlnAspValHisHis
(200)

Hincll   700
GCTGCTGATAATCAGTTAACAGATCCGAACGTCATGTATGCATTTCATTTTTATGCAGGAACACATGGACAAAATTTACGAGACCAAGTA
AlaAlaAspAsnGlnLeuThrAspProAsnValMetTyrAlaPheHisPheTyrAlaGlyThrHisGlyGlnAsnLeuArgAspGlnVal

800
GATTATGCATTAGATCAAGGAGCAGCAATATTTGTTAGTGAATGGGGAACGAGTGAAGCTACTGGTGATGGCGGCGTGTTTTTAGATGAA
AspTyrAlaLeuAspGlnGlyAlaAlaIlePheValSerGluTrpGlyThrSerGluAlaThrGlyAspGlyGlyValPheLeuAspGlu

900
GCACAAGTGTGGATTGACTTTATGGATGAAAGAAATTTAAGCTGGGCAAACTGGTCTCTAACGCACAAAGATGAGTCATCTGCGGCGTTA
AlaGlnValTrpIleAspPheMetAspGluArgAsnLeuSerTrpAlaAsnTrpSerLeuThrHisLysAspGluSerSerAlaAlaLeu
(300)

ATGCCAGGTGCAAGCCCAACTGGTGGGTGGACAGAGGCTGAACTATCTCCATCTGGGACATTTGTGAGGGAAAAAATAAGAGAGTCAGCA
MetProGlyAlaSerProThrGlyGlyTrpThrGluAlaGluLeuSerProSerGlyThrPheValArgGluLysIleArgGluSerAla

1000   BamHI
ACAACACCACCTAGTGATCCAACACCACCATCTGATCCAGATCCAGGTGAACCAGAACCAGATCCAGGTGAACCGGATCCAACGCCACCA
ThrThrProProSerAspProThrProProSerAspProAspProGlyGluProGluProAspProGlyGluProAspProThrProPro

1100   BamHI   HaeIII
AGTGATCCAGGAGATTATCCGGCATGGGATCCAAATACAATTTATACAGATGAAATTGTGTACCATAACGGCCAGCTATGGCAAGCAAAA
SerAspProGlyAspTyrProAlaTrpAspProAsnThrIleTyrThrAspGluIleValTyrHisAsnGlyGlnLeuTrpGlnAlaLys

1200   1227
TGGTGGACGCAAAATCAAGAGCCAGGCGACCCATACGGTCCGTGGGAACCACTCAATTAACGATATAATGATAGAAATTTACTAATGATA
TrpTrpThrGlnAsnGlnGluProGlyAspProTyrGlyProTrpGluProLeuAsn***
(400)   (409)

TAAGGAGAATGCCAAGAGTCTAAATTGGACGAITTGGCATICTCATTTTACATATTAGAGTTGAAATACTTTCCGCTATTCGTCTGCAATA
HpaI/HincII
AGTTAACCCATACTTAGTAGTAGTTTCCAATCTGTAAAACACTTTTAAGAAAAGGTTTTCTTACGAATCATTTACATGCAAATTAATAAA

HindIII
TTGTGAGACATTAGTGTAAAACTTTTTATGTAGTAGTAAAGCTT

Fig. 7.32 (a) pNK1 (*CelB* gene) (p.208). (b) pNK2 (*CelA* gene).
Nucleotide sequences of the cellulase genes carried by the 2.1-kb *Hind* III fragment in pNK1 and the 1.6-kb *EcoR* I-*Hind* III fragment in pNK2 and the primary structure of their products. The nucleotides are numbered by taking initiation code A as 1. The amino acids are numbered taking the initiation methionine as (1). The Shine-Dalgarno (SD) sequences are indicated by broken lines. The putative transcription terminator sequences are indicated by heavy arrows under the nucleotide sequences. The major direct repeat sequences in the pNK1-encoded cellulase gene are boxed. The short direct repeats (Pro-Asp-Pro-Gly-Glu) are underlined with light arrows. Asterisks indicate stop codons.

(Reproduced with permission from Fukumori *et al.*, *J. Bacteriol.*, **168**, 480 (1986))

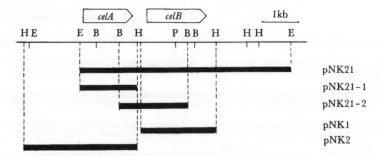

Fig. 7.33   Restriction map of the cloned *Eco*R I fragment of pNK 21 and of the subcloned fragments pNK 21-1 and pNK 21-2, and localization of the *celA* and
celB genes (arrows). B: *Bam*H I; H: *Hin*d III; P: *Pst* I.
(Reproduced with permission from *FEMS Symposium* (eds. Aubert, J.-P., *et
al.*), **48**, 207 (1987))

F.   Molecular Cloning of the *celC* Gene

Bacillus sp. No. N-4 produced two more alkaline cellulases having pH optima for
enzyme action at 9-10. We tried to clone these cellulase genes in addition to the *celA* and
celB genes. A plasmid pNK3 containing a 5.6 kb *Hin*d III fragment was obtained
(Fukumori *et al.*, 1989). The cellulase produced by *E. coli* carrying pNK3 had an optimum
pH of 9-10 for cellulase activity. The cellulase gene of pNK3 was subcloned in a functional state on a 3.5 kb *Xba* I-*Hin*d III fragment. Nucleotide sequence of the fragment
showed an open reading frame starting from a GTG as an initiation codon located 12 bp
downstream of ATG. Although the *Xba* I-*Hin*d III fragment encoded a protein that
expressed cellulase activity, no stop codon could be found within this fragment, but the
open reading frame should be stopped in the pBR322. The *celC* protein had a molecular
weight of about 100,000. The deduced amino acid sequence exhibited strong homology
with cellulases from *Bacillus* species, especially the *celF* from *Bacillus* sp. No. 1139. The *celC*
had an optimum pH for enzyme action in the range of pH 9 to 10, which is similar to that
of *celF*.

7.4.2   A Cellulase from Alkaliphilic *Bacillus* sp. No. 1139

A.   Isolation of *Bacillus* sp. No. 1139

Isolation procedures were almost same as those for *Bacillus* sp. No. N-4. The alkaliphilic *Bacillus* sp. No. 1139 was selected from about 1,200 colonies (Fukumori, Kudo and
Horikoshi, 1985). The characteristics of the isolate are listed in Table 7.26. It was essentially similar to *Bacillus firmus* in taxonomic characteristics, except for growth pH and ability to grow anaerobically. Neither Avicel nor filter paper was hydrolyzed at all.

Bacillus sp. No. 1139 was grown with shaking on a rotary shaker at 37°C for 3 days in
modified Horikoshi-II CMC medium. The culture broth was centrifuged at 5,000 × g for
20 min and the supernatant fluid was used as a crude cellulase preparation. The crude
cellulase (carboxylmethylcellulase, CMCase) in the culture broth was precipitated with
ammonium sulfate at 80% saturation, redissolved in water and dialyzed against 0.05M Tris-
HCl buffer (pH 8.0) containing 0.05 M NaCl. The dialysate was applied onto a DEAE-
Toyopearl ion-exchange column equilibrated with the same buffer described above. The

Table 7.26   Characteristics of isolate No. 1139

Shape	Rod	Growth temperature:	
Size (μm)	0.6–0.8 × 3–4	Maximum	45 ℃
Pigment production on agar slants	Negative	Optimum	30–37 ℃
Motility	Positive	Optimum pH for growth	8.8–10.0
Gram stain	Positive	Growth with NaCl:	
Endospores	Formed; not swollen	5% (w/v)	Positive
Catalase	Positive	7% (w/v)	Negative
Oxidase	Positive	Hydrolysis of starch	Positive
Anaerobic growth	Positive	Reduction of NO_3^- to NO_2^-	Positive
Voges-Proskauer reaction	Negative	Formation of indole	Negative
		mol% G + C	38.8

(Reproduced with permission from Fukumori *et al.*, *J. Gen. Microbiol.*, **131**, 3341 (1985))

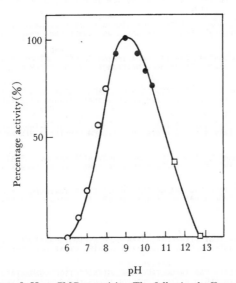

Fig. 7.34   Effect of pH on CMCase activity. The following buffer systems were used: ○, 0.05 M-sodium phosphate; ●, 0.05 M-glycine/NaCl/NaOH. Other conditions were the same as for the standard assay. One hundred percent activity was taken as 150 mU.
(Reproduced with permission from Fukumori *et al.*, *J. Gen. Microbiol.*, **131**, 3343 (1985))

enzyme was eluted with a linear gradient of 0.05–1.0 M NaCl in the same buffer. The CMCase was eluted at about 0.3 M NaCl and purified further by passing through a Toyopearl HW-55F column. The purified CMCase was electrophoretically homogeneous, with an estimated molecular weight of 92,000 (SDS-PAGE). The isoelectric point was pH 3.1. The CMCase was most active at pH 9.0 and still retained some activity at pH 10.5 (Fig. 7.34). The enzyme was stable over the range of pH 6–11 (24 h at 4 ℃ and up to 40 ℃ for 10 min). The enzyme hydrolyzed cellotriose or cellotetraose but not cellobiose. Cellotriose was converted to cellobiose, which was the main product. The ratio between

```
                                                                          -202(HpaI/HincII)
                                                                          GTTAACCTTGTGCTATATGCCG

ATTTAGGAAGGGGGTAGATTGAGTCAAGTAGTCATAATTTAGATAACTTATAAGTTGTTGAGAAGCAGGAGAGAATCTGGGTTACTCACA


AGTTTTTTAAAACATTATCGAAAGCACTTTCGGTTATGCTTATGAATTTAGCTATTTGATTCAATTACTTTAATAATTTTAGGAGGTAAT
                                                                                  SD
1
ATGATGTTAAGAAAGAAAACAAAGCAGTTGATTTCTTCCATTCTTATTTTAGTTTTACTTCTATCTTTATTTCCGACAGCTCTTGCAGCA
MetMetLeuArgLysLysThrLysGlnLeuIleSerSerIleLeuIleLeuValLeuLeuLeuSerLeuPheProThrAlaLeuAlaAla
(-30)
     100
GAAGGAAACACTCGTGAAGACAATTTTAAACATTTATTAGGTAATGACAATGTTAAACGCCCTTCTGAGGCTGGCGCATTACAATTACAA
GluGlyAsnThrArgGluAspAsnPheLysHisLeuLeuGlyAsnAspAsnValLysArgProSerGluAlaGlyAlaLeuGlnLeuGln
(1)
              200
GAAGTCGATGGACAAATGACATTAGTAGATCAACATGGAGAAAAAATTCAATTACGTGGAATGAGTACACACGGATTACAATGGTTTCCT
GluValAspGlyGlnMetThrLeuValAspGlnHisGlyGluLysIleGlnLeuArgGlyMetSerThrHisGlyLeuGlnTrpPhePro

   BglII       300
GAGATCTTGAATGATAACGCATACAAAGCTCTTGCTAACGATTGGGAATCAAATATGATTCGTCTAGCTATGTATGTCGGTGAAAATGGC
GluIleLeuAsnAspAsnAlaTyrLysAlaLeuAlaAsnAspTrpGluSerAsnMetIleArgLeuAlaMetTyrValGlyGluAsnGly

                           400        BglII
TATGCTTCAAATCCAGAGTTAATTAAAAAGCAGAGTCATTAAAGGAATAGATCTTGCTATTGAAAATGACATGTATGTCATCGTTGATTGG
TyrAlaSerAsnProGluLeuIleLysSerArgValIleLysGlyIleAspLeuAlaIleGluAsnAspMetTyrValIleValAspTrp
                           (100)
                                        500
CATGTACATGCACCTGGTGATCCTAGAGATCCCGTTTACGCTGGAGCAGAAGATTTCTTTAGAGATATTGCAGCATTATATCCTAACAAT
HisValHisAlaProGlyAspProArgAspProValTyrAlaGlyAlaGluAspPhePheArgAspIleAlaAlaLeuTyrProAsnAsn

                                              600
CCACACATTATTTATGAGTTAGCGAATGAGCCAAGTAGTAACAATAATGGTGGAGCTGGGATTCCAAATAATGAAGAAGGTTGGAATGCG
ProHisIleIleTyrGluLeuAlaAsnGluProSerSerAsnAsnAsnGlyGlyAlaGlyIleProAsnAsnGluGluGlyTrpAsnAla

                                                      700
GTAAAAGAATACGCTGATCCAATTGTAGAAATGTTACGTGATAGCGGGAACGCAGATGACAATATTATCATTGTGGGTAGTCCAAACTGG
ValLysGluTyrAlaAspProIleValGluMetLeuArgAspSerGlyAsnAlaAspAspAsnIleIleIleValGlySerProAsnTrp
                                                      (200)
               PvuII                    AccI                        800
AGTCAGCGTCCTGACTTAGCAGCTGATAATCCAATTGATGATCACCATCACAATGTATACTGTTCACTTCTACACTGGTTCACATGCTGCT
SerGlnArgProAspLeuAlaAlaAspAsnProIleAspAspHisHisThrMetTyrThrValHisPheTyrThrGlySerHisAlaAla

                                                                   900
TCAACTGAAAGCTATCCGCCTGAAACTCCTAACTCTGAAAGAGGAAACGTAATGAGTAACACTCGTTATGCGTTAGAAAACGGAGTAGCA
SerThrGluSerTyrProProGluThrProAsnSerGluArgGlyAsnValMetSerAsnThrArgTyrAlaLeuGluAsnGlyValAla

GTATTTGCAACAGAGTGGGGAACTAGCCAAGCAAATGGAGATGGTGGTCCTTACTTTGATGAAGCAGATGTATGGATTGAGTTTTTAAAT
ValPheAlaThrGluTrpGlyThrSerGlnAlaAsnGlyAspGlyGlyProTyrPheAspGluAlaAspValTrpIleGluPheLeuAsn
                                                                           (300)

    1000
GAAAACAACATTAGCTGGGCTAACTGGTCTTTAACGAATAAAAATGAAGTATCTGGTGCATTTACACCATTCGAGTTAGGTAAGTCTAAC
GluAsnAsnIleSerTrpAlaAsnTrpSerLeuThrAsnLysAsnGluValSerGlyAlaPheThrProPheGluLeuGlyLysSerAsn

         1100               KpnI/BanI
GCAACAAGTCTTGACCCAGGGCCAGACCAAGTATGGGTACCAGAAGAGTTAAGTCTTTCTGGAGAATATGTACGTGCTCGTATTAAAGGT
AlaThrSerLeuAspProGlyProAspGlnValTrpValProGluGluLeuSerLeuSerGlyGluTyrValArgAlaArgIleLysGly

                     1200
GTGAACTATGAGCCAATCGACCGTACAAAATACACGAAAGTACTTTGGGACTTTAATGATGGAACGAAGCAAGGATTTGGAGTGAATGGA
ValAsnTyrGluProIleAspArgThrLysTyrThrLysValLeuTrpAspPheAsnAspGlyThrLysGlnGlyPheGlyValAsnGly

                          1300
GATTCTCCAGTTGAAGTAGTAGTTATTGAGAATGAAGCGGGCGCTTTAAAACTTTCAGGATTAGATGCAAGTAATGATGTTTCTGAAGGT
AspSerProValGluAspValValIleGluAsnGluAlaGlyAlaLeuLysLeuSerGlyLeuAspAlaSerAsnAspValSerGluGly
                          (400)
                                1400
AATTACTGGGCTAATGCTCGTCTTTCTGCCGACGGTTGGGGAAAAAGTGTTGATATTTTAGGTGCTGAAAAACTTACTATGGATGTGATT
AsnTyrTrpAlaAsnAlaArgLeuSerAlaAspGlyTrpGlyLysSerValAspIleLeuGlyAlaGluLysLeuThrMetAspValIle

                                      1500
GTTGATGAGCCGACCACGGTATCAATTGCTGCAATTCCACAAGGGCCATCAGCCAATTGGGTTAATCCAAATCGTGCAATTAAGGTTGAG
ValAspGluProThrThrValSerIleAlaAlaIleProGlnGlyProSerAlaAsnTrpValAsnProAsnArgAlaIleLysValGlu

                                        PvuII       1600
CCAACTAATTTCGTACCGTTAGAGGATAAGTTTAAAGCGGAATTAACTATTAACTTCAGCTGACTCTCCATCGTTAGAAGCTATTGCGATG
ProThrAsnPheValProLeuGluAspLysPheLysAlaGluLeuThrIleThrSerAlaAspSerProSerLeuGluAlaIleAlaMet
                                                    (500)
                                              1700
CATGCTGAAAATAACAACATCAACAACATCATTCTTTTTGTAGGAACTGAAGGTGCTGATGTTATCTATTTAGATAACATTAAAGTAATT
HisAlaGluAsnAsnAsnIleAsnAsnIleIleLeuPheValGlyThrGluGlyAlaAspValIleTyrLeuAspAsnIleLysValIle
```

Fig. 7.35   (Continued)

```
                                                                                   1800
GGAACAGAAGTTGAAATTCCAGTTGTTCATGATCCAAAAGGAGAAGCTGTTCTTCCTTCTGTTTTTGAAGACGGTACACGTCAAGGTTGG
GlyThrGluValGluIleProValValHisAspProLysGlyGluAlaValLeuProSerValPheGluAspGlyThrArgGlnGlyTrp

GACTGGGCTGGAGAGTCTGGTGTGAAAACAGCTTTAACAATTGAAGAAGCAAACGGTTCTAACGCGTTATCATGGGAATTTGGATACCCA
AspTrpAlaGlyGluSerGlyValLysThrAlaLeuThrIleGluGluAlaAsnGlySerAsnAlaLeuSerTrpGluPheGlyTyrPro
                                                                                   (600)
    1900
GAAGTAAAACCTAGTGATAACTGGGCAACAGCTCCACGTTTAGATTTCTGGAAATCTGACTTGGTTCGCGGTGAAAATGATTATGTAACT
GluValLysProSerAspAsnTrpAlaThrAlaProArgLeuAspPheTrpLysSerAspLeuValArgGlyGluAsnAspTyrValThr

         2000
TTTGATTTCTATCTAGATCCAGTTCGTGCAACAGAAGGCGCAATGAATATCAATTTAGTATTCCAGCCACCTACTAACGGGTATTGGGTA
PheAspPheTyrLeuAspProValArgAlaThrGluGlyAlaMetAsnIleAsnLeuValPheGlnProProThrAsnGlyTyrTrpVal

            AccI       2100
CAAGCACCAAAAACGTATACGATTAACTTTGATGAATTAGAGGAACCGAATCAAGTAAATGGTTTATATCACTATGAAGTGAAAATTAAC
GlnAlaProLysThrTyrThrIleAsnPheAspGluLeuGluGluProAsnGlnValAsnGlyLeuTyrHisTyrGluValLysIleAsn

                     2200
GTAAGAGATATTACAAACATTCAAGATGACACGTTACTACGTAACATGATGATCATTTTTGCAGATGTAGAAAGTGACTTTGCAGGGAGA
ValArgAspIleThrAsnIleGlnAspAspThrLeuLeuArgAsnMetMetIleIleIlePheAlaAspValGluSerAspPheAlaGlyArg
             (700)
                     2300
GTCTTTGTAGATAATGTTCGTTTTGAGGGGGCTGCTACTACTGAGCCGGTTGAACCAGAGCCAGTTGATCCTGGCGAAGAGACGCCGCCT
ValPheValAspAsnValArgPheGluGlyAlaAlaThrThrGluProValGluProGluProValAspProGlyGluGluThrProPro,
                      2400
GTCGATGAGAAGGAAGCGAAAACAGAACAAAAAGAAGCAGAGAAAGAAGAGAAAGAAGAGTAAAAGAAGAAAAGAAAGAAGCTAAAGAAG
ValAspGluLysGluAlaLysThrGluGlnLysGluAlaGluLysGluLysGluLys***
                                                         (770)
AAAAGAAAGCAATCAAAAATGAGGCTACGAAAAAATAATCTAATAAACTAGTTATAGGGTTATCTAAAGGTCTGATGCAGATCTTTTAGA

TAACCTTTTTTTGCATAACTGGACATAGAATGGTTATTAAAGAAAGCACGGTGTTTATACGATATTAAAAGGTAGCGATTTTAATTGAAA

CCTTTAATAATGTCGTGTGATAGAATGATGAAGTAATTTAAGAGGGGGGAAACGAAGTGAAAACGGAAATTTCTAGTACAACAAAAACAG
         HindIII
ACCAACAAATACTGCAAGCTT
```

Fig. 7.35 Nucleotide sequence of the cellulase gene from *Bacillus* sp. No. 1139. The DNA sequence of the coding strand is given from 5' to 3', numbered from nucleotide 1 at the putative initiation site. The proposed ribosomal binding site (SD) is underlined with a dashed line. The predicted amino acid sequence is given below the DNA sequence. The amino acids are numbered taking the NH_2-terminal amino acid of the mature protein as 1. Underlined amino acids have been determined by automated Edman sequencing of the purified cellulase. The hairpin loop of the putative rho-independent terminator site is underlined, and the poly-T region is overlined.
(Reproduced with permission from Fukumori *et al.*, *J. Gen. Microbiol.*, **132**, 2332 (1986))

glucose and cellobiose was about 0.1. This result indicates that the enzyme has *trans* glucosidase activity. To obtain further information we cloned the alkaliphilic *Bacillus* sp. No. 1139 CMCase gene in *E. coli* HB101 using the plasmid pBR322. Details are given in the following section.

B. Cloning of the Alkaline Cellulase Gene of *Bacillus* sp. No. 1139

Fukumori *et al.* (1986) cloned the cellulase gene of *Bacillus* sp. No. 1139 to gain further information. *Bacillus* sp. No. 1139 was grown aerobically to the early stationary phase at 37 °C in CMC medium and bacterial DNA was digested with *Hin*d III at 37 °C for 1 h (plasmid DNA) or for 14 h (chromosomal DNA). After digestion, 1 μg of plasmid and 3 μg of bacterial chromosomal DNA were mixed and ligated with T_4 ligase overnight at 15°C. This ligated mixture was used for transformation experiments. The ligated mixtures described above were introduced into *E. coli* HB101 and about 10^4 Ap^r Tc^s transformants per mg DNA obtained. A plasmid, pFK1, was obtained from a transformant showing shallow craters around the colony. The plasmid contained a 4.6 kb *Hin*d III fragment. Southern hybridization experiments showed that pFK1 hybridized to a 4.6 kb of *Hin*d III fragment of *Bacillus* sp. No. 1139 DNA, but no sequences complementary to pFK1 were detected in *E. coli* DNA fragments.

Subcloning experiments revealed that a 2.9 kb *Hin*c II-*Hin*d III fragment was necessary for cellulase production. The DNA sequence of this fragment was determined by the dideoxy chain termination method. The DNA sequence and the deduced amino acid sequence are shown in Fig. 7.35. Underlined amino acids were determined by automated Edman sequencing of the purified cellulase of *Bacillus* sp. No. 1139. A putative ribosome binding site, an AGGAGG sequence that was highly complementary to the 3′ end of *B. subtilis* 16S ribosomal RNA, was observed upstream of the open reading frame.

E. coli HB101 (pFK1) was grown aerobically in LB broth for 24 h at 37°C. The extracellular, periplasmic and cellular cellulase activities were 66, 402 and 116 mU per ml of culture, respectively. The synthesis of cellulase in *E. coli* was constitutive and no effect of CMC supplementation was observed. The optimum pH for the cellulase activity of *E. coli* (pFK1) was 9.0, as was observed for *Bacillus* sp. No. 1139. Periplasmic cellulase gave a line of precipitation that fused with that for cellulase from *Bacillus* sp. No. 1139 in the Ouchterlony double-diffusion test and showed a molecular weight of 94,000 by the SDS-PAGE method. This molecular weight was slightly higher than the 92,000 estimated from the open reading frame. This difference may be due to processing of the protein. It is possible that the cellulase secreted into the periplasmic space of *E. coli* may be processed at a different site(s) during the secretion process.

C. Truncation Analysis of an Alkaline Cellulase from Alkaliphilic *Bacillus* sp. No.1139

It is of interest to examine whether or not the entire sequence is indispensable for cellulose activity. In order to study this point, variously sized truncation products of the alkaline cellulase were constructed using an insertional terminator cartridge (Fukumori, Kudo and Horikoshi, 1987). Plasmid pFKl was digested with *Hin*c III, ligated with T_4 DNA ligase, and the ligation mixture introduced into *E. coli* C600. pFK3 was isolated from cellulase-positive transformants. pFK3 lacked a 0.7 kb *Hin*c II fragment of pFK1, and the 3.5 kb *Hin*c II fragment was inserted in the direction opposite to that of pFK1. pFK4 was constructed by removing a 1.7 kb *Hin*d III fragment of pFK3. To analyze the role of the C-terminal region of the alkaline cellulase an omega fragment was used to mutate the DNA fragment at specific restriction endonuclease sites. This yielded the plasmids pFK4-*Mlu* I, pFK4-*Sph* I, pFK4-*Kpn* I, and pFK4-*Sca* I with the omega fragment within the alkaline cellulase gene. *E. coli* C600 carrying each plasmid was cultured in LB-broth for 24 h at 37°C and the cellulase activities expressed in *E. coli* were assayed. As shown in Fig. 7.36, *E. coli* carrying pFK4, pFK4-*Mlu* I, pFK4-*Sph* I and pFK4-*Sca* I exhibited cellulase activity showing similar pH optima. However, the protein encoded by the pFK4-*Kpn* I showed no enzyme activity, although the protein corresponding to pFK4-*Kpn* I was detected by the maxicell experiment. Molecular weights of the modified proteins were as follows. The pFK encod-

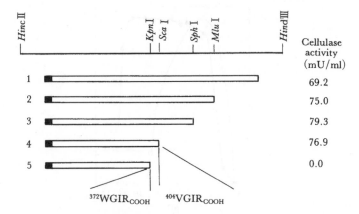

Fig. 7.36 Construction of pFK4 derivatives and cellulase activity of their products. 1, pFK4; 2, pFK4-*Mlu* I; 3, pFK4-*Sph* I; 4, pFK4-*Sca* I; 5, pFK4-*Kpn* I. The putative length of the genes is indicated by open boxes. The C-terminal amino acid sequence of pFK4-*Sca* I and pFK4-*Kpn* I proteins is indicated by one-letter abbreviations. Amino acids are numbered starting at the N-terminal methionine. Cellulase activity is expressed as mU/ml of culture medium. (Reproduced with permission from *FEMS Symposium* (eds. Aubert, J.-P., *et al.*), **43**, 212 (1987))

ed protein (control) was 92,000; pFK4-*Sca* I, 46,000 and pFK4-*Kpn* I, 43,000. The cellulase of alkaliphilic *Bacillus* sp. No. 1139 consists of 770 amino acids and has a molecular weight of 92,000. However, a truncated cellulase (molecular weight of 46,000, pFK4-*Sca* I protein) has enzyme activity and pH optimum similar to those of the original protein. It is of interest that the large C-terminal part of the cellulase is not necessary for the enzyme activity. Furthermore, when 32 amino acids were deleted from the pFK4-*Sca* I protein, no enzyme activity was observed. This region may include some part of the active site or may be important for maintaining the structure of the cellulase molecule.

D. Chimeric Cellulases between *Bacillus subtilis* and *Bacillus* sp. N-4

Nakamura *et al.* constructed many chimeric cellulases from *Bacillus subtilis* and *Bacillus* sp. N-4 enzyme genes to better understand the results described above (Nakamura, Fukumori *et al.*, 1991). The amino acid sequences of cellulase from *Bacillus subtilis* (BSC) and that from an alkaliphilic *Bacillus* sp. N-4(NK1) show significant homology in most parts except for the C-terminal portions. Despite the high homology, the pH activity profiles of the two enzymes are quite different; BSC has its optimum pH at 6–6.5, whereas NK1 is active over a broad pH range from 6 to 10.5 (Fig. 7.37). In order to identify the structural features which determine such pH activity profiles, chimeric cellulases between BSC and NK1 were constructed using four restriction sites commonly present within the homologous coding sequences, and were present within the homologous coding sequences, and were produced in *E. coli* (Fig. 7.38). The chimeric cellulases showed various chromatographic behaviors, reflecting the origins of their C-terminal regions. The pH activity profiles of the chimeric enzymes in the alkaline range could be classified into either the BSC or NK1 type mainly depending on the origin of the fifth C-terminal region (Fig. 7.39). In

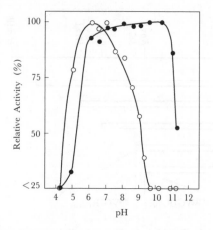

Fig. 7.37 pH activity profiles of BSC and NK1. ○, BSC; ●, NK1.
(Reproduced with permission from Nakamura *et al.*, *J. Biol. Chem.*, **266**, 1580 (1991))

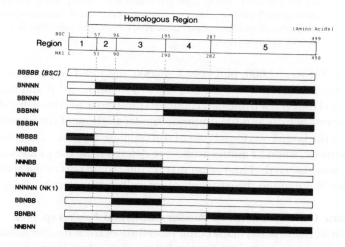

Fig. 7.38 Structures of the chimeric cellulases. *Open bars* and *black boxes* show the regions derived from BSC and NK1, respectively. Amino acid positions which punctuate protein regions are also indicated.
(Reproduced with permission from Nakamura *et al.*, *J. Biol. Chem.*, **266**, 1581 (1991))

the acidic range the profile was determined only by the fourth enzyme region from the N-terminus. Only a limited number of amino acids in the fourth region may affect the deprotonation of catalytic residues of cellulases and modulate the catalytic activity in the acidic pH values.

E. pH Activity Profile Depends on the Substrates Used
Hitomi *et al.* (1994) conducted research to determine pH activity profiles. A neutral

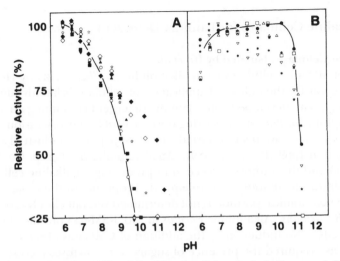

Fig. 7.39   pH activity profiles of the chimeric cellulases in the basic range. A, BSC-type chimeras; B, NK1-type chimeras. ○, BBBBB; ▲, NBBBB; ■, NNBBB; ▼, NNNBB; ◆, NNNNB; ●, NNNNN; △, BNNNN; □, BBNNN; ▽, BBBNN; ◇, BBBBN: ★, NNBNN; *, BBNBN; ☆, BBNBB. Only the profiles of BSC and NK1 are indicated with *lines* in A and B, respectively.
(Reproduced with permission from Nakamura *et al., J. Biol. Chem.*, **266**, 1581 (1991))

cellulase (BSC) from *Bacillus subtilis* and an alkaline cellulase (NKl) from alkaliphilic *Bacillus* sp. N-4 show significant amino acid sequence homology as shown above. Despite the high homology, the pH activity profiles of the two enzymes for carboxymethly cellulose (CMC) hydrolysis are quite different; BSC shows a sharp optimum pH at 6, whereas NKl shows its full activity in a broad range, from pH 6 to 10.5. For elucidation of the reason for the difference in their pH activity profiles their activities were examined at various pHs using a series of cellooligosaccharides and their derivatives, cellotetraose (G4), cellopentaose (G5), cellohexaose (G6), cellopentaitol (G5OH), and cellohexaitol (G6OH), as substrates. The optimum pH of BSC was around 6 for all the cellooligosaccharides examined. On the other hand, the optimum pH of NKl varied depending on the substrate, *i.e.*, a sharp optimum at pH 6 with G4 and G5OH, and a broad optimum of pH 6 to 10.5 with G5, G6, and G6OH. Comparison of the kinetic parameters of the two cellulases at pH 7 and 9 using G6OH as the substrate revealed that NKl showed similar values at both pHs, while BSC showed a greatly increased K_m value for this substrate at pH 9. In addition, NKl showed a greatly increased K_m value for G5OH hydrolysis at pH 9. Both enzymes cleaved these substrates at the same position, which suggests the same productive binding mode of these substrates for both enzymes. All these observations suggest that the reduced enzyme activity of BSC in the alkaline pH range can be attributed to a decrease in the affinity of a subsite for the third glucose moiety from the scissile site of these substrates.

7.4.3 Alkaline Cellulases for Laundry Detergent

A. Alkaline Cellulases Isolated by Ito *et al.*

Discovery of alkaline cellulases from alkaliphilic *Bacillus* strains by Horikoshi *et al.* (1984) paved the way for the industrial application of cellulase as laundry detergent additives. Kao Corp., Japan, developed a new laundry detergent containing an alkaline cellulase of the alkaliphilic *Bacillus* strain. During the course of their work various types of alkaline cellulases (K-series) have been reported. The first paper on industrial alkaline cellulase was published in 1989 (Ito *et al.*, 1989). An alkaliphilic *Bacillus* sp. No. KSM-635 was isolated from soil and the authors succeeded in producing an alkaline cellulase as laundry detergent additive in an industrial scale plant. Properties of the isolate are shown in Table 7.27. The taxonomical position is not determined yet, but closely resembles that of *B. pumilus* and *B. circulans*. Alkaliphilic *Bacillus* sp. No. KSM-635 produced CMCases, but did not utilize CMC. No induction by the addition of CMC was observed. Higher production of CMCase required the presence of sugars (*e.g.*, fructose, sucrose, maltose and

Table 7.27   Properties of *Bactillus* sp. No. KSM-635

Morphology		Hydrolysis of starch	−
Form	Rods	Hydrolysis of casein	−
Size (μm)	0.5–1.2 × 1.5–4.0	Utilization of citrate	−
Motility	+	Utilization of inorganic	
Gram staining positive		nitrogen	+
spores (μm)	0.7–1.2 × 1.0–2.0	Formation of pigment	−
Central to terminal		Urease	−
acid-fastness	−	Oxidase	Unclear
		Catalase	+
Characteristics of culture		Temperature for growth	20°–45 ℃
On nutrient agar plates :		pH for growth	8–11
Circular colonies, flat		Behavior on oxygen	Aerobic
surface, white or yellow,		Anaerobic growth	−
semi-transparent glass			
Gelatin	Not liquefied	Utilization of sugars	
		L-Arabinose	+
Litmus milk	Not liquefied ;	D-Xylose	+
	Color unchanged	D-Ribose	+
		D-Glucose	+
In nutrient broth with7% NaCl :		D-Fructose	+
+		D-Mannose	+
		D-Galactose	−
Biochemical properties		Sucrose	+
Reduction nitrate to nitrite	+	Lactose	−
Denitrification	−	Maltose	+
MR test	Color unchanged	Trehalose	+
VP test	+	Raffinose	−
Formation of indole	−	D-Solbitron	−
Formation of hydrogen sulfide	−	D-Mannitol	+
		Insitol	+
		Glycerol	+

(Reproduced with permission from Ito *et al.*, *Agric. Biol. Chem.*, **53**, 1277 (1989))

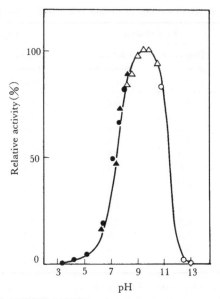

Fig. 7.40   Effect of pH on activity.
(Reproduced with permission from Ito *et al.*, *Agric. Biol. Chem.*, **53**, 1278 (1989))

mannitol) and complex nitrogen, with an initial pH of 8-9. The CMCase, partially purified by precipitation with ethanol, showed an optimum pH for activity of 9.5 (Fig. 7.40)
and an optimum temperature of 40 °C in glycine-NaOH buffer. It showed strong activity
toward CMC, but very little activity to *p*-nitrophenyl-β-glucoside or cellulosic substrates
showing high crystallinity. Activity towards *p*-nitrophenyl-β-cellobioside was detected at
pH 7, but it amounted to less than 2% of maximum CMCase activity. Characteristically,
the alkaline CMCase activity was not affected by various laundry components such as surfactants, chelating agents and alkaline proteinases, and the addition of fructose, sucrose
or mannitol increased enzyme production. Glucose, ribose, glycerol and maltose gave relatively lower production. Xylose strongly repressed enzyme production, as shown in Table
7.28. The alkaline cellulase barely hydrolyzed cotton, Avicel, swollen cellulose and cellobiose. Other properties are summarized in Table 7.29. It is of interest that the molecular weight of the enzyme varied (100,000 to 600,000) by cultivation condition such as
time, medium and temperature (Ito, personal communication). Polyacrylamide gel electrophoresis revealed that the cellulase in the culture broth was in the multi-enzyme form
as reported previously (Horikoshi *et al.*, 1984).

Yoshimatsu *et al.* (1990) isolated and purified two enzymes from the culture fluid of
Bacillus KSM-635. Two CMCases (E-H and E-L) were purified to homogeneity from a culture filtrate of the alkaliphilic *Bacillus* sp. KSM-635 by chromatography on DEAE-
Toyopearl 650S and gel filtration on Bio-Gel A-0.5 M. The purified CMCases both contained approx. 2-3% (w/w) glucosamine. Molecular masses deduced from SDS-PAGE
were 130 kDa for E-H and 103 kDa for E-L. The pH optima of the enzymes were both
about 9.5, and their optimum temperatures were around 40 °C. Activities of these enzymes
were inhibited by Hg^{2+}, Cu^{2+}, Fe^{2+} and Fe^{3+}, but sulfhydryl inhibitors, such as *N*-ethyl-

Table 7.28　Effect of carbon sources on alkaline cellulase KSM-635 production

Carbon source	Experiment 1		Experiment 2	
	Degree of growth (A_{600})	CMCase produced (units/ml)	Degree of growth (A_{600})	CMCase produced (units/ml)
None	3.3	0.8	4.0	1.3
Cellulose powder	—	—	Nm[†]	1.4
Avicel	Nm	0.8	Nm	1.5
CMC	3.3	1.5	4.0	1.8
Cellobiose	3.3	1.6	4.1	2.2
Xylose	4.9	0.1	6.1	0.1
Ribose	2.9	1.2	3.7	1.5
Glucose	6.0	1.0	5.5	1.4
Fructose	6.0	3.0	7.5	2.7
Sucrose	4.8	2.4	5.8	3.2
Maltose	3.9	4.8	4.5	2.5
Glycerol	5.1	1.4	5.9	2.9
Mannitol	4.0	3.8	4.7	2.1

[†] Not measurable, because of the nature of the substrate.

Bacillus sp. No. KSM-635 was cultivated at 30 °C for 3 days in duplicate flasks, in which was placed 50 ml of NY-medium containing 1% (w/v) of a carbon source. Assays were performed at 40 °C in glycine–NaOH buffer, pH 9.5.

(Reproduced with permission from Ito *et al., Agric. Biol. Chem.*, **53**, 1278 (1989))

Table 7.29　Properties of alkaline cellulase produced by alkaliphilic *Bacillus* sp. No. KSM-635

Property	Alkaline cellulase KSM-635
Molecular weight	100,000–600,000
Optimum pH	9.5
Stable pH	5.0–12.0 (30 °C, 60 min)
Optimum temperature	40 °C
Stability	Stable in EDTA or EGTA
	LAS, AS, ES, AOS, Soup
	Pronase, Alcanase
	Savinase
Isoelectric point	Lower than pH 4
Inhibitors	*p*-chloromercuribenzoate
	1 mM metals such as Hg^{2+}, Pb^{2+},
	Cd^{2+} Cu^{2+}, etc. did not inhibit
	enzymatic activity

maleimide, monoiodoaceate and 4-chlormercuribenzoate, had either no effect or a slight inhibitory effect. *N*-Bromosuccionimide was strongly inhibitory, suggesting that a tryptophan residue is essential for the activity of the CMCases. In addition, the activities of these enzymes were stimulated by Co^{2+}, and they required Mg^{2+}, Ca^{2+}, Mn^{2+} or Co^{2+} for stabilization. Both enzymes efficiently hydrolyzed CMC and lichenan, but crystalline cellulosic substrates, curdlan, laminarin and 4-nitrophenyl-*β*-glucopyranoside were hydrolyzed very little, if at all. 4-Nitrophenyl-*β*-cellobioside was hydrolyzed by both enzymes, and their

Table 7.30 Substrate specificity of CMCase

Assays were performed at 30 °C in 0.1 M-glycine/NaOH buffer (pH 9.5) with 0.86 µg E-H or with 0.63 µg E-L under the standard reaction conditions. Each value is the mean of three determinations, and activity of E-H against CMC was taken as 100%.

	E-H activity	E-L activity
	Rel. act. (%)	Rel. act. (%)
CMC	100	186
Activity	tr	tr
Filter paper	tr	tr
H_3PO_4-swollen cellulose	tr	tr
NaOH-swollen cellulose	tr	tr
Lichenan	84	181
Curdlan	0	0
Laminarin	0	0
Cellobiose	0	0
NPC[†] at pH 9.5	1.3	2.6
at pH 7.0	3.4	6.0
NPG[†] at pH 9.5	0	0
at pH 7.0	0	0

tr: Very low but not negligible activity (< 1.1 units/mg).

[†] Assays were performed with 4 µmol substrate in 0.5 ml 0.1 M-glycine/NaOH buffer (pH 9.5) or 0.1 M-phosphate buffer (pH 7.0), with 0.86 µg E-H or 0.63 µg E-L as enzyme.

hydrolysis rates were higher at neutral pH than at alkaline pH (Table 7.30).

Besides KSM-635 enzyme, Shikata *et al.* (1990) isolated three strains, alkaliphilic *Bacillus*, KSM-19, KSM-64 and KSM-520, producing alkaline cellulases for laundry detergents. Their activities were not inhibited at all by metal ions or various components of laundry products such as surfactants, chelating agents and proteinases. The enzyme preparations showed strong activities towards CMC, the optimum pH being 8.5–9.5 and the optimum temperature about 50 °C. Maximum growth of the isolates was observed at an initial pH of 8.5–9.5; slight growth occurred at neutral pH. Production of the alkaline cellulases required the presence of CMC, and it was controlled by a mechanism involving catabolite repression and induction. KSM-19 and KSM-64 produced alkaline lichenan-hydrolyzing enzymes. Neutrophilic *Bacillus* strain KSM-522 that could produce an alkaline carboxymethyl cellulase with broad pH activity profile (pH 5 to 10) was isolated in neutral media (Kawai *et al.*, 1988). However, the properties and yield of the enzyme were not suitable for industrial applications. Three endo-β-1,4-glucanases (CMCase E-1, E-II and E-III) from *Bacillus* sp. KSM-522 were purified to homogeneity by repeated chromatography on DEAE Bio-Gel A. The molecular weights estimated by SDS polyacrylamide gel electrophoresis were 78,000 for E-I, and 61,000 for E-II and E-III. The optimum pH values for activity and isoelectric points were around 6 and 4.4 for E-I, and 7–10 and 3.5 for E-II and E-III. These enzymes hydrolyzed CMC in a similar random fashion. The alkaline CMCases, E-II and E-III, were not distinguishable from each other (Okoshi *et al.*, 1990).

B. Cloning of Alkaline Cellulase Genes of *Bacillus* sp. KSM Series

Ozaki *et al.* (1990) cloned the alkaline cellulase gene of their industrial strain, *Bacillus* sp. KSM-635, in *E. coli.* A gene for alkaline cellulase from the alkaliphilic *Bacillus* sp. KSM-635 was cloned into the *Hin*d III site of pBR322 and expressed in *E. coli* HB101. Although the recombinant plasmid contained two *Hin*d III inserts of 2.6 kb and 4.0 kb, the inserts were found to be contiguous in the *Bacillus* genome by hybridization analysis. The nucleotide sequence of a 2.4-kb region, which was indispensable for the production of cellulase, and the flanking, 1.1-kb region, were determined. There was an open reading frame (ORF) of 2823 bp in the 3498 bp sequence determined, which encoded 941 amino acid residues. Two putative ribosome binding sites and a σ43-type, promoter-like sequence were found upstream from an initiation codon in the ORF. As exhibited in Fig. 7.41, the deduced amino-terminal sequence resembles the signal peptide of extracellular proteins. The amino acid sequence of the enzyme exhibited 72% homology with the alkaline cellulase of *Bacillus* sp. No. 1139. A region of amino acids, 249 to 568, of the deduced amino acid sequence of the cellulase from this organism is homologous with those of alkaline and neutral enzymes of other microorganisms, but nine amino acid residues were found to be conserved only in the alkaline enzymes. Furthermore, KSM-64 and KSM-330 cellulase genes that were sequenced and expressed in *E. coli* were cloned (Ozaki , Sumitomo and Ito, 1991; Sumitomo *et al.*, 1992).

Ozaki *et al.* (1995) reported that the N-terminal of the cellulase of *Bacillus* sp. KSM-635 was essentially exchangeable. Part of a 2,4-kb DNA fragment that encoded the amino-terminal 584 residues (65 kDa) of the alkaline cellulase from *Bacillus* sp. KSM-635 (941 amino acid residues ; 105 kDa) was spontaneously deleted during subcloning of the fragment. The remaining 1.1-kb insert of the deleted plasmid encoded amino acids from Ala 228 to Leu 584 of the enzyme. However, *Escherichia coli* HB101 cells harboring this plasmid produced an active endoglucanase. After addition of a termination codon, TAA, immediately downstream of the codon for Leu 584, the 1.1-kb fragment was inserted into an expression vector, pHSP64. The resultant plasmid was introduced into *Bacillus subtilis* ISW1214 for extracellular production of the truncated endoglucanase. Amino-terminal sequencing of the enzyme showed that the enzyme consisted of 7 amino acid residues encoded by the vector and 357 amino acid residues encoded by the truncated gene, with a molecular mass of 40.2 kDa. The purified enzyme was very active against carboxy methyl-cellulose and its pH and temperature profiles were almost identical to those of the enzyme produced by *Bacillus* sp. KSM-635.

C.   Other Alkaline Cellulases

Park , Horinouchi and Beppu (1991) and Damude *et al.* (1993) studied a semi-alkaline cellulase produced by the alkaliphilic *Streptomyces* strain KSM-9. Sequencing of the enzyme showed that the cellulase was a typical member of family B.

In order to obtain cellulases that improve the washing effect of laundry detergent additives, Dasilva *et al.* (1993) reported two alkaliphilic microorganisms, *Bacillus* sp. B38-2 and *Streptomyces* sp. S36-2, which were isolated from soil and compost by incubating samples in enrichment culture medium containing CMC and Na_2CO_3 at pH 9.6. It was found that they secrete a constitutive extracellular alkaline carboxymethyl cellulase (CMCase) in high quantity. The maximum enzyme activity was observed between 48 h and 72 h at 30 °C for the *Streptomyces* and between 72 h and 96 h at 35 °C for the *Bacillus.* The optimum pH and temperature of the crude enzyme activities ranged from 6.0 to 7.0 at 55 °C for the *Streptomyces* and 7.0 to 8.0 at 60 °C for the *Bacillus.* Two crude CMCases activities were ther-

alkaline neutral   alkaline neutral   alkaline neutral

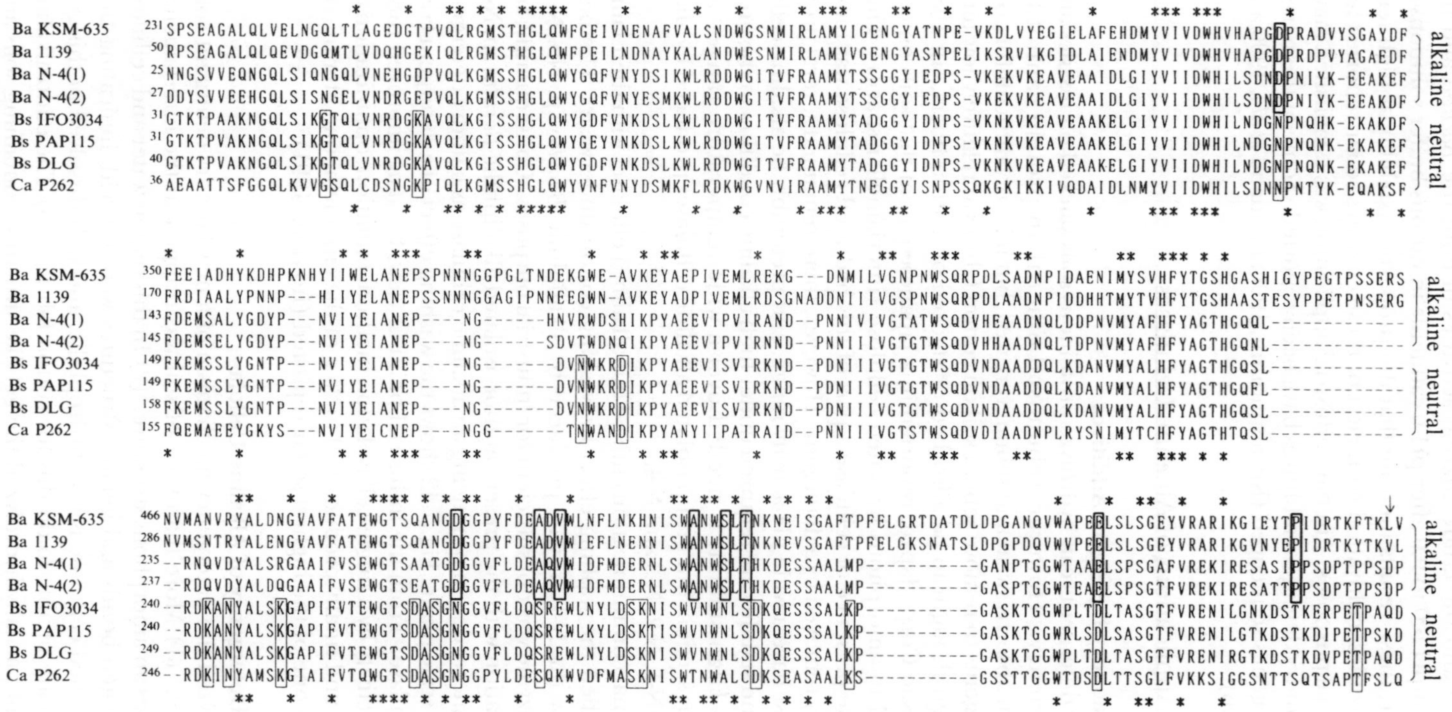

Ba KSM-635
Ba 1139
Ba N-4(1)
Ba N-4(2)
Bs IFO3034
Bs PAP115
Bs DLG
Ca P262

Fig. 7.41 Comparison of the amino acid sequence of alkaline cellulase from *Bacillus* sp. KSM-635 with those of alkaline and neutral enzymes from other microorganisms. Amino acid residues conserved in all the cellulases are indicated by asterisks. Amino acid residues conserved only in alkaline cellulases or in neutral cellulases are boxed with bold or thin lines, respectively. The vertical arrow indicates the carboxy-terminus (Leu-584) of the truncated cellulase encoded by the 2.4 kb fragment. Ba, *Bacillus* sp. (alkaliphilic); Bs, *Bacillus subtilis*; Ca, *Clostridium acetobutylicum*. (Reproduced with permission from Ozaki *et al.*, *J. Gen. Microbiol*, **136**, 1132 (1990))

mostable at 45°C for 1 h and the crude enzyme activities of both the *Bacillus* and the *Streptomyces* were stable at pH 5.0 to 9.0 after pH treatments in various buffer solutions at 30 °C for 24 h. However, these results indicate that the properties (stability at high temperatures, pH-activity profiles, cultivation cost, etc.) of these enzymes so far reported are not sufficient for industrial purposes. After an industrial application was established many alkaline cellulases as laundry detergent additives have been reported. Some of them have been purified and their genes sequenced (Khyamihorani, 1996; Landaud *et al*., 1996 and Sancheztorres *et al*., 1996).

7.4.4   Industrial Applications of Alkaline Cellulases

A.   Cellulases as Laundry Detergent Additives

Protease in laundry detergents can hydrolyze proteinous soils stacked on textiles and improve washing efficiency. However, sebum soil other than protein cannot be hydrolyzed at all. Cotton absorbs sweat and sebum very well, but it is very difficult for conventional laundry detergents to remove sebum soil on cotton fabrics at relatively low temperatures. Soil had been thought to be adsorbed or stacked on cotton fibers. In 1981 Murata *et al.* (1988) of Kao Corp., Japan, reported in a Japanese patent that soils on cotton fibers were trapped in amorphous hydrated cellulose (interlamellar space). They demonstrated that in the presence of cellulase a part of this hydrated cellulose was modified and soils were easily removed by a detergent. However, there remained several problems. (1) Cellulases must have a wide pH activity range, and preferably be alkaline cellulases. (2) High activity and stability are required under high alkaline condition in the presence of detergent. (3) High stability is needed over a wide temperature range. As described in the previous section, our group isolated alkaline cellulases produced by alkaliphilic *Bacillus* sp. Nos. N-1, N-2, N-3 and N-4.

Saito and Ito (personal communication) mixed our alkaline cellulases with their laundry detergents and studied the washing effect by washing cotton underwear. The best results were obtained by the N-1 enzyme produced by alkaliphilic *Bacillus* sp. No. N-1. However, the yield of enzyme was not sufficient for industrial purposes. Consequently, Ito *et al.*, (1989) in collaboration with the present author, isolated alkaliphilic *Bacillus* sp. No. KSM-635 from soil and succeeded in producing an alkaline cellulase as laundry detergent additive in an industrial scale plant. Fig. 7.42 shows the washing effect of a detergent containing the alkaline cellulase. From the industrial point of view, Ito *et al.* improved their strain to increase productivity by a single-cell isolation method, mutation, optimization of culture conditions, etc., and finally produced 20 to 25 g of the enzyme in one liter of culture broth. This yield is one of the highest so far reported (Ito *et al.*, 1991).

B.   Mechanism of Detergency of Alkaline Cellulase

Murata (1988) analyzed the detergency mechanism of a new laundry detergent containing alkaline cellulase. (1) More than 90% of soil was trapped in the interlamellar space of cotton, which has an amorphous structure. (2) In this region water and cellulose molecules formed hydrogen bonds and gel-like structures were formed, indicating that the structure of the modified cellulose is very hydrophilic, and soils dissolved in this gel; conventional detergents are unable to do this for cotton fabrics. (3) The endo-alkaline cellulase in the detergent partially hydrolyzed the gel-like structure containing soils and

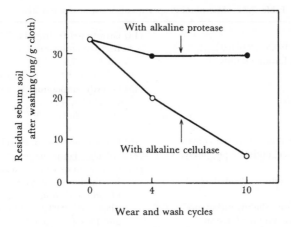

Fig. 7.42 Detergency effect of alkaline cellulase on naturally soiled cotton underwear in a controlled laundry test.

the detergent could pull out soils from cotton fiber. This mechanism is hypothetical but quite reasonable because after many washings practically no change in strength or loss of weight was observed and only alkaline CMCase exhibited good experimental results. Several laundry detergents containing alkaline cellulase are now commercially available. In Japan such cellulase laundry detergents enjoy approximately 40% of the market. This was the first large scale industrial application of cellulase.

C. Alkaline Cellulase for Cellulosic Waste

About two decades ago many Japanese microbiologists looked for alkaline cellulases that hydrolyze native cellulose to treat human excrement in Japan. During the treatment, the pH value became 8 to 9 due to the ammonia produced. No cellulolytic enzyme reported could hydrolyze the cellulose in human excrement.

1. Isolation of No. 212 enzyme

After two years systematic screening, we isolated alkaliphilic *Aeromonas* sp. No. 212 (ATCC 31085) producing a cellulase that hydrolyzed cellulose in human feces (unpublished data). The strain isolated has the microbial properties shown in Table 7.31. This strain is also alkaliphilic because growth was very poor at neutral pH value. During cultivation the pH of the medium should be adjusted to a value of about 8 to 9 by the addition of sodium carbonate, otherwise no enzyme was produced at all. The cellulase was produced as follows. 90 ml of a culture medium containing 0.5% ammonium sulfate, 1.5% pulp flock, 0.02% glucose, 0.1% yeast extract, 0.02% $MgSO_4 \cdot 7H_2O$, 0.2% K_2HPO_4 was charged in a 500-ml flask and sterilized at 120 °C for 20 min. Then 10 ml of separately sterilized aqueous solution containing 7% $NaHCO_3$ was added to the medium. The strain was inoculated on the culture medium, and cultivated with shaking at 37 °C. After 72 h, the culture fluid was centrifuged and the crude enzyme solution was obtained. The enzyme thus obtained had an activity of 350–500 units/ml.

Activity of the enzyme was assayed as follows. An aqueous suspension containing 1.5% micro-crystalline cellulose was used as the substrate. The mixture (2 ml of phosphate

Table 7.31   Microbiol Properties of *Aeromenas* sp. No. 212

Morphology
(1) Rod or short rod (0.4–0.5 × 2.0–6.0 μm)
(2) One polar flagellum having motility
(3) No spores
(4) Gram : Negative
(5) Acid fastness : Negative

Growth conditions
(1) Bouillon-agar plate culture : Circular convex colony, smooth surface, undulate periphery, wetty gloss, viscous
(2) Bouillon-agar streak culture : Beaded, scanty growth
(3) Bouillon liquid culture : Slightly turbid, no growth on the surface, no precipitation
(4) Bouillon-gelatin stab culture : Not liquefied
(5) Litmus milk : Slightly acidic, not peptonized, not coagulated

Biochemical properties
(1) Nitrate reduction : Positive (in medium containing bacto-tryptone, meat extract, and potassium nitrate)
(2) Denitrification : Positive
(3) MR test : Negative
(4) V-P test : Positive
(5) Formation of indole : Negative
(6) Formation of hydrogen sulfide : Negative
(7) Hydrolysis of starch : Positive
(8) Utilization of citric acid : Negative
(9) Utilization of inorganic nitrogen source : Neither nitrate nor ammonium salt is utilized
(10) Formation of pigment : Negative
(11) Urease : Negative
(12) Oxidase : Positive
(13) Catalase : Positive

(14) Growth pH (cultured at 37 ℃ for 3 days in 1% bacto-tryptone water) :

6.2	7.1	8.1	9.0	10.0
−	+	+	+	+

(15) Growth temperature (1% bacto-tryptone water of pH of 7.5) :

10 ℃	20 ℃	30 ℃	34 ℃	37 ℃	42 ℃
−	+	+	+	++	+

(16) Behavior to oxygen : Aerobic
(17) O-F test : Fermentative
(18) Formation of acids from saccharides (cultured in a medium containing 2 g/l of peptone, 5 g/l of sodium chloride, 0.3 g/l of dipotassium phosphate, 0.03 g/l of bromocresol purple, and 15 g/l of agar) :

Saccharide	Formation of acid
L-Arabinose	+
Trehalose	+
D-Xylose	+
D-Sorbitol	+
D-Glucose	+
D-Mannitol	+
D-Mannose	+
Inositol	−
D-Fructose	+
Glycerine	−
Maltose	+
Starch	+
Sucrose	+
D-Galactose	+
Lactose	+

(19) Formation of gases from saccharides : Negative.

(Reproduced with permission from Horikoshi and Akiba, *Alkalophilic Microorganisms,* p.159, Springer-Verlag: Japan Scientific Societies Press (1982))

buffer, pH 7.0, 1ml of the substrate, and 1 ml of the enzyme solution) was incubated at 37 ℃ for 60 min. The reaction was stopped by the addition of 1N sulfuric acid (1 ml) and the mixture was filtered through filter paper (Toyo Filter Paper No. 131). Then 1 ml of 5% phenol was added to 1 ml of the filtrate, and 5 ml of concentrated sulfuric acid was poured directly into the mixture. The resulting mixture was agitated, cooled in water for 10 min and subjected to calorimetric determination at 485 nm. When a saccharide corresponding to 1 mg of glucose was released in 1 ml of the filtrate under the above conditions, the enzyme was defined as having one unit of activity.

The enzyme is a mixture of at least six enzymes. The crude preparation has the following properties. Optimum pH was approximately 6.0 and the enzyme had 50% of the maximum activity at pH values in the range 4.5 to 8.5 (Fig. 7.43). The enzyme was only very slightly inactivated at a pH value of 5 to 10 (40 °C for 30 min incubation). The enzyme exhibited a high activity at temperatures in the range from 45 to 55 °C and was very thermostable. When the enzyme was heated at 60 °C for 10 min at pH 8.0, it still maintained 70% activity. The enzyme acts on cellulose, microcrystalline cellulose, filter paper, swollen cellulose, absorbent cotton, and CMC. Furthermore, the crude preparation contains three xylanases as described in section 7.5.

2. Degradation of Fecal Cellulose

Small-scale test: A quantity of 15 g of toilet paper, microcrystalline cellulose, cellulose powder of fecal cellulose in 1 liter liquid (pH 8.0) was tested as a substrate. The weight loss of the substrate solid was measured, and results are shown in Table 7.32.

Bench-scale test: A quantity of 9.5 l of excrement-digested sludge was poured into each of four brown glass excrement digestion tanks (diameter 20 cm, height 32 cm, effective volume 10 l), with temperature maintained at 37 °C. To forcibly form scum in each

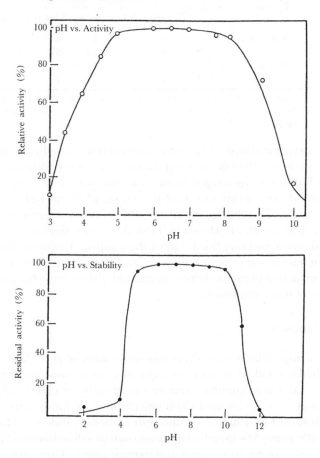

Fig. 7.43 Effects of pH on the cellulase from *Aeromonas* sp. No. 212.
(Reproduced with permission from Horikoshi and Akiba, *Alkalophilic Microorganisms*, p. 161, Springer-Verlag: Japan Scientific Societies Press (1982))

Table 7.32 Substrate-decomposing activity of cellulase of
Aeromonas sp. No. 212 in excrement–digested liquid

Substrate	Weight loss (%)	
	0 h	20 h
Toilet paper	0	15.0
Microcrystalline cellulose	0	7.5
Cellulose powder	0	5.0
Excrement cellulose	0	15.0

(Reproduced with permission from Horikoshi and Akiba,
Alkalophilic Microorganisms, p.161, Springer-Verlag: Japan
Scientific Societies Press (1982))

Table 7.33 Change in amount of accumulated scum

Enzyme	Tank	Thickness (mm) of scum				
		1 day	5 days	9 days	13 days	17 days
Not added	1	9	18	31	40	45
	2	9	18	29	37	40
Added	3	10	13	14	15	15
	4	7	9	14	12	14

(Reproduced with permission from Horikoshi and Akiba *Alkalophilic
Microorganisms*, p.162, Springer-Verlag: Japan Scientific Societies Press
(1982))

tank, 400 ml of excrement and 100 g of the same excrement cellulose as above were added
to each tank once a day. In order to keep the tank contents constant, 500 ml of the excrement-digested liquid was discharged from each tank once a day. Five grams of cellulase
powder was added once a day to two of the tanks, while no enzyme was added to the
remaining two tanks. The change in the amount of accumulated scum was observed with
the naked eye. The results are shown in Table 7.33. The thickness of scum in the cellu-
lase-added tank was about one-third that of the control. Cellulase No. 212 decomposed
the cellulose in human excrement quite well, and reduced the thickness of scum.

Although no actual plant has been constructed, this enzyme has potential application
in various types of waste treatment.

7.5 Xylanases

Xylan occurring in land plants is composed mainly of β-1,4-linked D-xylose units to
which is attached a small proportion of D-glucose, L-arabinose and D-glucuronic acid, as
shown in Fig. 7.44. Consequently, xylan can refer to three classifications of polysaccharide: (1) pentosan, (2) glycan, and (3) hemicellulose. Xylan is principally a polymer of
pentose and is found as a major component of hemicellulose. The activity of these
enzymes is usually assayed by measuring the amount of xylose liberated from the substrate.
Other types of xylan occur in seaweed and marine algae. These are composed of β-1,3-
linked D-xylose units that are hydrolyzed only by β-1,3-xylanase.

Several xylanase genes have been cloned in *E. coli* and genetically studied. Recently,
enzymatic debleaching in pulp-milling industries has been extensively investigated using

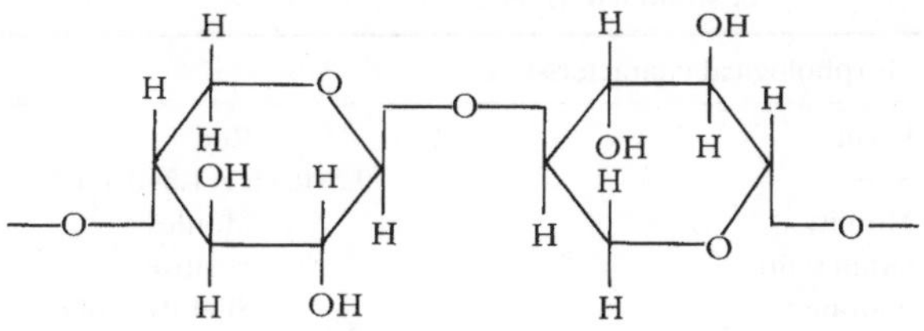

β-1,4-Xylan

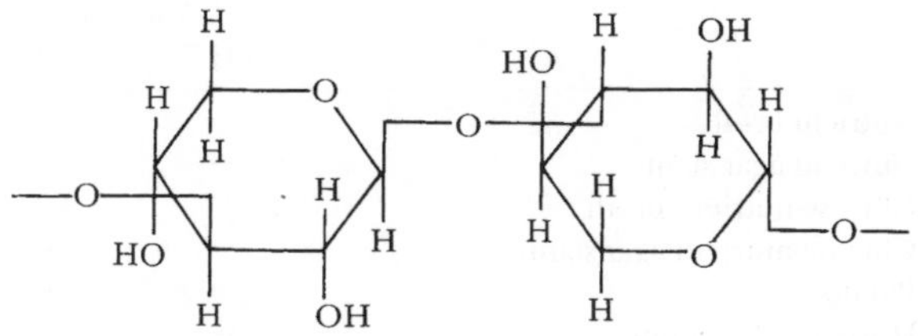

β-1,3-Xylan

Fig. 7.44 Structures of β-1,4- and β-1,3-xylan.
(Reproduced with permission from Horikoshi and Akiba, *Alkalophilic Microorganisms*, p.118, Springer-Verlag: Japan Scientific Societies Press (1989))

thermostable alkaline xylanases. This is a potential application of xylanase in industrial-scale plants, although the stability of xylanase is not sufficient in actual paper-milling plants.

7.5.1 Xylanases of Mesophilic Alkaliphilic *Bacillus* Strains
The first paper on xylanase of alkaliphilic bacteria was published in 1973 by Horikoshi and Atsukawa (1973*b*). The enzyme producer was alkaliphilic *Bacillus* sp. No. C-59-2.

A. Isolation of Xylan-degrading Alkaliphilic *Bacillus* sp. No. C-59-2
A small amount of soil was suspended in sterilized water and spread on alkaline xylan agar plates (Horikoshi-I medium with glucose substituted for 5 g of xylan). The plates were incubated at 37 °C for three days. From about 300 colonies on the plates, the colonies isolated were inoculated into a bran medium: wheat bran, 50 g; polypeptone, 10 g; KH_2PO_4, 1 g; $MgSO_4 \cdot 7H_2O$, 0.2 g; Difco yeast extract, 5 g and Na_2CO_3, 10 g in 1 liter of water. After three days cultivation under continuous shaking at 37 °C, xylanase activity of the culture broth was measured at pH 8.0. Several xylanase-positive colonies were picked and one strain, designated alkaliphilic *Bacillus* sp. No. C-59-2, was selected as a potent producer of xylanase. The isolate was an aerobic, spore-forming, gram-positive, motile, and rod-shaped bacterium. Table 7.34 summarizes the morphological and cultural characteristics of the isolate. The optimum enzyme production by strain No. C-59-2 was greatly affected by carbonate ions in the medium. As seen in Table 7.35, *Bacillus* sp. No. C-59-2 produced xylanase in a medium containing Na_2CO_3, K_2CO_3, or $NaHCO_3$. Maximum pro-

Table 7.34 Morphological cultural and biochemical characteristics of strain No. C-59-2

1. Morphological characteristics	
Form	Rod
Size	0.3–0.4 μ × 1.5–2.5 μ
Motility	Motile
Gram stain	Positive
Sporangia	Slightly swollen
Spore	0.5–0.7 μ × 1.0–1.2 μ

2. Cultural characteristics

	Growth at	
	pH 7	pH 10.3
Nutrient broth	−	+
Nutrient agar slant	−	+
Glucose-nutrient broth	−	++
Glucose-nutrient agar slant	−	++
Potato	−	++
Horikoshi I-medium	−	++
Horikoshi II-medium	−	++
Glucose-nitrate agar slant	−	±
Glucose-asparagine agar slant	−	+
Anaerobic growth in glucose broth	−	+
Anaerobic production of gas from nitrate	−	−
Horikoshi I-medium containing 7% NaCl	−	−

3. Biochemical characteristics

Hydrolysis of genatine and casein	Very weak
Hydrolysis of starch	Positive
Utilization of citrate	Utilized
Utilization of ammonium salts	Utilized
Reduction of nitrate to nitrite	Negative
Voges-Proskauer test	Negative

4. pH and temperature

pH for growth in Horikoshi I-medium pH 7.5 to pH 11
Temperature for growth in 15 ~ 42 ℃ at pH 10.3
Horikoshi I-medium

pH in the Horikoshi I-medium was adjusted by adding HCl of NaOH.
− indicates no growth ; ±, poor growth ; +, normal growth ;
++ abundant growth.
(Reproduced with permission from Horikoshi and Atsukawa, *Agric. Biol. Chem.*, **37**, 2098 (1973))

duction of the enzyme was achieved with medium containing 1% NaCO₃. The xylanase of *Bacillus* sp. No. C-59-2 was purified by precipitation with 70% saturated $(NH_4)_2SO_4$, column chromatography on CM-cellulose and hydroxyl apatite followed by gel filtration on Sephadex G-75. The homogeneity of the purified xylanase was examined by ultracentrifugation and isoelectric focusing. The sedimentation constant was 3.5S; the isoelectric point was pH 6.3. The purified enzyme exhibited a broad optimum pH range from 6.0 to

Table 7.35   Effect of carbonate salts on xylanase production by alkaliphilic *Bacillus* sp. No. C-59-2

Salts added (%)		Initial pH	Xylanase[2] (units/ml)
None		7.2	ND
NaCl	1.0	10.5[1]	ND
KCL	1.0	10.5[1]	ND
Na_2CO_3	0.5	9.7	2.8
	1.0	10.2	5.1
$NaHCO_3$	1.0	9.0	2.0
	2.0	9.3	3.1
K_2CO_3	1.0	10.2	4.1

[1] Adjusted with NaOH.
[2] ND, not detected.
(Reproduced with permission from Horikoshi and Akiba *Alkalophilic Microorganisms*, p.120, Springer-Verlag: Japan Scientific Societies Press (1982))

Table 7.36   Properties of bacterial xylanases (mesophiles)

Property	*Bacillus* sp. No. C-59-2	*B. subtilis* C-2	*Streptomyces xylophagus*	*Bacillus* sp. No. C-11
Optimum pH	6.0–8.0	6.0–6.2	6.2	7.0
Optimum temperature, °C	60	37–40	55–60	—
Sedimentation constant	3.5			
Isoelectric point	6.3			
Stable pH	7.0–7.5	5.0–7.0	5.3–7.3	5.5–9.0
Stable temperature, °C	Up to 60	Up to 45	Up to 40	
Predominant product	$X_2 > X_3 > X_4$	Ara, X_2, X_3	X, X_2	X, X_2, X_3
Max. hydrolysis rate, %	40	38		

8.0. Maximum hydrolysis of xylan was about 40% at either pH 6.0 or pH 9.0.

The optimum pH of xylanase from another xylanase producer, *Bacillus* sp. No. C-11, was 7.0, and approximately 37% of the activity remains even at pH 10.0. The properties of the enzymes are summarized in Table 7.36 with those of other bacterial enzymes for comparison. Xylanases from alkaliphilic strains retain considerably higher activity in the alkaline pH range than do other bacterial xylanases. Therefore, *Bacillus* sp. No. C-11 was applied to the biological treatment of rayon waste with alkaline pH and contained hemicellulose. It was found that several alkaliphilic strains such as *Aeromonas* sp. No. 212 (Ohkoshi *et al.*, 1985), *Bacillus* sp. No. C-125 (Honda *et al.*, 1985), etc. produced multi-component xylanases. The author carefully checked through his old experiment notebooks and found that an elution profile on a Sephadex G-75 column exhibited a minor peak of xylanase activity which eluted in early fractions. However, since he did not have any information on the multiplicity of xylanase at that time, he ignored this minor component (approximately 1/10 of the major component), and no further work on this minor component has been done.

B.   Isolation of Another Xylanase Producer, *Bacillus* sp. No. C-125

As isolation method analogous to that described above was applied. From about 3,000

Table 7.37 Cultural and biochemical characteristics of the
alkaliphilic *Bacillus* sp. No. C-125

	pH 7	pH 10
Nutrient broth	+	+
Nutrient	+	+
Glucose-nutrient broth	+ −	+
Glucose-nutrient agar	+ −	+
Tyrosin agar	+ −	+
Horikoshi-I medium	+ −	+
Glucose-nitrate agar	+	+
Glucose-asparagine agar	−	+
Anaerobic growth in Horikoshi-I medium	−	+ −
Horikoshi-I medium 5% Nacl	+	+
Utilization of citrate		
Koser medium	−	+
Christensen medium	−	+
Utilization of propionate	−	+
Reduction of nitrate	−	+ −
Voges-Proskauer test	−	−
Urease test	−	+ −
Hydrolysis of starch	+	+
Hydrolysis of casein	+	+
Max. pH for growth	10.5	
Max. temperature for growth	55	
GC, %	43.7	

Morphological characteristics of the alkaliphilic *Bacillus* sp. C-125

Form	Rods
Size (μm)	0.5–0.7 × 3.0–4.0
Motility	+
Gram strain	+
Sportangia	Definitely swollen
Spore	Subterminal
	Ellipsoidal
Spore size (μm)	0.6–0.8 × 1.0–1.2

colonies on the plates, xylanase-positive colonies were picked and one strain, designated alkaliphilic *Bacillus* sp. No. C-125, was selected as a potent producer of xylanase. The isolated sp. No. C-125 was an aerobic, definitely swollen spore-forming (ellipsoidal, subterminal), gram-positive, motile, and rod-shaped bacterium (0.5–0.7 × 3.0–4.0 μm). The C + G content of the DNA was 43.7% (Table 7.37).

C. Purification of Xylanases of Strain No. C-125 and Their Properties

Strain No. C-125 was grown aerobically for three days at 37 °C in the following medium: wheat bran, 10 g; polypeptone, 5 g; Difco yeast extract, 5 g; K_2HPO_4, 1 g; $MgSO_4 \cdot 7H_2O$, 0.2 g; $NaCO_3$, 10 g; made up to a volume of 1 liter with water. The culture broth was centrifuged to remove the cells and the supernatant fluid was used as crude xylanase. Besides the medium described above, bran could be substituted by xylan (Honda *et al.*, 1985). Ikura and Horikoshi (1987*c*) found that the addition of 0.5% glycine to the xylan

Table 7. 38 Purification of xylanases

Steps	Volume (ml)	Activity (units)	Protein (mg)	Specific activity (units/mg protein)	Recovery (%)
Supernatant	1000	830	1080	0.71	100
Ammonium sulfate					
precipitate	11	472	92.4	5.15	6.9
Biogel P-30					
Xylanase A [†]	54	78	65.2	1.2	9.4
Xylanase N [†]	55	318	12.9	24.7	38.3
DE-52					
Xylanase A [†]	110	55	13.3	4.1	6.6
Xylanase N [†]	—	—	—	—	—
Sephadex G-75					
Xylanase A [†]	7	31	1.9	16.3	3.7
Xylanase N [†]	11	220	9.2	23.9	26.5

[†] Activity was measured at pH 7.0.

(Reproduced with permission from Honda *et al.*, *Can. J Microbiol.*, **31**, 539, (1985))

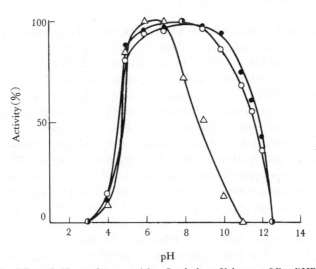

Fig. 7.45 Effect of pH on xylanase activity. Symbols: ○, Xylanase of *E. coli* HB101 carrying pCX311; ●, Xylanase A of alkaliphilic *Bacillus* sp. No. C-125; △, Xylanase N of alkaliphilic *Bacillus* sp. No. C-125.
(Reproduced with permission from Honda *et al.*, *J. Bacteriol.*, **161**, 785 (1985))

medium increased the enzyme yield by about 1.8-fold. But protease production was strongly repressed, while the cell mass was not changed. From an economical standpoint, wheat bran is recommended in larger fermentation vessels.

Xylanases were purified by an ammonium sulfate precipitation, Biogel P-30 gel filtration, DEAE-cellulose chromatography, and Sephadex G-75 gel filtration. Table 7.38 summarizes the results of the purification of xylanases from the culture fluid. Purified samples were analyzed by SDS-PAGE to determine molecular weight. Xylanase A had a molecular weight of 43,000 and that of xylanase N was 16,000. The Ouchterlony test between xylanase A and xylanase N indicated that xylanase A is immunologically different from xylanase N. Xylanase N was most active at pH 6–7 and xylanase A was most active at a pH range of 6 to 10 and has some activity at pH 12, as shown in Fig. 7.45.

Stability of the enzymes was also investigated in buffer solutions of various pH values. The mixtures were incubated at 4 °C for 24 h and the residual activities measured. Xylanase A was most stable in the pH range of 4 to 12, and xylanase N was most stable in the range of pH 5 to 12. Xylanase A and N were stable up to 50 °C at pH 7.0 (10 min incubation). The maximum degree of hydrolysis of xylan with the xylanases was about 25%. No significant difference between the two xylanases was observed. The end products were oligosaccharides, such as xylobiose, xylotriose, xylotetraose, and higher oligosaccharides. Both xylanases had *trans*-xylosidation activities. Properties of xylanase A were somewhat similar to those of xylanase W1-II and W2-II of thermophilic *Bacillus* strains isolated by Okazaki *et al.* (1985), *e.g.* molecular weight and their broad pH activity curves. However, their end products were different: no xylose was found in the hydrolysate of xylanase A. Xylanase A belongs to the family F group.

7.5.2 Alkaline Xylanases from Thermophilic Alkaliphilic *Bacillus* Strains

Since the discovery of alkaline xylanases, many reports on thermostable alkaline xylanases have been published. Four *Bacillus* strains (W1, W2, W3 and W4) which produce xylanase were isolated from soils by using xylan medium (Horikoshi and Atsukawa, 1973*b*; Okazaki *et al.*, 1984). The distinctive feature of the four organisms was that they grew at higher temperatures than reported for other xylan-degrading alkaliphilic bacteria at that time. The maximum temperature for growth was approximately 55 °C.

The cells were gram-positive, spore-forming, not acid-fast, aerobic and motile rods. The size of the cells was 2.0–5.0 μm × 0.13–0.5 μm. Strains W2 and W4 often showed filamentous cells. The spores were oval and central with swollen sporangia. The colonies of strains W1 and W3 were translucent, and those of W2 and W4 dry opaque. Gelatin was liquefied. Litmus milk was peptonized by strains W1 and W3, and clotted by strains W2 and W4. The strains grew well in nutrient broth containing 12% NaCl. The range of pH for growth of these strains was on the alkaline side, from 8.0 to 11.0; no growth was observed below pH 7.0. The four xylanases were produced in media containing xylan or xylose but not in media containing glucose. Therefore, the production of these enzymes was not constitutive. The strains were grown aerobically in 300-ml flasks containing 100 ml of xylan medium (pH 10.0) at 45 °C for 48 h on a rotary shaker, and xylanases of culture supernatant were studied. The pH optimum for enzyme action of strains W1 and W3 was 6.0 and for strains W2 and W4 it was between 6 and 7, as shown in Fig. 7.46. The enzymes were stable between pH 4.5 and 10.5 at 45 °C for 1 h. The optimum temperature of xylanases of W1 and W3 was 65 °C and that of W2 and W4 70 °C. The enzymes were sta-

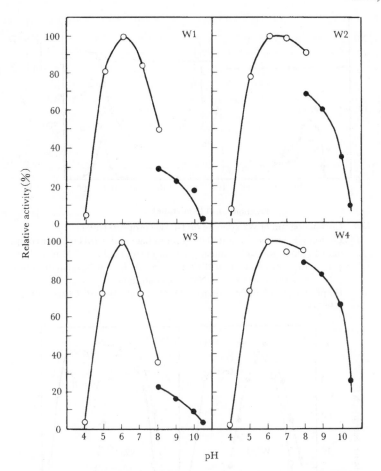

Fig. 7.46 Optimum pH of xylanases from strains W1, W2, W3 and W4. The pH was adjusted with the following buffer system: ○ McIlvaine buffer (pH 4-8); ● McIlvaine-NaOH buffer (pH 8-10.5).
(Reproduced with permission from Okazaki *et al.*, *Appl. Microbiol. Biotechnol.*, **19**, 338 (1984))

ble up to 60 °C and the addition of 5 mM CaCl$_2$ had no effect on their thermal stability. All xylanases hydrolyzed xylan to yield xylose and xylobiose. The degree of hydrolysis of xylan was about 70% after 24 h incubation.

As representatives of the enzymes, two strains, *Bacillus* sp. No. W1 (JCM2888) and W2 (JCM2889), were cultured in a 10-liter jar fermenter containing Horikoshi-II medium without starch supplemented with 1% xylan at 45 °C for 20 h. The xylanases from strains W1 and W2 were separated into two active components (W1-I and W1-II; W2-I and W2-II) on DEAE-Toyopearl 650M. As shown in Fig. 7.47, W2-I was not adsorbed on DEAE-Toyopearl 650M but W2-I was and eluted at a concentration of about 0.4 M NaCl. DEAE-Toyopearl 650M column chromatography of W1 xylanases gave the same chromatogram. The activities of these purified enzymes were measured at various pHs in McIlvain buffer. The results are shown in Fig. 7.48. The pH activity curves of components I and II were

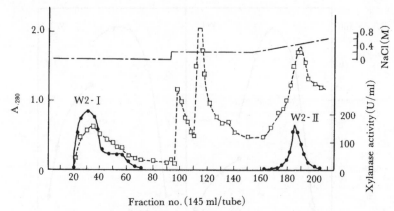

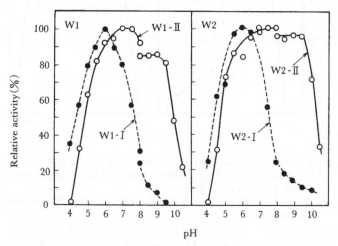

Fig. 7.47 Chromatogram of W2 xylanase on DEAE-Toyopearl 650M.
Elution was carried out with 0.05 M phosphate buffer (pH 6.2) for component I (W2-I) and with a linear gradient of from 0.2 to 0.6 M NaCl in the
same buffer for component II (W2-II). Symbols: ——●——, xylanase activity;
----□----, A_{280} (protein); —·—, NaCl concentration.
(Reproduced with permission from Okazaki *et al., Agric. Biol. Chem.*, **49**, 2035
(1985))

Fig. 7.48 Effect of pH on xylanase activity.
pH was obtained with the following buffer systems: McIlvaine buffer, pH
4.0–8.0; McIlvaine-NaOH buffer, pH 8.0 – 10.5. Symbols: ——●——, component I; ——○——, component II.
(Reproduced with permission from Okazaki *et al., Agric. Biol. Chem.*, **49**, 2036
(1985))

entirely different. W1-I and W2-I had sharp pH activity curves with optimum activity at pH
6.0, but W1-II and W2-II had very broad pH activity curves (pH 6–9) and were still active
even at pH 10.0. The four enzymes were stable between pH 4.5 and 10.0 (45 °C for 60 min
incubation). Transxylosidation activity was observed in these enzymes because the
enzymes yielded higher production of xylotetraose and/or xylopentaoase from xylotriose
and xylotetraose. Further properties are shown in Table 7.39.

Table 7.39 Characteristics of xylanases from thermophilic alkaliphilic *Bacillus* strains

| Characteristic | Xylanase of *Bacillus* : | | | |
| | Strain No. W1 | | Strain No. W2 | |
	I	II	I	II
Optimum pH	6.0	7.0–9.0	6.0	7.0–9.5
Optimum temperature (°C)	65	70	65	70
Isoelectric point (pH)	8.5	3.6	8.3	3.7
Molecular weight	21,500	49,500	22,500	50,000
K_m (mg-xykab/mol)	4.5	0.95	4.0	0.57
Hydrolysis product[†]	X_2, X_3, X_4, X_5,	X_1, X_2, X_3, X_5,	X_2, X_3, X_4, X_5,	X_1, X_2, X_3, X_5,
Inhibitor (at 5 mM)	$HgCl_2$	$HgCl_2$	$HgCl_2$	$HgCl_2$

[†] X_1, Xylose ; X_2, Xylobios ; X_3, Xylotriose ; X_4 , Xylotetraose ; X_5, Xylopentaose

Park *et al.* (1992) isolated a xylanase from alkaliphilic *Bacillus* sp. YC-335. The molecular weight of the enzyme was 40,000, the optimum pH for activity was 6.0, and the optimum temperature 55 °C. Dey (1992) then isolated from soil an alkaliphilic thermophilic *Bacillus* sp. (NCIM59) that produced two types of cellulase-free xylanase at pH 10 and 50 °C. Their molecular weights were 35,000 and 15,800, respectively. They had similar properties and both were stable at 50 °C at pH 7.0 for 4 days incubation. Subsequently, a *Bacillus* strain with xylanase activity has been isolated (Blanco and Pastor, 1993). Maximum xylanase production was achieved when the strain was cultured in media supplemented with birchwood xylan or rice straw; production was repressed by glucose and xylose. The optimal temperature and pH for xylanase activity were 45–50 °C and 5.5–7.5, respectively. However, crude xylanase was highly stable in a wide range of pH values, retaining 100 % of the activity after 24 h of incubation at 37 °C in buffer at pH 10.0. Analysis by polyacrylamide gel electrophoresis and zymogram techniques showed four xylanase activity bands with apparent molecular masses of 32, 48, 61, and 66 kDa. The most active of them (molecular mass 32 kDa) apparently corresponded to a xylanase with an isoelectric point (p*I*) of 9.3. Carboxymethylcellulase activity was detected only in the band of 48 kDa.

Furthermore, Khasin *et al.* reported that alkaliphilic *Bacillus stearothermophilus* T-6 produced an extracellular xylanase that was shown to optimally bleach pulp at pH 9 and 65 °C (Khasin, Alchanati and Shoham, 1993). The enzyme was purified and concentrated in a single adsorption step onto a cation exchanger and is made of a single polypeptide with a molecular weight of 43,000. Xylanase T-6 is an endoxylanase that completely degrades xylan to xylose and xylobiose. The p*I*s of the purified protein were 9 and 7 under native and denaturing conditions, respectively. The optimum activity was at pH 6.5; however, 60 % of the activity was still retained at pH 10. At 65 °C and pH 7, the enzyme was stable for more than 10 h; at 65 °C and pH 9, the half-life of the enzyme was approximately 6 h. The enzyme had no apparent requirement for cofactors, and its activity was strongly inhibited by Zn^{2+}, Cd^{2+}, and Hg^{2+}. The N-terminal sequence of the first 45 amino acids of the enzyme showed high homology with the N-terminal region of xylanase A from the alkaliphilic *Bacillus* sp. strain C-125.

Nakamura *et al.* reported that an alkaliphilic *Bacillus* sp. strain, 41M-1, isolated from soil produced multiple xylanases extracellularly. One of these xylanases was purified to homogeneity by ammonium sulfate fractionation and anion-exchange chromatography (Nakamura *et al.*, 1993*a*, *b*). The molecular weight of this enzyme (xylanase J) was 36 kDa, and the isoelectric point was p*I* 5.3. Xylanase J was most active at pH 9.0. The optimum temperature for the activity at pH 9.0 was around 50 °C. The enzyme was stable up to 55 °C at pH 9.0 for 30 min. Xylanase J was completely inhibited by Hg^{2+} ion and N-bromosuccinimide. The predominant products of xylan hydrolysate were xylobiose, xylotriose, and higher oligosaccharides, indicating that the enzyme was an endoxylanase. Xylanase J showed high sequence homology with the xylanases from *Bacillus pumilus* and *Clostridium acetobutylicum* in the N-terminal region.

An alkaliphilic and thermophilic *Bacillus* sp. strain TAR-1 was isolated from soil (Nakamura, Nakai *et al.*, 1994). The xylanase was most active at a pH range of 5.0 to 9.5 at 50 °C, as shown in Fig. 7.49. Optimum temperature of the crude xylanase preparation was 75 °C at pH 7.0 and 70 °C at pH 9.0 (Fig. 7.50). Zymogram analysis of the culture supernatant showed that the molecular mass of the xylanase was 40 kDa and the isoelectric point was p*I* 4.1. The predominant products of xylan hydrolysate were xylobiose, xylotriose, and higher oligosaccharides, indicating that the enzyme was an endoxylanase. Production of the thermophilic alkaline xylanase was induced by xylan and xylose, but was repressed in the presence of glucose. The xylanases described above acted on neither crystalline cellulose nor cellulose, indicating a possible application of the enzyme in biobleaching processes.

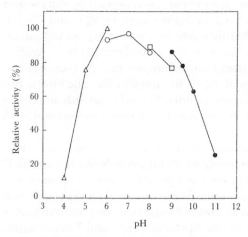

Fig. 7.49 Effect of reaction pH on activity.

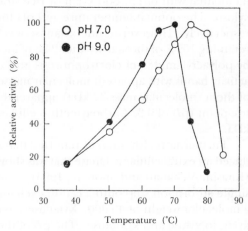

Fig. 7.50 Effect of temperature on activity.

7.5.3   Xylanases of Other Alkaliphiles

A.   Xylanases of an Alkaliphilic *Aeromonas* sp.

Ohkoshi *et al.* (1985) have isolated an alkaliphilic *Aeromonas* sp. No. 212 (ATCC31085) from soil. This species was found to produce three types of xylanases in a medium containing $NaHCO_3$ (0.5% w/v). The isolated strain was grown aerobically for 48 h at 37 °C in a xylan medium. The culture broth was centrifuged at 6,000 × g for 10 min and the supernatant fluid was used as the crude xylanase. The purification procedure for xylanases is summarized in Table 7.40. DEAE-cellulose ion-exchanger equilibrated with 0.5% $NaHCO_3$ was added directly to the crude xylanase and the mixture was stirred overnight. The enzyme which was not adsorbed on DEAE-cellulose was precipitated with 70% saturation of ammonium sulfate. The precipitate was dissolved in water and dialyzed against 50 mM phosphate buffer (pH 7.0). The dialysate was loaded onto a Sephacryl S-200 column equilibrated with 50 mM phosphate buffer (pH 7.0) containing 0.1 M NaCl; the column was eluted with the same buffer. Two activity peaks were detected, xylanases M and S. Xylanase that was adsorbed on a DEAE-cellulose column was eluted with 50 mM phosphate buffer (pH 6.0) containing 1 M NaCl. The fractions containing xylanase were pooled then dialyzed against 50 mM phosphate buffer (pH 6.0). The dialysate was loaded onto a DEAE-cellulose column equilibrated with 50 mM phosphate buffer (pH 6.0). The enzyme, xylanase L, was eluted with a linear gradient of NaCl (0–1.0 M). Purified samples were subjected to SDS-PAGE to determine their molecular weights. Xylanases L, M and S had molecular weights of 145,000, 37,000 and 23,000, respectively.

Xylanase L was most active at pH 7–8. Xylanases M and S were most active at pH 6–8 and pH 5–7, respectively. The effect of pH on stability was studied in buffer solutions of varying pH values. The mixtures were incubated at 40 °C for 1 h and the residual activities

Table 7.40   Purification of xylanases

Step	Xylanase	Volume (ml)	Activity[1] (units)	Protein[2] (mg)	Specific activity (U/mg protein)
Crude enzyme	L, M, S	5,000	9,750.0	10,800.0	0.9
First DEAE-52 through fraction	M, S	5,000	9,000.0	7,600.0	1.2
Ammonium sulfate precipitate	M, S	240	5,052.5	285.5	17.7
CM-52	M, S	230	2,867.0	23.5	122.0
Sephacryl S-200	M	53	625.4	5.3	118.0
	S	36	234.3	3.2	73.2
First DEAE-52	L	180	82.8	1,152.0	0.1
Second DEAE-52	L	5	36.0	0.5	72.0
Sephacryl S-200	L	68	26.4	0.4	66.0

[1] The xylanase assay was carried out by the method of Horikoshi and Atsukawa (1973*b*). One unit of enzyme activity is defined as the amount of enzyme which liberates 1 mg of reducing sugar expressed as xylose per minute at 40 °C in 10 min. Xylan purified from purum was purchased from Fluka A.G. Chem. Fabrik and the xylan solubilized by sonication was used as the substrate.

[2] Protein concentration was measured with bovine serum albumin as a standard.

(Reproduced with permission from Ohkoshi *et al.*, *Agric. Biol. Chem.*, **49**, 3037 (1985))

measured. Xylanase L was stable in the range of pH 5 to 12, xylanase S from pH 3 to 10 and xylanase M from pH 5 to 10. No significant differences among the three xylanases were observed except for molecular weight. The products in the hydrolysates after a short period of incubation were oligosaccharides such as xylobiose, xylotriose, xylotetraose and higher oligosaccharides. No xylose was detected after 24 h incubation. The enzymes were not activated by Ca^{2+} or Mg^{2+}. There are two possibilities as to how these xylanases are produced. (1) Three xylanase genes directly produce three enzyme molecules. (2) Proteolytic enzymes process the protein molecules during cultivation. To clarify this point, cloning experiments were carried out.

B. Xylanase of Alkalitolerant Fungi

An alkalitolerant *Cephalosporium* (NCL 87.11.9) strain capable of rapid growth, and xylanase secretion over a wide pH range (pH 4-10) has been isolated from soil samples (Bansod *et al.*, 1993). This is the first report of an extracellular fungal xylanase which is active and stable at high alkaline pH (8-9.5). The culture filtrate did not show any significant cellulase activity. Gel filtration studies indicated two peaks of xylanase activities corresponding to molecular weights of 70,000 and 30,000 in a proportion of 1 : 9.

Kang *et al.* (1996) also isolated two different xylanases from alkaliphilic *Cephalosporium* sp. strain RYM-202. Two different xylanases, CX-I and CX-II, have been purified to homogeneity. The enzymes had similar pH (7.5 to 8.0) and temperature (50 °C) optima and were stable over a wide pH range of 5.5 to 12.0. Both enzymes were shown to be cellulase-free endoxylanases with transglycosidation activity. Another alkalitolerant fungus, *Aspergillus fischeri* Fxn1, secreted an extracellular cellulase-free xylanase in the pH range 5-10 (Raj and Chandra, 1995 and 1996). The optimum pH of xylanase was 6.0-6.5 and it was stable in a wide range of pH 5-9.5. The optimum temperature was 60 °C and it was stable up to 55 °C. Half-life at 50 and 55 °C was 240 and 40 min, respectively. This enzyme released reducing sugars from pulp at pH 9.0 and 40 °C.

C. Xylanases of Anaerobic Alkaliphiles

Three thermophilic and strictly anaerobic *Dictyoglomus* strains were isolated from different environments: the type strain *D. thermophilum* from a hot spring in Japan; strain B1 from a paper-pulp factory in Finland; and strain B4a from a thermal pool in Iceland (Mathrani and Ahring, 1992). The xylanases of all the isolates had a broad range of activity (pH 5.5 to 9.0). Optimum temperature ranged between 80° and 90 °C.

7.5.4 Genetics of Alkaline Xylanases

There are many xylanase sequences available in the literature and they can be grouped into two families on the basis of primary structure similarity. The first family, termed family F, is composed of high molecular weight xylanases and also features some cellulolytic enzymes such as the enzyme of *Cellulomonas fimi*. The other family (G) comprises small molecular weight xylanases exemplified by *Bacillus pumilus*. No significant homology could be found between the two families, even around the catalytic residues.

A. Cloning of Xylanase Gene of *Bacillus* sp. No. C-125

The first cloning of the xylanase gene of alkaliphilic *Bacillus* sp. No. C-125 was carried out by Honda, Kudo and Horikoshi (1985*a*, *b*). The ratio of xylanase N to A was approximately 6 to 1, and this ratio could not be changed by varying culture conditions. For hyperproduction, the gene of xylanase A in *E. coli* was cloned using pBR322. It was striking that *E. coli* carrying the xylanase A gene produced xylanase A in the culture broth.

Alkaliphilic *Bacillus* sp. No. C-125 was aerobically cultured in Horikoshi-I medium to the early stationary phase at 37 °C. The chromosomal DNA was digested with *Hin*d III and ligated with T$_4$ ligase overnight at room temperature. This ligated DNA mixture was used for the transformation of *E. coli* HB101. Xylanase activity was detected on the plates because a clear zone formed around a colony producing xylanase on an LB-agar plate containing 0.5% xylan. A plasmid, pCX311, was obtained from the transformants (Ap^r, xylanase) which contained two *Hin*d III fragments (2.6 and 2.0 kb). Two *Hin*d III fragments were essential for the expression of the xylanase gene. We demonstrated from the sequence of the pCX311 plasmid that the 2.0 kb fragment had a promoter, and the xylanase gene was detected in the fragment of 2.6 kb.

E. coli HB101 carrying pCX311 was aerobically grown in LB-broth for 15 h at 37 °C. The xylanase activities in the extracellular, periplasmic and intracellular fractions were assayed, and 5.0 U, 0.7 U and 1.4 U of total activity (7.1 U) were observed, respectively. Xylanase production was completely inhibited by the addition of glucose (final concentration: 5 g/l), although cell growth was not inhibited. Addition of a small amount of glucose (1 g/l) did not inhibit the production of xylanase; however, more than 80% of the xylanase produced was detected only within the cells (Table 7.41). Furthermore, extracellular xylanase was produced only in the presence of Na$^+$, Li$^+$ or K$^+$. These ions can be replaced by sucrose. These observations suggest that the xylanase expressed in *E. coli* carrying pCX311 was probably regulated by osmotic pressure.

Xylanase produced by *E. coli* HB101 (pCX311) in the culture was purified in the same manner as that of xylanases of alkaliphilic *Bacillus* sp. No. C-125 described above. The molecular weight of pCX311-encoded enzymes, as determined by SDS-PAGE, was about

Table 7.41   Effects of carbohydrates

Medium	h	Extracellular	Periplasm	Intracellular	Total
LB-medium	15	5.0	0.7	1.4	7.1
	20	5.8	1.6	0.5	7.9
LB-medium	15	0	0	0	0
+ glucose	20	0	0	0	0
LB-medium	15	0.7	7.2		7.9
+ glucose (1 g/l)	20	1.3	6.7		8.0
LB-medium	15	3.8	1.6	1.8	7.2
+ glycerol	20	5.0	3.0	3.8	11.8
LB-medium	15	5.4	1.9	0.6	7.9
+ xylan	20	5.4	1.9	0.8	8.1
LB-medium	15	5.2	1.4	1.8	8.4
+ wheat bran	20	6.4	0.7	1.0	8.1
LB-medium + wheat	15	5.0	1.8	2.0	8.8
bran + glycerol	20	5.2	2.3	4.0	11.5
LB-medium + wheat	15	0	0	0	0
bran + glucose	20	0	0	0	0

The carbohydrates tested were added (5 g/l) to the LB-medium.
The values represent units of xylanase activity per 10 ml of medium.

43,000, which was the same as that of xylanase A. The results of the Ouchterlony double-diffusion test indicated that xylanase A gave lines of precipitation which fused completely with that for xylanase purified from *E. coli* HB101 carrying pCX311. Fig. 7.45 shows the effect of pH on xylanase activity. The pH was adjusted with McIlvaine buffer (pH 3–8) or 0.05 M glycine-NaOH buffer (pH 9–13). No significant difference was observed between the pCX311-borne xylanase and xylanase A, both of which have enough enzyme activities at pH 11.0. These xylanases were stable at 4 °C for 24 h in the range of pH 5 to 12 but inactivated at pH 4.0 and 12.5.

Hamamoto *et al.* (1987) and Hamamoto and Horikoshi (1987) analyzed the nucleotide sequence of the fragment showing a single open reading frame. A protein of molecular weight 45,000 translated from this open reading frame of 1,188 bp. The sequence of the N-terminal amino acids of the extracellular form of xylanase A, determined with a peptide sequence, NH$_2$-Ala-Gln-Gly-Pro-Pro-Lys-Ser-Gly-Val-Phe, was identical to that deduced from the DNA sequence. The molecular weight of the matured xylanase A as calculated from the DNA sequence was 42,479, in good agreement with the molecular weight of 43,000 for the extracellular xylanase A produced by *E.coli* carrying plasmid pCX311 coding for xylanase A. But the promoter for alkaliphilic *Bacillus* sp. No. C-125 did not work well in the *E. coli* HB101 system. Plasmid pCX311 contains two *Hind* III fragments from alkaliphilic *Bacillus* DNA. The 2.6-kb fragment contained the xylanase A gene and no open reading frame could be detected in the 2.1-kb fragment. However, in the absence of the 2.1-kb fragment no xylanase A activity could be expressed in *E. coli* using pBR322. A plasmid pYEJCX3, which contained the 2.7-kb *Hind* III fragment in the synthetic promoter vector pYEJ001, was able to express xylanase A activity in *E. coli.* As shown in Fig. 7.51, subcloning and CAT cartridge insertion experiments indicated a promoter for *E. coli* located in between the *Hae* III-nucleotide number 988 fragment of the 2.1-kb fragment. Xylanase A should be a secretable protein from *E. coli* but further details are not yet available.

E. coli YK561 (an alkaline phosphatase constitutive mutant producing large amounts of alkaline phosphatase) carrying pCX311 was cultured for 15 h in LB-broth at 37 °C. As shown in Table 7.42, essentially no enzymatic activity was detected in the culture broth of *E. coli* YK561 or *E. coli* YK561 carrying pBR322. About 75% of the alkaline phosphatase was trapped in the periplasmic space and almost all β-galactosidase activity was detected in the cellular fraction. On the other hand, about 70% of the total β-lactamase and 80% of the total xylanase activities were found in the extracellular fraction of *E. coli* carrying pCX311.

However, about 85% of the β-galactosidase (which is a typical cellular enzyme) and most of the alkaline phosphatase (typical periplasmic enzyme) was detected in the cellular fraction and periplasmic fraction, respectively. Such selective secretion was observed not only in *E. coli* YK561 but also in *E. coli* HB101. These results indicate that the outer membrane of *E. coli* was transformed into a permeable form by the introduction of pCX311 or YEJCX3 having a xylanase A gene (not a β-lactamase gene) into the cells. It is noteworthy that not only xylanase but β-lactamase was also secreted into the culture broth. It is therefore highly probable that the secretion process is different from that of *E. coli* carrying a pEAP series. The amino acid sequence of these proteins showed no remarkable characteristic or homology with those of hemolysin or the outer membrane proteins. It seems reasonable to assume that the mature proteins have unknown properties which

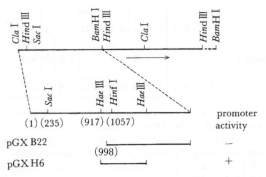

Fig. 7.51 Restriction map of pCX311.
Plasmid pCX311 contains two *Hin*d III fragments from alkaliphilic *Bacillus* DNA on pBR322.
Vector plasmid pBR322 is indicated by the dotted line. The thick arrow indicates the xylanase A gene and the direction of transcription. The 2.1 kb *Hin*d III fragment is enlarged. The numbers in parentheses denote the nucleotide numbers from the distal *Hin*d III site at position 1. The DNA fragments contained by the subcloned plasmids and the promoter activity are also shown: (+) means promoter positive and (−) promoter negative. (Reproduced with permission from Hamamoto and Horikoshi, *Agric. Biol. Chem.*, **51**, 3111 (1987))

Table 7.42 Distribution of enzymes in *E. coli* YK561 carrying plasmids

Plasmid	Enzyme	Activity in the following fractions		
		Extracellular (%)	Periplasmic (%)	Cellular (%)
None	β-Gal	0.04 (2)	0.01 (0)	2.44 (98)
	APase	0.10 (7)	1.05 (73)	0.29 (20)
	β-Lactamase	0.00 (0)	0.00 (0)	0.00 (0)
	Xylanase	0.00 (0)	0.00 (0)	0.00 (0)
pBR322	β-Gal	0.03 (1)	0.01 (0)	2.36 (99)
	APase	0.11 (9)	0.85 (66)	0.32 (25)
	β-Lactamase	2.60 (13)	17.50 (85)	0.38 (2)
	Xylanase	0.00 (0)	0.00 (0)	0.00 (0)
pCX311	β-Gal	0.07 (3)	0.02 (1)	2.32 (96)
	APase	0.11 (7)	1.07 (73)	0.29 (20)
	β-Lactamase	4.70 (68)	1.60 (23)	0.60 (9)
	Xylanase	0.43 (80)	0.11 (20)	0.00 (0)

(Reproduced with permission from Honda *et al.*, *Agric. Biol. Chem.*, **49**, 3012 (1985))

allow specific secretion through the outer membrane of *E. coli*.

B. Cloning of Xylanase Genes of Other Alkaliphilic *Bacillus* Strains

Recently, Nakamura *et al.* (1993*b*) cloned an alkaline xylanase J which was expressed in *E. coli*, in *Bacillus subtilis*, and *Bacillus* sp. C-125. Xylanase J showed high sequence homology with the xylanases from *Bacillus pumilus*. More than 90% of the xylanase pro-

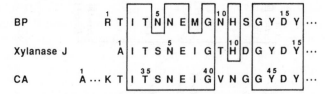

Fig. 7.52 N-terminal amino acid sequence of xylanase J compared with those of other microbial xylanases. BP, a xylanase from *B. pumilus*; CA, xylanase B from *C. acetobutylicum*. Identical amino acids are boxed. The number above each amino acid sequence corresponds to the mature enzyme.
(Reproduced with permission from Nakamura *et al.*, *Appl. Environ. Microbiol.*, **59**, 2315 (1993*b*))

duced was located in the periplasmic space of *E. coli*. The alkaline xylanase expressed in these bacilli was secreted from cells. Amino acid sequence analysis revealed that the xylanase J N-terminal exhibited high homology with family G xylanases such as *Bacillus pumilus* and *Clostridium acetobutylicum*. However, the C-terminal (125 amino acid residues) that was a characteristic amino acid sequence had a sequence different from those of family G xylanases (Fig. 7.52).

Shendye and Rao (1993*a*; Shendye, Gaikaiwari and Rao, 1994) also cloned xylanase genes of an alkaliphilic thermophilic *Bacillus* sp. NCIM59 and expressed them in *E. coli*. Expression in *B. subtilis* was also exhibited. However, there were different xylan hydrolysis products from NCIM59 and the host *B. subtilis*, and the two plasmid-borne xylanases appeared to have different modes of action. Total xylanase expression in the transformants was 6-fold higher than in the host. There was no significant enhancement in the expression of recombinant xylanases by adding xylan to the growth medium. From an industrial point of view, they integrated xylanase gene into the chromosomal gene in an alkaliphilic thermophilic *Bacillus* sp. (NCIM59) to enhance the production of xylanase (Shendye and Rao, 1993*b*).

C. Cloning and Expression of a Xylanase Gene of *Aeromonas* sp. No.212

Alkaliphilic *Aeromonas* sp. No. 212 was grown aerobically to the early stationary phase at 37 °C in Horikoshi-I medium and chromosomal DNA was purified. A xylanase gene was essentially cloned by a method similar to that for cloning the *Bacillus* sp. No. C-125 xylanase gene. *E. coli* HB101 carrying the pAX1 plasmid was grown aerobically in LB-broth for 20 h, and xylanase activity was detected in sonically disrupted cells. The xylanase activity expressed in *E. coli* was about 80 times higher than that of xylanase L produced by alkaliphilic *Aeromonas* sp. No. 212. All xylanase activity was detected in either periplasmic or cytoplasmic fractions and no extracellular xylanase was detected.

The synthesis of xylanase in *E. coli* carrying pAX1 was constitutive and no effect of supplement of xylan was observed. The molecular weight of the xylanase purified by ion-exchange chromatography followed by gel filtration was determined to be 135,000 by SDS-PAGE. The xylanase of *E. coli* carrying pAX1 exhibited the same pH activity curve as that of xylanase L (pH optimum 7.0-8.0). The end products in the hydrolysate of xylan with the xylanase of *E. coli* (pAX1) were oligosaccharides such as xylobiose, xylotriose, xylotetraose and high oligosaccharides. No xylose was detected as a major end product. The Ouchterlony double-diffusion test indicated that only xylanase L gave a line of precipitation which fused with that for xylanase purified from *E. coli* HB101 carrying pAX1. These results showed that the gene of xylanase L was located in the *Hin*d III fragment of 6 kb, (Kudo, Ohkoshi and Horikoshi, 1985), although no nucleotide sequence of this gene has yet been determined. Using the secretion vector pEAP37, we were able to produce xylanase L in the culture broth of *E. coli* HB101 (see p. 114).

7.5.5 Applications
Although actual industrial application has not yet been established, the following potential applications are under consideration.

A. Biological Degradation of Rayon Waste
Ikura and Horikoshi (1977) reported that alkaliphilic *Bacillus* sp. No. C-11 could utilize rayon waste well under alkaline conditions. The optimum initial pH for alkaliphilic *Bacillus* sp. No. C-11 was about 11.0, although after 24-h incubation the pH of the medium decreased to 9.5 due to the organic acids produced. No further investigation has been reported.

B. Biobleaching
In the course of the pulping process, residual lignin in degraded by chlorine-based chemicals or other oxidizing agents and extracted by alkaline solutions in successive steps to increase pulp brightness. Lignin is thought to be linked to hemicellulose in pulps. Under alkaline pulping conditions, xylan is first dissolved then reprecipitated into the fibers. It has been suggested that this type of xylan is present in fibers in different forms: loosely bound in the pulp, crystallized onto the cellulose fibers or chemically inked to cellulose. Cleavage of hemicellulose bonds near the points of attachment between lignin and hemicellulose may also facilitate the extraction of residual lignin. The addition of xylanase in bleaching aims to decrease chlorine consumption. Lignin content, usually expressed as a kappa number, was reduced in pine and birch Kraft pulps by 25% and 33%, respectively, using xylanases (Viikari *et al.*, 1994). From an industrial point of view, alkaline xylanases are very good candidates for biological bleaching under alkaline conditions.

Nakamura *et al.* reported that their xylanases of alkaliphilic Bacilli were available for biological bleaching (Nakamura *et al.*, 1994). Khasin, Alchanati and Shoham (1993) also isolated *Bacillus stearothermophilus* T-6, producer of an alkaline xylanase that was shown to optimally bleach pulp at pH 9 and 65 °C. Kulkarni and Rao (1996) isolated a xylanase from alkaliphilic thermophilic *Bacillus* sp. NCIM59 and tested biobleaching of bagasse pulp. The biotreatment resulted in a 2 unit decrease in the K number without altering the strength properties of the pulp. However, these reported xylanases are not stable enough for actual biological bleaching in industrial-scale plants. Therefore, the isolation and production of more xylanases stable at higher pH and temperature is highly desirable.

7.6 Mannan-degrading Enzymes

β-Mannan is a kind of hemicellulose contained in higher plants such as konjac, guargum, locust bean and copra, and it easily dissolves in alkaline water. Mannan-degrading enzymes of neutrophilic bacteria, actinomyces, and fungi have been studied. However, no mannan-degrading enzyme that hydrolyzes under alkaline condition has been discovered.

7.6.1   Isolation and Properties of Mannan-degrading Microorganisms (Akino, Nakamura and Horikoshi, 1987; 1988*a, b*)

A small amount of soil was spread on agar plates containing 1% β-mannan from larch wood, 1% polypeptone, 0.2% yeast extract, 0.1% KH_2PO_4, 0.02% $MgSO_4 \cdot 7H_2O$ and 0.5% sodium carbonate. The plates were incubated at 37 °C for 48–72 h. Strain AM-001 with a large clear zone around the colony was selected as the enzyme producer. The isolate grew at temperatures from 20 °C to 45 °C, with an optimum at 37 °C in the medium described above. The pH range for growth was from pH 7.5 to 11.5 with the optimum at pH 8.5 to 9.5. The bacterium was aerobic, motile and gram-variable; the rod-shaped cells (0.6–0.8 $\mu m \times 3.0$–6.0 μm) had peritrichous flagella and terminal swollen sporangia containing oval spores (1–1.2 $\mu m \times 1.5$–2.0 μm). The taxonomical characteristics of this alkaliphilic *Bacillus* strain were almost the same as those of *Bacillus circulans* except for the pH range for growth.

Bacillus sp. No. AM-001 was cultivated aerobically under various conditions and activities of β-mannanase in the culture broth and β-mannosidase extracted from cells treated with 0.1% Triton X-100. Both enzymes formed when the bacterium was grown under alkaline conditions, and the optimum concentration was 0.5% Na_2CO_3 or 0.5–1.0% $NaHCO_3$. Various carbohydrates were also tested and the best carbohydrate for enzyme production was konjac powder (1% w/v). The optimum cultivation temperature for enzyme production was 31 °C for β-mannosidase and 37 °C for β-mannanase in the production medium composed of 1% konjac powder, 0.2% yeast extract, 2% polypeptone, 0.1% K_2HPO_4, 0.02% $MgSO_4 \cdot 7H_2O$ and 0.5% sodium carbonate. The crude enzyme preparation showed optimum pH and temperature for β-mannosidase of 7.0 and 55 °C and for β-mannanase 9.0 and 65 °C, respectively.

7.6.2   Purification of β-Mannanase and Its Properties (Akino *et al.*, 1988*a*)

Three extracellular β-mannanases (M-I, M-II and M-III) were purified by ammonium sulfate precipitation (80% saturation) followed by chromatography on a DEAE-Toyopearl 650 M column (4.6 × 35 cm) equilibrated with 0.01 M phosphate buffer (pH 7.0) and by a hydroxyapatite column (1.6 × 25 cm). As shown in Fig. 7.53, two active fractions (Fractions 1 and 2) were detected. Each fraction was applied onto a Sephacryl S-200 column (2.6 × 90 cm) equilibrated with 0.01 M phosphate buffer (pH 7.0) containing 0.1 M NaCl and eluted with the same buffer. Mannanase-I and -II were isolated from fraction 1 and mannanase-III from fraction 2. Polyacrylamide electrophoresis revealed that these three β-mannanases were electrophoretically homogeneous. The molecular weight estimated by SDS-PAGE was 58,500 for M-I, 59,500 for M-II and 42,000 for M-III. As shown in Fig. 7.54, β-mannanases, M-I and M-II were most active at pH 9.0, and M-III demonstrated optimum enzyme action at pH 8.5. Although M-III enzyme was relatively more stable than the others, there were no significant differences among these three mannanases except for molecular weight. These enzymes hydrolyzed β-1,4-mannooligosaccharides larger than mannotriose and the major components in the digest were di-, tri- and tetrasaccharides.

7.6.3 β-Mannosidase

β-Mannosidase hydrolyzes the β-mannosidic linkage in various β-1,4-mannans and

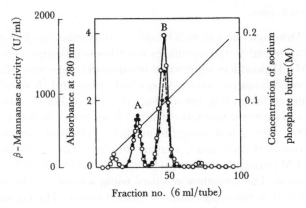

Fig. 7.53 Chromatography of the β-mannanases with hydroxyapatite column. Symbols: (●), β-mannanases activity; (○), A_{280} (Protein); (—), sodium phosphate buffer concentration.
(Reproduced with permission from Akino *et al.*, *Agric. Biol. Chem.*, **52**, 775 (1988))

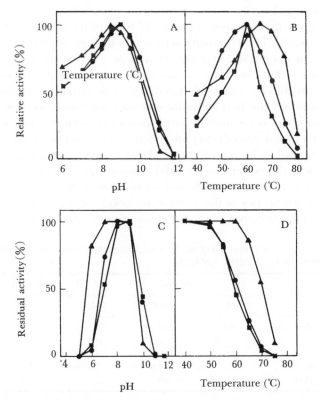

Fig. 7.54 Effects of pH and temperature on enzyme activity and stability.
The enzyme activities were measured by the standard assay methods of various pHs and temperatures. The buffers used were Carmody buffer (A) and 50 mM of Good's buffer (CHES, pH 9.0) (B). The enzyme solutions were incubated at various pHs in Carmody buffer for 30 min at 50° or 60 ℃ (C), and at various temperatures for 30 min in 50 mM Good's buffer (CHES, pH 9.0) (D). The residual activities were measured under the standard assay conditions. Symbols: (●), β-mannanase M-I; (■), β-mannanase M-II; (▲), β-mannanase M-III.
(Reproduced with permission from Akino *et al.*, *Agric. Biol. Chem.*, **52**, 776 (1988))

yields D-mannan. Although many studies have been performed on β-mannosidases of animal tissues, few microbial β-mannosidases have been investigated. Our alkaliphilic *Bacillus* sp. No. AM-001 produces significant amounts of three extracellular β-mannanases and a cell-associated β-mannosidase (Akino, Nakamura and Horikoshi, 1988b).

The cells (cultured for 24 h in the medium described above) were collected by centrifugation and the β-mannosidase was extracted from the cells in the presence of 0.1% (w/v) Triton X-100. After 2 h shaking the insoluble cells were centrifuged off and the supernatant fluid was used as the crude enzyme solution. The crude enzyme solution was dialyzed against 0.1 M phosphate buffer (pH 7.0) followed by DEAE Toyopearl 650 M column chromatography. The enzyme was eluted using a linear gradient of NaCl concentration from 0 to 0.5 M in the same buffer described above. The enzyme fraction which was dialyzed against the same buffer was subjected to a D-mannose Sepharose 6B affinity column. The enzyme was eluted by phosphate buffer containing 5% (w/v) D-mannose and finally purified by passing through a Sephadex G150 column. Table 7.43 summarizes the purification process of the enzyme. The enzyme was electrophoretically homogeneous and its molecular weight was estimated to be 94,000. Isoelectric focusing showed the isoelectric point to be 5.5.

The purified enzyme was most active at pH 6.0 and stable at 40 °C for 30 min in the pH range of 6.5 to 8.0. The enzymatic activity was strongly inhibited by the addition of metal ions and some chemicals but not by D-mannose, as shown in Table 7.44. Among the synthetic substrates so far tested, the enzyme did not hydrolyze any except p-nitrophenyl-β-D-mannopyranoside (K_m value was 1.3 mM). The enzyme hydrolyzed β-1,4-mannooligosaccharides and the best substrate was mannotetraose, as shown in Table 7.45, although long incubation resulted in weak transferase activity. The molecular weight (94,000) was smaller than that of other microbial β-mannosidases such as *Aspergillus oryzae* (120,000–135,000) and *Tremella fuciformis* (160,000–200,000). The final product from the oligosaccharides with the enzyme was mainly D-mannose and the enzymatic activity was not inhibited by D-mannose or mannose-derivatives (D-mannosamine, D-mannonic acid and D-mannitol). These properties of the enzyme are good for the production of D-mannose from β-mannan in the presence of the β-mannanases described above.

Table 7.43 Comparison of some properties of the purified β-mannanases from alkaliphilic *Bacillus* sp. No. AM-001 and *E. coli* JM101 (pMAH3).

| | AM-001 | | | E. coli (pMAH3) | |
	M-I	M-II	M-III	A	B
Optimum temperature (°C)	60	60	65	60	65
Optimum pH	9.0	9.0	8.5	9.0	8.5
Thermal stability (°C)	50	50	60	50	60
pH stability 8.0–9.0	8.0–9.0	8.0–9.0	7.0–9.0	8.0–9.0	7.0–9.0
Molecular weight (kDa)	58	59	42	58	43

The enzyme activities were measured by the method described in the text under various pH and temperature conditions.

The buffers used were Carmody buffer and 50 mM of Good's buffer (CHES, pH 9.0). The enzyme solutions were incubated at various pHs in Carmody buffer for 30 min at 50 or 60 °C, and at various temperatures for 30 min in 50 mM Good's buffer (CHES, pH 9.0).

The residual activities were measured by the method reported previously (Akino *et al.*, 1988b).

Table 7.44 Effects of various reagents on β-mannosidase activity

Reagent		Residual activity (%)
None		
D-Mannoasamine HCl	20 mM	100
D-Mannonic acid-γ-lactone	20 mM	100
N-Acetyl glucosamine	20 mM	100
D-Mannose	20 mM	100
D-Mannitol	20 mM	100
Nojirimycin	0.0025%	91
Deoxynojirimycin	0.0025%	100
PMSF	0.1 mM	99
ICH$_2$COOH	1.0 mM	98
2-ME	1.0 mM	95
Cysteine	1.0 mM	92
PCMB	0.1 mM	0
N-Bromosuccinimide	1.0 mM	0
SDS	0.1%	0
DBS	0.1%	0
Ag$^+$		0
Cd^{2+}		0
Cu^{2+}		0
Zn^{2+}		0
Hg^{2+}		0
Fe^{2+}		5
Fe^{3+}	1.0 mM	100
Pb^{2+}		94
Co^{2+}		96
Mg^{2+}		100
Sn^{2+}		100
Mn^{2+}		100

PMSF, phenylmethanesulfonyl fluoride; PCMB, sodium p-chloro-mercuribenzoate; 2-ME, 2-mercaptoethanol; SDS, sodium dodecyl sulfate; DBS, N-dodecylbenzene sulfonate.
(Reproduced with permission from Akino *et al.*, *Agric. Biol. Chem.*, **52**, 1462 (1988))

Table 7.45 Relative hydrolysis rates of β-1,4-D-manno-oligosaccharides by β-mannosidase

Substrate	Activity (μmol/min·mg)	Relative activity (%)
Mannobiose	1.92	34
Mannotriose	3.88	68
Mannotetraose	5.68	100
Mannopentaose	4.48	79

(Reproduced with permission from Akino *et al.*, *Agric. Biol. Chem.*, **52**, 1463 (1988))

7.6.4 Molecular Cloning of Mannan-degrading Enzymes

As described in the previous section, this strain produced significant amounts of three extracellular β-mannanases and a cell-associated β-mannosidase. The three β-mannanases differed in several enzymatic properties, including optimum pH for enzyme action, optimum temperature, pH stability, thermal stability, isoelectric point and molecular weight. To elucidate the genetic basis for the production of multiple forms we tried to clone the β-mannanase genes of this strain into *E. coli* using pUC 19 as a vector (Akino, Kato and Horikoshi, 1989*a*).

Alkaliphilic *Bacillus* sp. No. AM-001 was aerobically grown to the early stationary phase at 37°C in alkaline medium (pH 9.0) consisting of 1% konjac mannan. Total chromosomal DNA was digested with *Hin*d III restriction enzyme and 2 to 4 kb DNA fractions of chromosomal DNA were collected by 1% agarose gel electrophoresis. The digested plasmid pHSG398 was successively dephosphorylated with bacterial alkaline phosphates. After digestion 1 μg of plasmid and 5 μg of chromosomal DNA were mixed and ligated with T_4 DNA ligase for 30 min at 16°C. The ligation mixtures described above were used to transform *E. coli* JM101. Transformants having mannanase activity could be detected directly on the LB-konjac plates containing ampicillin (50 μg/ml). Of 5×10^4 colonies one was surrounded by a large transparent halo and this colony showed β-mannanase activity in liquid cultivation using LB medium. A plasmid harbored by the transformant was designated pMAH1. After subcloning with *Xba* I treatment a plasmid, pMAH3, containing a 2.5 kb DNA fragment of *Bacillus* sp. No. AM-001 was isolated.

7.6.5 Properties of a Mannanase Produced by *E. coli* JM101 Carrying pMAH3

E. coli JM101 (pMAH3) was grown aerobically in LB broth for 24 h at 37°C. The β-mannanase activity was located mainly in the periplasmic (53%) and intracellular (43%) fractions. No induction of the enzyme could be detected in the presence or absence of mannan in media such as konjac, locust bean and copra. The periplasmic β-mannanase was purified by affinity chromatography and two active fractions (β-mannanase A and B) were separated. Molecular weight was 58,000 for A and 43,000 for B. As shown in Table 7.43, mannanase A has enzymatic properties similar to those of β-mannanase M-I and M-II of *Bacillus* sp. No. AM-001, and mannanase B properties are similar to those of β-mannanase M-III. Furthermore, the Ouchterlony double diffusion test showed that these five enzymes gave fused precipitation lines. However, N-terminal amino acid sequences of

Table 7.46 N-Terminal sequences of β-mannanases

Alkaliphilic *Bacillus* sp. No. AM-001

		1	2	3	4	5	6	7	8	9	10	11	12
β-mannanase	M-I	Asn	Gly	Ala	Ala	Leu	Ser	Asn	Pro	Asn	Ala	Asn	Gln
	M-II	Asn	Gly	Ala	Ala	Leu	Ser	Asn	Pro	Asn	Ala	Asn	Gln
	M-III	Asn	Gly	Ala	Ala	Leu	Ser	Asn	Pro	Asn	Ala	Asn	Gln

E. coli (pMAH3)

		1	2	3	4	5	6	7	8	9	10	11	12
β-mannanase	A	Ser	Glu	Ala	Asn	Gly	Ala	Ala	Leu	Ser	Asn	Pro	Asn
	B	Ser	Glu	Ala	Asn	Gly	Ala	Ala	Leu	Ser	Asn	Pro	Asn

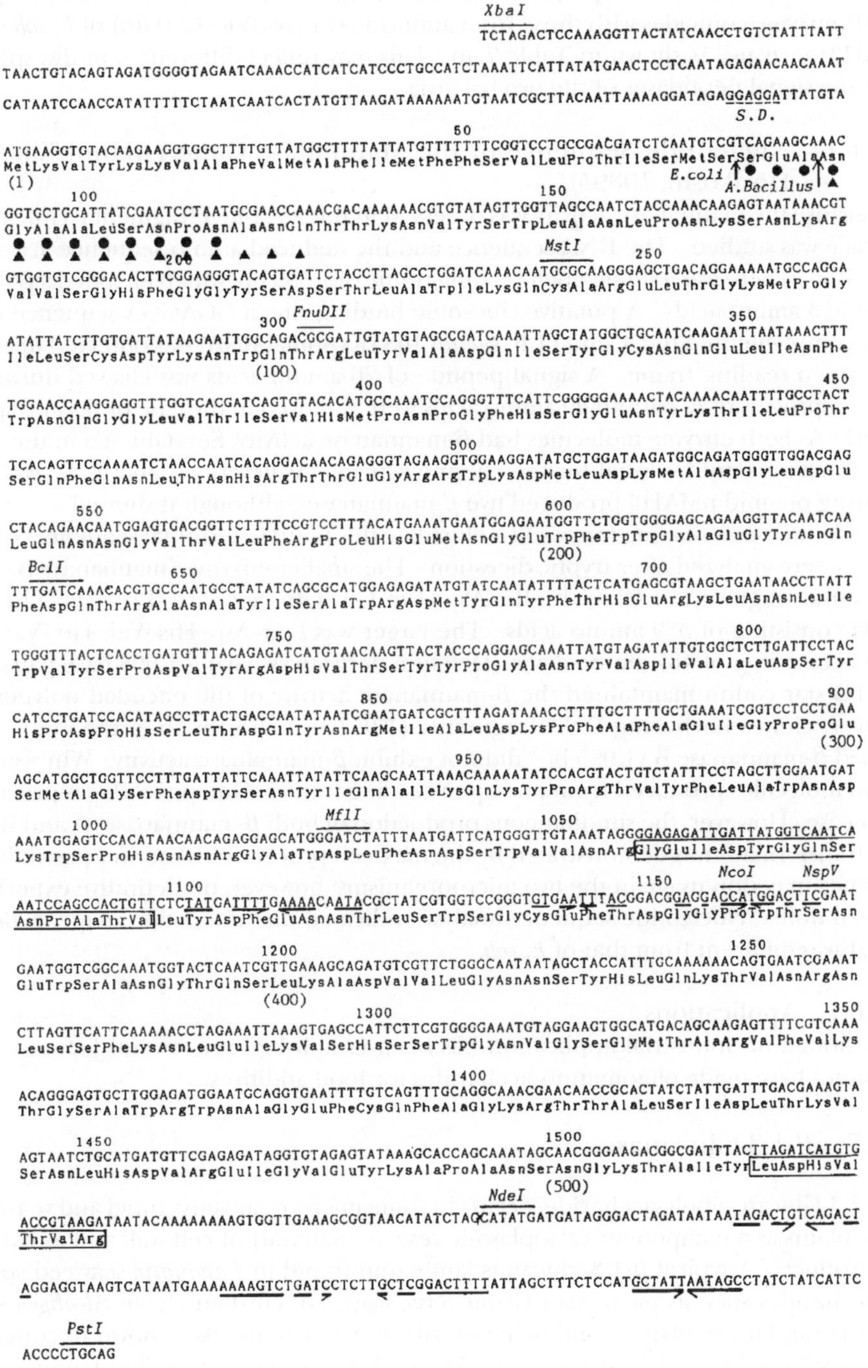

                                              XbaI
                                          TCTAGACTCCAAAGGTTACTATCAACCTGTCTATTTATT

TAACTGTACAGTAGATGGGGTAGAATCAAACCATCATCATCCCTGCCATCTAAATTCATTATATGAACTCCTCAATAGAGAACAACAAAT

CATAATCCAACCATATTTTTCTAATCAATCACTATGTTAAGATAAAAAATGTAATCGCTTACAATTAAAAGGATAGAGGAGGATTATGTA
                                                                            _S.D._
            50
ATGAAGGTGTACAAGAAGGTGGCTTTTGTTATGGCTTTTATTATGTTTTTTTCGGTCCTGCCGACGATCTCAATGTCGTCAGAAGCAAAC
MetLysValTyrLysLysValAlaPheValMetAlaPheIleMetPhePheSerValLeuProThrIleSerMetSerSerGluAlaAsn
(1)   _E.coli_ ↑● ● ●↑↑
            100   _A.Bacillus_ ▲ ▲
GGTGCTGCATTATCGAATCCTAATGCGAACCAAACGACAAAAAACGTGTATAGTTGGTTAGCCAATCTACCAAACAAGAGTAATAAACGT
GlyAlaAlaLeuSerAsnProAsnAlaAsnGlnThrThrLysAsnValTyrSerTrpLeuAlaAsnLeuProAsnLysSerAsnLysArg
▲▲●●●●●●● ▲ ▲ ▲
           ▲ ▲ ▲ 200   _MstI_   250
GTGGTGTCGGGACACTTCGGAGGGTACAGTGATTCTACCTTAGCCTGGATCAAACAATGCGCAAGGGAGCTGACAGGAAAAATGCCAGGA
ValValSerGlyHisPheGlyGlyTyrSerAspSerThrLeuAlaTrpIleLysGlnCysAlaArgGluLeuThrGlyLysMetProGly
                      300   _FnuDII_   350
ATATTATCTTGTGATTATAAGAATTGGCAGACGCGATTGTATGTAGCCGATCAAATTAGCTATGGCTGCAATCAAGATATTAATAAACTTT
IleLeuSerCysAspTyrLysAsnTrpGlnThrArgLeuTyrValAlaAspGlnIleSerTyrGlyCysAsnGlnGluLeuIleAsnPhe
                                                    (100)
            400   450
TGGAACCAAGGAGGTTTGGTCACGATCAGTGTACACATGCCAAATCCAGGGTTTCATTCGGGGGGAAAACTACAAAACAATTTTGCCTACT
TrpAsnGlnGlyGlyLeuValThrIleSerValHisMetProAsnProGlyPheHisSerGlyGluAsnTyrLysThrIleLeuProThr
            500
TCACAGTTCCAAAATCTAACCAATCACAGGACAACAGAGGGTAGAAGGTGGAAGGATATGCTGGATAAGATGGCAGATGGGTTGGACGAG
SerGlnPheGlnAsnLeuThrAsnHisArgThrThrGluGlyArgArgTrpLysAspMetLeuAspLysMetAlaAspGlyLeuAspGlu
    550   600
CTACAGAACAATGGAGTGACGGTTCTTTTCCGTCCTTTACATGAAATGAATGGAGAATGGTTCTGGTGGGGAGCAGAAGGTTACAATCAA
LeuGlnAsnAsnGlyValThrValLeuPheArgProLeuHisGluMetAsnGlyGluTrpPheTrpTrpGlyAlaGluGlyTyrAsnGln
    BclI   650   700
TTTGATCAAACACGTGCCAATGCCTATATCAGCGCATGGAGAGATATGTATCAATATTTTACTCATGAGCGTAAGCTGAATAACCTTATT
PheAspGlnThrArgAlaAsnAlaTyrIleSerAlaTrpArgAspMetTyrGlnTyrPheThrHisGluArgLysLeuAsnAsnLeuIle
                        750   800
TGGGTTTACTCACCTGATGTTTACAGAGATCATGTAACCAAGTTACTACCCAGGAGCAAATTATGTAGATATTGTGGCTCTTGATTCCTAC
TrpValTyrSerProAspValTyrArgAspHisValThrSerTyrTyrProGlyAlaAsnTyrValAspIleValAlaLeuAspSerTyr
                                    850   900
CATCCTGATCCACATAGCCTTACTGACCAATATAATCGAATGATCGCTTTAGATAAACCTTTTGCTTTTGCTGAAATCGGTCCTCCTGAA
HisProAspProHisSerLeuThrAspGlnTyrAsnArgMetIleAlaLeuAspLysProPheAlaPheAlaGluIleGlyProProGlu
                                                                        (300)
        950
AGCATGGCTGGTTCCTTTGATTATTCAAATTATATTCAAGCAATTAAACAAAAATATCCACGTACTGTCTATTTCCTAGCTTGGAATGAT
SerMetAlaGlySerPheAspTyrSerAsnTyrIleGlnAlaIleLysGlnLysTyrProArgThrValTyrPheLeuAlaTrpAsnAsp
    1000   _MflI_   1050
AAATGGAGTCCACATAACAACAGAGGAGCATGGGATCTATTTAATGATTCATGGGTTGTAAATAGGGGAGAGATTGATTATGGTCAATCA
LysTrpSerProHisAsnAsnArgGlyAlaTrpAspLeuPheAsnAspSerTrpValValAsnArg GlyGluIleAspTyrGlyGlnSer
                                                            1150   _NcoI_   _NspV_
    1100   GTGAATTTACGGACGGAGGACCATGGACTTCGAAT
AATCCAGCCACTGTTCTC TATGATTTTGAAAAACA ATACGCTATCGTGGTCCGGGT GTGAATTTACGGACGGAGGACCATGGACTTCGAAT
AsnProAlaThrVal LeuTyrAspPheGluAsnAsnThrLeuSerTrpSerGlyCysGluPheThrAspGlyValProTrpProThrSerAsn
                1200   1250
GAATGGTCGGCAAATGGTACTCAATCGTTGAAAGCAGATGTCGTTCTGGGCAATAATAGCTACCATTTGCAAAAAACAGTGAATCGAAAT
GluTrpSerAlaAsnGlyThrGlnSerLeuLysAlaAspValValLeuGlyAsnAsnSerTyrHisLeuGlnLysThrValAsnArgAsn
            (400)   1300
CTTAGTTCATTCAAAAAACCTAGAAATTAAAGTGAGCCATTCTTCGTGGGGAAATGTAGGAAGTGGCATGACAGCAAGAGTTTTCGTCAAA
LeuSerSerPheLysAsnLeuGluIleLysValSerHisSerSerTrpGlyAsnValGlySerGlyMetThrAlaArgValPheValLys
                                    1400
ACAGGGAGTGCTTGGAGATGGAATGCAGGTGAATTTTGTCAGTTTGCAGGCAAACGAACAACCGCACTATCTATTGATTTGACGAAAGTA
ThrGlySerAlaTrpArgTrpAsnAlaGlyGluPheCysGlnPheAlaGlyLysArgThrThrAlaLeuSerIleAspLeuThrLysVal
    1450   1500
AGTAATCTGCATGATGTTCGAGAGATAGGTGTAGAGTATAAAGCACCAGCAAATAGCAACGGGAAGACGGCCGATTTAC TTAGATCATGTG
SerAsnLeuHisAspValArgGluIleGlyValGluTyrLysAlaProAlaAsnSerAsnGlyLysThrAlaIleTyr LeuAspHisVal
                                                        (500)   _NdeI_
ACCGTAAGATAATACAAAAAAAAGTGGTTGAAAGCGGTAACATATCTAC CATATGATGATAGGGACTAGATAATAAT AGACTGTCAGACT
ThrValArg

AGGAGGTAAGTCATAATGAAAAAAAAGTCTGATCCTCTTGCTCGGACTTTTATTAGCTTTCTCCATGCTATTAATAGCCTATCTATCATTC

PstI
ACCCCTGCAG

Fig. 7.55 DNA sequence.

these five mannanases determined by an automatic amino acid sequencer revealed that the N-terminal amino acid sequence from amino acid I (Asn) to 9 (Gln) of *Bacillus* sp. No. AM-001 enzyme coincides with those from amino acid 4 (Asn) to 12 (Gln) of *E. coli* JM101 (pMAH3) enzyme, as shown in Table 7.46. This may reflect differences in the specificities of the signal peptidases of the two bacteria.

7.6.6   Nucleotide Sequence of the Mannanase Gene (Akino, Kato and Horikoshi, 1989*b*)

Because the mannanase activity was observed in the *Xba*-I-*pst* I fragment its nucleotide sequence was studied. The DNA sequence and the deduced amino sequence are shown in Fig. 7.55. There was a single open reading frame of 1,539 bp, which encoded a polypeptide of 513 amino acids. A putative ribosome binding site, a GGAGGA sequence which highly complemented the 3′ end of *B. subtilis* 16S ribosomal RNA, was observed upstream of the open reading frame. A signal peptide of 26 amino acids was cleaved during the secretion process of *E. coli* and 29 amino acids were removed in the case of *Bacillus* sp. No. AM-001. As both enzyme molecules had β-mannanase activity, Ser–Glu–Ala in the N-terminal fraction is not essential for the enzyme activity. As described above, *E. coli* JM101 harboring plasmid pMAH3 produced two β-mannanases, although it showed no capacity to encode two open reading frames. To elucidate this point C-terminal sequences of both enzymes were analyzed after tryptic digestion. The smaller enzyme (mannanase B) had a C-terminal fragment of Gly–Glu–Ile–Asp–Tyr–Gly–Gln–Ser–Asn–Pro–Ala–Thr–Val–COOH consisting of 339 amino acids. The larger was Leu–Asp–His–Val–Thr–Val–Arg–COOH consisting of 487 amino acids. Deletion derivatives having 1,098 base pairs from the ATG star codon maintained the β-mannanase activity of the encoded polypeptide. However, clones harboring DNA fragments (1,051 bp) shorter than the gene which encoded β-mannanase B (1,095 bp) did not exhibit β-mannanase activity. Why were two protein molecules produced from one open reading frame? One possibility is processing by protease. However, the simultaneous production of both β- mannanases A and B in an *E. coli* transformant was demonstrated by the maxicell procedure. Another possibility is a difference in codon usage in the two microorganisms; however, no definitive experiment has been done. Nucleotide sequence analysis indicated that the codon usage in *Bacillus* AM-001 was different from that of *E. coli*.

7.6.7   Applications

Although no industrial application has been established yet, several Japanese food companies have made oligomannosaccharides for food additives.

7.7   β-1,3-Glucanase

β-1,3-Glucan, a polysaccharide, occurs in some microorganisms (fungi and yeast) and higher plants as a component (cytoplasmic reserve material) of cell wall and as extracellular products. A typical β-1,3-glucan is laminaran found in *Laminaria* seaweed species. Polysaccharides such as pachyman (from *Basidiomycetes*), curdlan (from *Alcaligenes* sp.), callose (from higher plants), and lichenan (from lichens) are also known to contain β-1,3-glucosyl residues. β-1,3-Glucanase [EC 3.2.1.39], sometimes called laminaranase, hydrolyzes the β-1,3-glucosidic linkage of these polysaccharides. The β-1,3-glucanases have been found in bacteria, fungi, algae, and higher plants.

In our earlier studies, semialkaliphilic *Bacillus circulans* was found to be a β-1,3-glu-canase producer with strong lytic activity on the cell walls and spore coats of many fungi such as *Aspergillus oryzae* (Horikoshi and Iida, 1958, 1959; Horikoshi, Koffler and Arima 1963; Horikoshi and Iida, 1964). Almost all β-1,3-glucanases so far reported have pH optimum on the acid side (pH 4 to 6), but have no activity at pH 8. Alkaline β-1,3-glucanases that we isolated (Horikoshi and Atsukawa, 1973*a*, 1975) are active at pH over 8.0. However, no enzyme with optimum for enzyme action at higher than pH 10 has been reported. Nogi and Horikoshi (1990) isolated an alkaliphilic *Bacillus* sp. No. AG-430, which produced a thermostable β-1,3-glucanase having an optimum pH for enzyme action at pH 9.5, and some of the enzymes have been cloned for the structure genes to analyze their genetic information.

7.7.1   β-1,3-Glucanases as Lytic Enzymes

Horikoshi and Iida (1958, 1959, 1964) reported that *Bacillus circulans* isolated from a contaminated culture vessel was able to strongly lyse the cell walls and spore coats of fungi during the cultivation of *Aspergillus oryzae*. β-1,3-Glucanase was isolated and purified from the culture fluid of the isolate (Horikoshi, Koffler and Arima, 1963). This was the first paper showing that β-1,3-glucanase is the key enzyme to lyse the cell walls of fungi, including yeasts. The author's old laboratory notes revealed a very interesting phenomenon: The isolate *Bacillus circulans* IAM 1165 showed very poor growth at pH 5.5 to 6.5; this value is not bad for conventional *Bacillus* strains. The best growth of *B. circulans* IAM 1165 was observed in the range of pH 8.0 to 8.5, and growth was detected at pH 9.0. Since at the time the author had no knowledge of alkaliphilic microorganisms, the highest pH value tested was 9.0. Furthermore, the bacterium was able to increase the pH value of the mix-culture broth: *Aspergillus oryzae* was inoculated into a culture medium of pH 6 and cultivated for 3 days at 30 °C. Then *B. circulans* IAM 1165 was inoculated and mix-cultured at 37 °C. During cultivation, the pH value gradually increased and reached pH 8.5 to 9.0 after 3 days mix-cultivation. Fungal cells were lysed by the lytic enzyme produced by *B. circulans* IAM 1165 and no fungal cells were observed by microscopy. The yield of the lytic enzyme was observed in medium of pH 8.5. Although these findings were very primitive, it is of interest that *B. circulans* isolated from a contaminated culture flask was a kind of semi-alkaliphilic *B. circulans*. The β-1,3-glucanase of *B. circulans* IAM1165 was purified and characterized as shown in Table 7.47.

Aono *et al.* (1992) reported that *Bacillus circulans* IAM1165 produced three major extracellular β-1,3-glucanases (molecular masses, 28, 42, and 91 kDa) during the stationary phase of growth. One of these extracellular enzymes was purified to homogeneity. The molecular mass was 91 kDa, and the p*I* was 4.3. The optimum temperature of the enzyme reaction was 70 °C when laminarin (a soluble β-1,3-glucan) was used as the substrate. The pH range of the enzyme was broad (pH 4.5 to 9.0), and the optimum pH was 6.5. The enzyme is an endo β-1,3-glucanase and has a random cleavage pattern (Aono, Sato *et al.*, 1992). The 28 and 42 kDa enzymes were recently purified to homogeneity from the culture supernatant in this study (Aono, Hammura *et al.*, 1995). The properties of these two enzymes were examined, together with those of the 91 kDa enzyme previously isolated. The enzymatic properties of the 28 and 42 kDa β-1,3-glucanases closely resemble each other. The enzymes belong to a category of endo type 1,3-β-D-glucan glu-

Table 7.47 Physiochemical properties of β-1,3-glucanases

Property	β-1,3-Glucanase			
	B. circulans	Bacillus sp. No. 221	Bacillus sp. No. K-12-5	Basidiomycete sp. No. QM806
Molecular weight	28,000	36,000	40,000	57,000
Sedimentation constant	3.0	3.2	3.6	3.7
Isoelectric point	5.4	41	3.5	ND
Optimum pH	5.8	8.0	6.0–8.0	4.6–6.0
Stabilization by Ca^{2+}	+	−	−	ND

ND, Not done
(Reproduced with permission from Horikoshi and Akiba, *Alkalophilic Microorganisms*, p.115, Springer-Verlag: Japan Scientific Societies Press (1982))

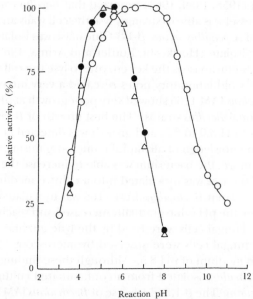

Fig. 7.56 Effect of reaction pH on the activities of β-1,3-glucanases. pH was adjusted with the following buffer systems: 0.1 M citric acid-NaH$_2$PO$_4$ (pH 3.0 to 3.5), 0.1 M citric acid-NaOH (pH 3.5 to 6.0), 0.1 M 2-(*N*-morpholino)ethanesulfonic acid (MES)-NaOH (pH 6.0 to 7.0), 0.1 M HEPES-NaOH (pH 6.5 to 8.2), 0.1 M *n*-tris(hydroxymethyl)methyl-3-aminopropanesulfonic acid (TAPS)-NaOH (pH 7.7 to 9.5), and 0.1 M monoethanolamine-HCl (pH 9.5 to 11.0). Activity was measured at 40 °C for 10 min. The level of activity of each enzyme at its optimum pH was defined as 100%. Symbols: ●, enzyme L; ▲, enzyme M; ○, enzyme H.
(Reproduced with permission from Aono *et al.*, *Appl. Environ. Microbiol.*, **58**, 523 (1992))

canohydrolases. The enzymes were active at pH 4.0 to 7.0 (Fig. 7.56). The optimum temperature of the reactions was 60 °C when laminarin (a soluble β-1,3-glucan) was used as the substrate at pH 7.0. The enzymes hydrolyzed barley glucan and lichenan (β-1,3-1,4-glucans) more effectively than laminarin. Of the three enzymes, the kDa enzyme lysed fungal cell walls the most effectively. Aono, Yamamoto *et al.* (1992) and Yamamoto, Aono and Horikoshi (1993) analyzed the structure of the 91 kDa β-1,3-glucanase gene of *Bacillus circulanas* IAM1165 and the properties of the enzyme accumulated in the periplasm of *E. coli* carrying the gene. The sequence showed an open reading frame of 877 amino acids, which corresponds to a precursor of the β-1,3-glucanase. The coding region of 2631 bp is flanked by a putative promoter and transcription terminator sequences. The signal peptide was considered to consist of 38 amino acids. The amino acid sequence of the mature enzyme composed of 839 amino acids showed high homology to that of the enzyme from *B. circulans* WL-12, although these enzymes are different in size. A catalytic domain of the enzyme was assumed to be the central region of the sequence based on comparison of amino acid sequences of β-1,3- or β-1,3-1,4-glucanases. Properties of the periplasmic enzyme produced in *E. coli* carrying the gene were identical with those of the extracellular enzyme produced by *B. circulans* IAM1165. β-1,3-Glucanase activity was detected in the cell walls of several yeasts and filamentous fungi, and it is assumed that its activity may play an important role in cell growth (Horikoshi and Grant, 1991). Therefore, Nakajima *et al.* tried to express a 91 kDa-β-1,3-glucanase gene in *Saccharomyces cerevisiae* under the control of the GAL1 gene promoter (Nakajima *et al.*, 1993). Yeast cells containing this fused gene produced active β-1,3-glucanase in the medium after a long period of incubation at low temperature. The enzyme produced by yeast was heterogeneous in size, and larger than the enzyme produced by *E. coli.*

7.7.2   β-1,3-Glucanase of Alkaliphilic *Bacillus* Strains

Most β-1,3-glucanases so far reported have pH optimum on the acid side but have no activity at pH 8. Alkaline β-1,3-glucanases, which are active at pH over 8, were isolated from the culture broths of alkaliphilic *Bacillus* sp. No. K-12-5 isolated from soil and *Bacillus* sp. No. 221, which has been found to be an alkaline protease producer (Horikoshi and Atsukawa, 1973*a*; 1975). The medium used for these two strains contains (per liter): pachyman, 20 g; yeast extract (Difco), 5.0 g; polypeptone, 5.0 g; K_2HPO_4, 1.0 g; $MgSO_4 \cdot 7H_2O$, 0.05 g; and sodium carbonate, 10.0 g. Pachyman powder should be defatted with ether for 20 h. The cultivations were conducted at 37 °C for 72 h under aerobic conditions. The production of the enzyme was greatly stimulated by the addition of carbonate or bicarbonate salts to the culture medium, as seen in Table 7.48. Difference of K^+ and Na^+ was not observed in this strain. The β-1,3-glucanase from both strains was purified by conventional procedures including column chromatography on DEAE-cellulose at pH 8.0, perspiration with 70% saturated ammonium sulfate, and gel filtration on a Sephadex G-100 or G-75. No. 221 enzyme was crystallized from the eluate from a Sephadex G-75 column with 30% saturated ammonium sulfate. No. K-12-5 enzyme was finally purified by column chromatography on hydroxy apatite. Two active peaks of β-1,3-glucanase were detected: one of the peaks contained approximately 90% of the original activity and the enzyme was used in the following experiments. The enzymatic properties are summarized in Table 7.47 together with those from *B. circulans* IAM 1165 and *Basidiomycete* sp. QM 806 for comparison. The β-1,3-glucanases from strains No. K-12-5 and No. 221 have relatively broader pH activity curves towards laminaran than that of *B. circulans*, as shown in Fig. 7.57. No. K-12-5 and No. 221 enzymes were most stable at pH 7.0, 10 min heating at 65 °C,

Table 7.48   Effects of the addition of Na$_2$CO$_3$ or NaHCO$_3$ on the production of β-1,3-glucanase by alkaliphilic *Bacillus* sp. No. K-21-5

Salt added (%)		Initial pH	Cell growth (A$_{660}$)	Enzyme activity (units/ml)
None	—	7.2	0.1	ND
None	—	10.0[†]	0.8	ND
NaCl	1.0	10.5[†]	0.8	ND
KCl	1.0	10.5	0.8	ND
Na$_2$CO$_3$	0.5	9.7	1.1	1.5
	1.0	10.2	1.3	5.6
Na$_2$HCO$_3$	1.0	9.0	1.1	1.0
	2.0	9.3	1.3	4.0
K$_2$CO$_3$	1.0	10.2	1.2	3.1

ND, not detected.
[†] Adjusted with NaOH.
(Reproduced with permission from Horikoshi and Akiba, *Alkalophilic Microorganisms*, p.114, Springer-Verlag: Japan Scientific Societies Press (1982))

whereas the *B. circulans* enzyme was most stable at pH 6.0. The optimum temperature of No. K-12-5 and No. 221 was 65 °C and 80 °C respectively. As seen in Fig. 7.58, the presence of 10 mM Ca^{2+} causes an increase in the thermostability of *B. circulans* enzyme; however, no protective action on the two alkaline enzymes was detected. Another different feature of these enzymes was the action pattern on substrates (Table 7.49). The two alkaline enzymes hydrolyzed laminaran, laminaripentaose, and laminaritetraose, but did not (or only very slowly) hydrolyze laminaribiose and laminaribiose. The fungal cell walls prepared from *Mucor* sp., *Fusarium* sp., and *Aspergillus oryzae* were scarcely hydrolyzed by the two alkaline enzymes, but *B. circulans* and *Basidiomycete* enzymes were able to hydrolyze well. The results suggest that the lytic activity towards fungal cell walls may depend partly on the activity of the enzyme towards laminaritriose.

No enzyme with an optimum for enzyme action at pH 10 has been found. Moreover, all the enzymes so far reported were inactivated by 10 min incubation at 80 °C. Nogi and Horikoshi (1990) isolated *Bacillus* sp. No. AG-430 as a potent enzyme producer having higher enzyme activity at pH 9.0. The isolate grew in a temperature range of 20 to 50 °C; the optimum temperature for enzyme production was 33 °C. The bacterium was aerobic, motile, gram-positive and rod-shaped (0.4–0.5 μm × 2.0–3.0 μm) with peritrichous flagella. A terminal swollen sporangium containing oval spores (0.8–1.0 μm × 1.5–2.0 μm) was observed. The pH range for growth was from pH 6.5 to 11.0 with an optimum at pH 8.5–9.5. The bacterium was able to grow in the presence of 5% NaCl, but not 7.5% NaCl. Catalase and oxidase reactions were positive. Negative reactions were observed for reduction of nitrate to nitrite, hydrolysis of casein, gelatin or starch, citrate assimilation, formation of indole, or H$_2$S, and the Voges-Proskauer test. The GC content was estimated to be 52 mol%. Its taxonomical position has not yet been determined.

Bacillus sp. No. AG-430 was grown aerobically at 35 °C for three days in pachyman medium. The enzyme was purified by DEAE-Sepharose CL-6B column chromatography

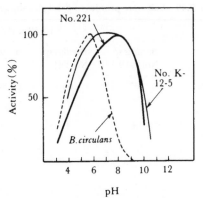

Fig. 7.57 pH activity profiles of β-1,3-glucanases.
(Reproduced with permission from Horikoshi and Akiba, *Alkalophilic Microorganisms*, p. 116, Springer-Verlag: Japan Scientific Societies Press (1982))

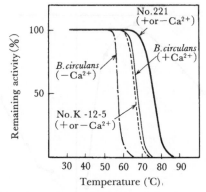

Fig. 7.58 Thermal stability of β-1,3-glucanases.
(Reproduced with permission from Horikoshi and Akiba, *Alkalophilic Microorganisms*, p. 116, Springer-Verlag: Japan Scientific Societies Press (1982))

followed by passage through a Sephadex G-75 column. Final purification was carried out by a hydroxyl apatite column. Molecular weight estimated by SDS-PAGE method was approximately 35,000 and isoelectric point was pH 3.8. The enzyme was most active at pH 9-10 (Fig. 7.59) and most stable in the range of pH 4 to 12 at 40 °C for 1 h incubation. The crude enzyme exhibited a broad pH activity curve, and the pH optimum was pH 8-9. Cloning experiments revealed the presence of a neutral β-1,3-glucanase in addition to an alkaline β-1,3-glucanase (unpublished data). The enzyme was most active at 65 °C in the presence of laminaran as the substrate. It is, however, very interesting that in the absence of the substrate the enzyme was very stable in 0.01 M phosphate buffer (pH 7.0), as shown in Fig. 7.60. About 50% of the original activity still remained after 50 min incubation in a boiling water bath. Since its thermal stability decreased in the presence of some ions such as Ca^{2+}, Mg^{2+}, etc., we suspect that the enzyme exhibits reversible thermal denatu-

Table 7.49   Substrate profiles of β-1,3-glucanases

Substrate and hydrolysis products	β-1,3-Glucanase			
	Bacillus sp. No. K-12-5	Bacillus sp. No. 221	B. circulans	Basidiomyscete sp. QM 806
Laminaran → G + L₂ + L₃, %	65	65	80	+
Laminaripentaose → G + L₂ + L₃	+	+	+	+
Laminaritetraose → G + L₃	+	+	+	+
Laminaritriose (L₃) → G + L₂	−	−	+	+
Laminaribiose (L₂) → 2G	−	−	−	+
cell walls from				
Mucor sp., %	0	0	1.2	
Fusarium sp., %	0.5	0	8.1	9.5
Aspergillus oryzae, %	2.5	0	7.2	8.2

(Reproduced with permission from Horikoshi and Akiba, *Alkalophilic Microorganisms*, p.117, Springer-Verlag: Japan Scientific Societies Press (1982))

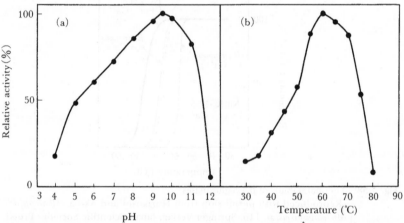

Fig. 7.59   pH activity curve and optimum temperature of the thermostable β-1,3-glu-canase.

ration. These results indicate that the β-1,3-glucanase of alkaliphilic *Bacillus* sp. No. AG-430 is the most thermostable β-1,3-glucanase so far reported. The end products in the enzymatic digest were glucose, laminaribiose and laminaritriose, and this enzyme is an endo-β-1,3-glucanase.

Louw, Reid and Watson (1993) cloned a *Bacillus brevis* gene coding for an endo-(1,3-1,4)-β-glucanase in *E. coli* and sequenced it. The open reading frame contains a sequence of 759 nucleotides encoding a polypeptide of 252 amino acid residues. The amino acid sequence of the β-glucanase gene showed only 50% similarity to previously published data from *Bacillus* endo-(1,3-1,4)-β-glucanases. The optimum temperature and pH for enzyme activity were 65-70 °C and 8-10, respectively. When held at 75 °C for 1 h, 75% residual activity was measured. The molecular mass was estimated to be about 29 kDa on SDS-

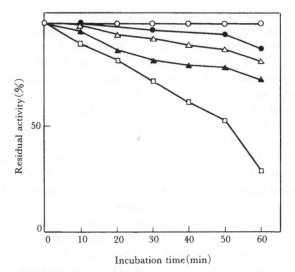

Fig. 7.60 Thermal stability of the enzyme.
The enzyme, dissolved in 0.01 M sodium phosphate buffer, pH 7.0, at a protein concentration of 0.1 mg/ml, was incubated at 60 °C (○), 70 °C (●), 80 °C
(△), 90 °C (▲), 100 °C (□).

PAGE and the enzyme was found to be resistant to SDS.

Tabernero *et al.* (1994) also reported cloning and DNA sequencing of *bgaA*, a gene
encoding an endo-(1,3-1,4)-β-glucanase. The gene *bgaA* encoding an alkaline endo-β-1,3-
1,4-glucanase (lichenase) from an alkaliphilic *Bacillus* sp. strain N-137 was cloned and
expressed from its own promoter in *Escherichia coli.* The nucleotide sequence of a 1,416-
bp DNA fragment containing *bgaA* was determined and revealed an open reading frame
of 828 nucleotides. The deduced protein sequence consists of 276 amino acids and has a
31-amino-acid putative signal peptide which is functional in *E. coli*, in which the *bgaA* protein is located mainly in the periplasmic space. The lichenase activity of *bgaA* is stable
between pH 6 and 12, it shows optimal activity at a temperature between 60 and 70 °C, and
it retains 65% of its activity after incubation at 70 °C for 1 h. This protein is similar to some
other lichenases from *Bacillus* species such as *B. amyloliquefaciens, B. brevis,
B. licheniformis, B. macerans, B. polymyxa,* and *B. subtilis.* However, it has a lysine-rich region
at the carboxy terminus which is not found in any other published lichenase sequence and
may be implicated in the unusual biochemical properties of this enzyme. The location of
the mRNA 5′ end was determined by primer extension and corresponds to nucleotide 235.
A typical *Bacillus* σA promoter precedes the transcription start site.

7.8   α-Glucosidases

In the starch industry, glucoamylase and α-glucosidase from bacteria and molds have
been used extensively for the production of glucose, and many investigators continue to
seek new starch-degrading enzymes which have different properties.

7.8.1 Isolation and Purification of α-Glucosidases

Kelly, O'Reilly and Fogarty (1983) reported that alkaliphilic *Bacillus* sp. No. A-59 (ATCC 21591) isolated by our research group produced three enzymes associated with degradation of starch: α-amylase reported by Yamamoto, Tanaka and Horikoshi (1972), pullulanase and α-glucosidase. The organism was grown in a medium containing 10 g soluble starch, 6 g peptone, 3 g yeast extract, 1 g K_2HPO_4, 0.04 g $MnSO_4 \cdot 4H_2O$, 0.20 g $MgCl_2 \cdot 7H_2O$, 3 g $Na_2CO_3 \cdot H_2O$ and 1000 ml water (final pH 9.7). Analysis of the growth curve of the organism *Bacillus* sp. No. A-59 relative to α-glucosidase production revealed that biomass rose rapidly and reached a peak after 18 h. The pH fell sharply from 9.7 to 6.6 after 15 h, and after 18 h rapid cell lysis took place. The level of α-glucosidase reached a maximum after 24 h. Optimum enzyme production was reached at initial pH 9.7 for α-glucosidase and pullulanase after 24 h growth (Fig. 7.61). However, the initial pH of the culture, 10.2, dropped to 9.0 after 48 h cultivation, but no trace of α-glucosidase activity could be detected. However, α-amylase and pullulanase activities were detected primarily in media having a pH between 9.7 and 10.4. It is of interest that the alkaliphilic *Bacillus* sp. No. A-59 produces a glucosidase with a pH optimum at 7.0 and that the enzyme had only 15% of optimum activity when assayed at pH 9.5 and was inactive at pH 10.0. The enzyme was substrate-specific for *p*-nitrophenyl-α-D-glucoside, maltose and maltotriose in that order. Almost all of the activity was located in the cell-free supernatant. Transferase activity was detected using maltose (4%, w/v) as substrate at pH 7.0 at 40 °C for 1 h. Analysis by thin-layer chromatography revealed a number of products the major one of which had an R_f value similar to that of isomaltose.

Another alkaliphilic *Bacillus* sp. NCIB 11203 (Kelly, Brennon and Fogarty, 1987), which was isolated from soil, produced several enzymes in the culture broth, including alkaline amylase, alkaline protease and alkaline phosphatase. In the culture broth two extracellular α-glucosidases were detected. The organism was grown in 250-ml Erlenmeyer flasks containing 50 ml of Horikoshi-II medium at 30 °C for 18 h. Cells were removed from the culture medium by centrifugation and the cell-free supernatant was used for purification of the extracellular enzymes. The activity of the crude extracellular α-glucosidase system was examined during cultivation. Essentially, in its extracellular nature and pattern of production the system is similar to that of Kelly's report on alkaliphilic *Bacillus* sp. No. A-59 (Kelly, O'Reilly and Fogarty, 1983). The enzyme was purified by fractionation with ammonium sulfate and chromatography on DEAE Biogel A. Two types of α-glucosidase were isolated (Table 7.50). The first of these hydrolyzed *p*-NPG preferentially and had minor activity on isomaltose and isomaltotriose; the second enzyme strongly hydrolyzed maltose and maltotriose and had some activity on *p*-NPG. Both enzymes had pH optimum at 7.0, which is distinct from enzymes of most other alkaliphilic bacilli but similar to that of *Bacillus* sp. No. A-59.

Shirokane, Arai and Uchida (1994) isolated a new enzyme, maltobionated α-D-glucohydrolase, and purified it to apparent homogeneity from a cell-free extract of alkaliphilic *Bacillus* sp. N-1053. The enzyme showed optimum activity at about pH 7.0 and was stable over the range of pH 6.0–9.5. The molecular weight was estimated to be 152,000 and 71,000 by HPLC gel filtration and SDS-polyacrylamide gel electrophoresis, respectively. The enzyme hydrolyzed maltobionate more effectively than disaccharides such as maltose and maltitol or trisaccharides such as maltotrionate, maltotriose and maltotriitol, but

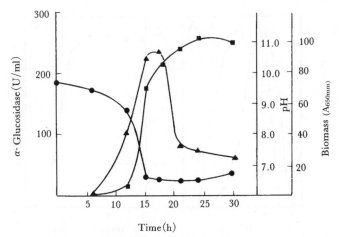

Time (h)

Fig. 7.61 Production of α-glucosidase by *Bacillus* sp. No. A-59. (■) α-glucosidase; (▲) biomass; (●) pH.
(Reproduced with permission from Kelly *et al.*, *FEMS Microbiol. Lett.*, **20**, 57 (1983))

Table 7.50 Substrate specificities of purified maltase and α-glucosidase of *Bacillus* sp. No. 11203

Substrate (0.25%)	Maltase fraction	α-Glucosidase fraction
Maltose	100	0
Maltotriose	64	0
Isomaltose	0	0.84
Isomaltotriose	0	2.40
p-NPG	17	100

No activity was detected with either system on sucrose, raffinose, melezitose, α-methyl-D-glucopyranoside or starch.

showed no activity toward polysaccharides such as amylose, amylopectin and soluble starch. The enzyme activity was almost completely inhibited by Hg^{2+}, Ag^+, iodine and *N*-bromosuccinimide, and also inhibited by *p*-nitrophenyl-α-D-glucoside, maltose and maltitol.

7.8.2 Isolation of *Bacillus* sp. No. F5 Enzyme and Cloning of the Enzyme Gene

From an industrial point of view, an alkaliphilic isolate, *Bacillus* sp. No. F5, which accumulated a significant amount of oligo-1,6-α-glucosidase (dextrin-6-α-D-glucanohydrolase; EC 3.2.1.10) in the cells was studied (Yamamoto and Horikoshi, 1987*a*, *b*, 1990*b*). The characteristic property of the enzyme was in the strong activity towards isomaltose, panose and a series of isomaltooligosaccharides. Little is known about an α-glucosidase like this capable of hydrolyzing isomaltose as a primary substrate. The culture medium of strain No. F5 contained 1% soluble starch (or maltose), 5% defatted soybean powder, 0.1% K_2HPO_4, 0.02% $MgSO_4 \cdot 7H_2O$, and 1% $NaHCO_3$ (initial pH 9.0). The accumulation of

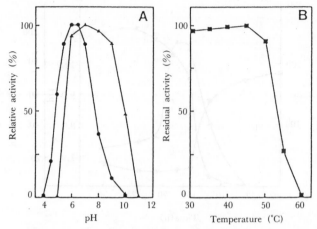

Fig. 7.62 Effect of pH (A) and temperature (B) on activity and stability. Symbols: optimum pH (●), pH stability (▲) and heat stability (■). The enzyme activities at various pH values were determined by the standard assay method. Residual activities were measured (A) after incubation at various pH values for 10 min at 50 °C, and (B) at various temperatures for 10 min at pH 7.0. Buffers used: 50 mM citrate-sodium citrate (pH 4.0-6.0), 50 mM Na_2HPO_4-NaH_2PO_4 (pH 6.0-8.0), 50 mM glycine-NaCl-NaOH (pH 8.0-10.0), and 50mM Na_2HPO_4-NaOH (pH 11.0-12.0).
(Reproduced with permission from Yamamoto and Horikoshi, *Denpun Kagaku*, **37**, 143 (1990))

enzyme was induced by maltose or soluble starch added to the culture medium. As shown in Fig. 7.62, the enzyme was most active at pH 6.0-6.5 and temperature of 45 °C. The molecular weight of the purified glucosidase was 60,000 by SDS-PAGE. Panose and isomaltotriose are the best substrates for this enzyme. The restricted substrate specificity indicated the assignment of the enzyme to be an oligo-1,6-glucosidase, but there were indications that it could be a new type of oligo-α-1,6-glucosidase. The enzymatic saccharification of starch is always accompanied by the formation of oligosaccharides such as isomaltose, panose, isopanose, etc. If these by-products are converted to glucose, the yield of glucose from starch should be improved. The apparent V_{max} and K_m values for isomaltose was 34 mmol/min/mg of protein, and 3.33 mM (Yamamoto and Horikoshi, 1990b). The nucleotide sequence of the alkaliphilic *Bacillus* oligo-1,6-glucosidase gene was determined by a conventional method (Yamamoto and Horikoshi, 1990a). The coding sequence on a 3.1 kb DNA fragment if pMC21 showed an open reading frame of 1,527 bp encoding 509 amino acids (Fig. 7.63). The amino acid sequence from position 2 to 11 corresponds exactly to the N-terminal sequence of the purified enzyme protein from strain F5. The molecular weight (60,142) of the protein derived from the 1,522 bp open reading frame coincided with the molecular weight (60,000) of the parent strain. However, the molecular weight of the purified oligo-1,6-α-glucosidase produced by *E. coli* HB101 carrying plasmid pMC21 showed 65,000 by SDS-PAGE. The enzymatic properties, pH and temperature optima, pH and thermal stability, and substrate specificity were almost the same as those obtained for alkaliphilic *Bacillus* strain F5.

Debranching enzymes such as pullulanase and isoamylase have been applied in the

```
        -50          -40          -30          -20          -10
CAATGTGCAGGTAGAGCAAAACAGCGTCGTTTTCAGTCTTTTAAAAGGAGAACCGCTTTC

TATGACAGTTTATGGAGAACAGATCGCACTGAACGGACGGTATGAAAGGAGGATGGCCCG
                                                      S D
                                                SacII   50
ATGAGCCAATGGTGGAAAGAGGCAGTCGTGTATCAAATTTACCCGCGGAGTTTTTATGAT
MetSerGlnTrpTrpLysGluAlaValValTyrGlnIleTyrProArgSerPheTyrAsp
    N-terminal

                                              100
TCCAACGGAGACGGCTTCGGTGATTTGCAGGGGGTGATTCAAAAGCTCGATTACATCAAA
SerAsnGlyAspGlyPheGlyAspLeuGlnGlyValIleGlnLysLeuAspTyrIleLys

                      150
AGGCTTGGAGCCGATGTCATATGGCTGTGTCCTGTGTTTGATTCCCCGCAGGATGACAAC
ArgLeuGlyAlaAspValIleTrpLeuCysProValPheAspSerProGlnAspAspAsn

                  200
GGCTACGACATCAGCGACTATAGAAGCATATACGAAAAGTTCGGGACGAACGATGATATG
GlyTyrAspIleSerAspTyrArgSerIleTyrGluLysPheGlyThrAsnAspAspMet

              250                                      300
TTTCAATTGATTGATGAAGTGCATAAACGCGGAATGAAAATCATTATGGATCTTGTCGTG
PheGlnLeuIleAspGluValHisLysArgGlyMetLysIleIleMetAspLeuValVal

                          350
AATCACTCGTCTGATGAACATGCCTGGTTTGCCGAAAGCCGCAAGTCAAAGGACAATCCT
AsnHisSerSerAspGluHisAlaTrpPheAlaGluSerArgLysSerLysAspAsnPro

                   BamHI       400
TACCGTGATTATTATTTTTGGAAGGATCCAAAGGCGGATGGTTCAGAGCCGAACAATTGG
TyrArgAspTyrTyrPheTrpLysAspProLysAlaAspGlySerGluProAsnAsnTrp

                   450
GGAGCGATCTTTTCAGGGCCGGCATGGAGCGCGATGAGCACAGCACAGTATTATCTTCAC
GlyAlaIlePheSerGlyProAlaTrpSerAlaMetSerThrAlaGlnTyrTyrLeuHis

          500
TACTTTTCGAAAAAGCAGCCTGACTTAAACTGGGAGAATGAAGCGGTTCGCCGCGAAGTA
TyrPheSerLysLysGlnProAspLeuAsnTrpGluAsnGluAlaValArgArgGluVal

      550                                          600
TACGATTTGATGACATTTTGGATGGACAGAGGCGTGGACGGCTGGCGGATGGACGTGATC
TyrAspLeuMetThrPheTrpMetAspArgGlyValAspGlyTrpArgMetAspValIle

              SalI                            650
GGTTCCATCTCCAAATTCGTCGACTTTCCCGATTATGAAACAGATGACAGCCGTCCATAT
GlySerIleSerLysPheValAspPheProAspTyrGluThrAspAspSerArgProTyr

                      700
GTCGTCGGCCGCTATCATTCGAACGGCCCCCGTCTGCACGAATTTATACAAGAGATGAAC
ValValGlyArgTyrHisSerAsnGlyProArgLeuHisGluPheIleGlnGluMetAsn

                  750
CGGGAAGTGCTGTCCCGTTACGACTGTATGACGGTCGGCGAAGCAGGCGGCTCAGATGTT
ArgGluValLeuSerArgTyrAspCysMetThrValGlyGluAlaGlyGlySerAspVal

         800BamHI
GAAGAAGCGAAAAAATACACGGATCCAAGCAGACATGAGCTGAACATGATTTTTACATTT
GluGluAlaLysLysTyrThrAspProSerArgHisGluLeuAsnMetIlePheThrPhe

    850                                      900
GAACATATGGATATTGATACCAAACAGCATTCTCCGAATGGGAAATGGCAGATGAAGCCG
GluHisMetAspIleAspThrLysGlnHisSerProAsnGlyLysTrpGlnMetLysPro
```

Fig. 7.63   (Continued)

```
                                                          950
TTTGACCCGATCGCTTTAAAAAAGACGATGACGAGGTGGCAGACCGCTTTAATGAATGTC
PheAspProIleAlaLeuLysLysThrMetThrArgTrpGlnThrAlaLeuMetAsnVal

                                      1000
GGCTGGAATACGCTTTATTTCGAAAATCATGATCAGCCGAGGGTCATATCCGCTGGGGCA
GlyTrpAsnThrLeuTyrPheGluAsnIleSerAspGlnProArgValIleSerAlaGlyAlaAla

            Sac I               1050
ATGACCCGCGAGCTCCGCAAACAATCGCGCCAAAGCATTTCCAACAGTTCTGCACGGCAT
MetThrArgGluLeuArgLysGlnSerArgGlnSerIleSerAsnSerSerAlaArgIlis

               1100
GAAGGGAACCCGTTCATTTATCAAGGTGAAGAAATCGGCATGACAAACAGCGAGATGCCG
GluGlyAsnProPheIleTyrGlnGlyGluGluIleGlyMetThrAsnSerGluMetPro

            1150                                          1200
CTCGAAATGTATGATGATCTCGAAATCAAAAACGCATACCGCGAACTTGTCATTGAAAAC
LeuGluMetTyrAspAspLeuGluIleIleLysAsnAlaTyrArgGluLeuValIleGluAsn

                                   1250
AAAACGATGACAGAGGAAGACTTTCGAAAAGCCGTGGCTAAAAAAGGGCGGGATCATGCG
LysThrMetThrGluGluAspPheArgLysAlaValAlaLysLysGlyArgAspHisAla

                     1300
AGAACGCCGATGCAATGGGATGACGGAAAATACGCGGGCTTTACGGATGGAGAGGCTTGG
ArgThrProMetGlnTrpAspAspGlyLysTyrAlaGlyPheThrAspGlyGluAlaTrp

               1350
CTGGCGGTAAATCCCCGCTATCAAGAAATCAATGTGAAAGAGTCGCTTGCCGATGAAGAT
LeuAlaValAsnProArgTyrGlnGluIleAsnValLysGluSerLeuAlaAspGluAsp

      1400
TCGATTTTCTATTATTATCAGAAGCTCATCGGGCTGCGCAAACAAAATAAAGTCATCGTG
SerIlePheTyrTyrTyrGlnLysLeuIleGlyLeuArgLysGlnAsnLysValIleVal

      1450              Xho I                             1500
TACGGAGACTATCGGCTGCTGCTCGAGGAAGATCCGCGAATCTTTGCGTATATCCGCGAA
TyrGlyAspTyrArgLeuLeuLeuGluGluAspProArgIlePheAlaTyrIleArgGlu

               1530
TATCGGGGCGAAAAGCTGTTAGTGCCGTGAATTTATCAGAAGAAAAGGCGCTTTTTCATC
TyrArgGlyGluLysLeuLeuValPro***              Inverted repeat
                (509)


TTCGCCGGAATTGCTTGAGGACAGATGGGATGTGCTGTTGTCTAACTACGCCCGAGAGCG
```

Fig. 7.63 Nucleotide sequence of the oligo-1,6-glucosidase gene from alkaliphilic *Bacillus* strain F5.

The DNA sequence of the coding strand is given from 5′ to 3′, numbered from nucleotide 1 at the putative initiation site. The proposed ribosomal binding site (SD) is underlined. The predicted amino acid sequence is given below the DNA sequence. The amino acids are numbered taking the NH$_2$-terminal amino acid of the mature protein as 1. Enclosed amino acids have been determined by automated Edman sequencing of the purified oligo-1, 6-glucosidase. The hairpin loop of the putative rho-independent terminator site is underlined.

(Reproduced with permission from Yamamoto and Horikoshi, *Denpun Kagaku*, **37**, 140 (1990))

starch industry but small amounts of by-products having α-1,6-linkage remain. The oligo-1,6-α-glucosidase of strain No. F5 is a good candidate for this purpose because the enzyme can hydrolyze the by-products.

7.9 Lipases

Lipases [EC3.1.1.3] occur in microorganisms, higher plants and animals. Much attention has been devoted in recent years to lipases synthesized by microorganisms because of their versatility and practical applications in industry and medicine. Commercially, certain lipases have been utilized as detergent additives. Some lipases exhibit fatty acid transfer activity. Oil processing using this property has been investigated especially in the food industry.

The pH optimum of microbial lipases is generally in the range of 5 to 8; a few exceptions are lipases from *Penicillium crustosum* and *M. lipolyticus*, which have pH optimum around 9.0 (Oi, Sawada and Satomura, 1967; Nagaoka and Yamada, 1969; 1973). Savitha and Ratledge reported an inducible, intracellular alkaliphilic lipase in *Aspergillus flavipes* that is not alkaliphile. The lipase produced had an optimum pH for activity of 8.8 and retain 30% of its activity at pH 10.0 (Savitha and Ratledge, 1992).

7.9.1 Alkaline Lipases

Attempts to isolate from soil samples alkaliphilic bacteria that are capable of producing alkaline lipases have been made. Strain No.865, which was identified as an *Achromobacter* species, produced a large amount of extracellular lipase in an alkaline medium (pH 10). The optimum pH of the purified lipase was 10.0, and the enzyme retains 50% of the maximum activity between pH 6.0 and 11.8. The general properties of lipases are summarized in Table 7.51. The lipase was significantly inhibited by incubation with either alkylbenzene sulfate or dodecyl benzene sulfonate (0.1%) for 30 min at 35 °C. Although the initial motivation for studying alkaline lipase was its application to detergents, this alkaline lipase was not good for this purpose.

Watanabe *et al.* (1977) conducted an extensive screening for alkaline lipase-produc-

Table 7.51 General properties of alkaline lipases from microorganisms

Property	*Achromobacter* sp. No. 865	*Pseudomonas nitoreducens* nov. var. *thermotolerans* 26.1B[†1]	*Ps. fragi* 22.39B[†1]	*Penicillium crustosum*[†2]	*Mucor lypolyticus*[†3] 7-3A	*Mucor lypolyticus*[†3] 7-3B
Optimum pH	10.0	9.5	9.5	9.0	9.0	8.0
Optimum temperature, °C	45	50	75–80		37	30–37
Stable pH	6.0–11.5	5.0–11.0	5.0–11.0	6.0–9.0	3.3–10.0	
Stable temperature, °C	Up to 50	Up to 70	Up to 70	Up to 45	Up to 30	
Molecular weight	>300,000			I : 29,000 II : 32,000 III : 11,000	25,400	29,000
Isoelectric point	4.2				9.7	10.2

[†1] Watanabe *et al.* (1977). [†2] Oi, Sawada and Satomura (1967). [†3] Nagaoka and Yamada (1969, 1973).
(Reproduced with permission from Horikoshi and Akiba, *Alkalophilic Microorganisms*, p.130, Springer-Verlag: Japan Scientific Societies Press (1982)).

ing microorganisms from soil and water samples. Of 1,606 strains isolated, two bacterial strains, 26.1B and 22.39B, were selected as potent producers of alkaline lipase. These were identified as *Pseudomonas nitroreducens* nov. var. *thermotolerans* and *Pseudomonas fragi*, respectively. The optimum pH of the two lipases was 9.5. Both enzymes were inhibited by bile salts such as sodium cholate, sodium deoxycholate, and sodium taurocholate at a concentration of 0.25%. On the other hand, No. 865 lipase was not inhibited by these bile salts. However, further work has not been reported.

Although the enzyme isolated was not pure lipase, Wang and Saha (1993) reported purificaiton and characterization of thermophilic and alkaliphilic tributyrin esterase from *Bacillus* strain A30-1 (ATCC 53841). An extracellular esterase [EC 3.1.1.1] from a thermophilic *Bacillus* A30-1 (ATCC 53841) was purified 139-fold to homogeneity by sodium chloride (6 M) treatment, ammonium sulfate fractionation (30-80%) and phenyl-Sepharose CL-6B column chromatography. The native enzyme was a single polypeptide chain with a molecular weight of about 65,000 and an isoelectric point at pH 4.8. The optimum pH for esterase activity was 9.0, and its pH stability range was 5.0-10.5. The optimum temperature for its activity was 60 °C. The esterase had a half-life of 28 h at 50 °C, 20 h at 60 °C and 16 h at 65 °C. It showed the highest activity on tributyrin, with little or no activity toward long-chain (12-20 carbon) fatty acid esters. The enzyme displayed $K(m)$ and $K(cat)$ values of 0.357 mM and 8365/min, respectively, for tributyrin hydrolysis at pH 9.0 and 60 °C. Cyclodextrin (α, β, and γ), Ca^{2+}, Co^{2+}, Mg^{2+} and Mn^{2+} enhanced the esterase activity, and Zn^{2+} and Fe^{2+} acted as inhibitors of the enzyme activity. The enzyme activity was not affected by ethylenediaminetetraacetic acid, *p*-chloromercuribenzoate and *N*-bromosuccinimide.

Bhushan (Bhushan and Hoondal, 1994; Bhushan *et al.*, 1994) found a lipase produced from an alkaliphilic yeast by solid state fermentation. The lipase-producing alkaliphilic yeast species was isolated, identified, classified and termed *Candida* BG-55. The lipase from this microorganism had temperature and pH optimum of 40 °C and 8.5, respectively, and was stable at 45 °C for 4 h. Enzyme activity was stimulated by Ni^{2+} and Ca^{2+} ions whereas Fe^{2+} and Fe^{3+} ions inhibited the activity.

Lin *et al.* (1995, 1996) reported alkaline lipase of *Pseudomonas pseudoalcaligenes* F-111. An extracellular alkaline lipase of alkaliphilic *Pseudomonas pseudoalcaligenes* F-111 which was isolated from soil using a rhodamine B agar plate wiht Na_2CO_3 was purified to homogeneity. The apparent molecular weight determined by sodium dodecyl sulfate-polyacrylamide gel electrophoresis was 32,000, and the isoelectric point was 7.3. With *p*-nitrophenyl esters as its substrates, the enzyme shows preference for C-12 acyl and C-14 acyl groups. The enzyme was stable in the pH range of 6 to 10, which coincides with the optimum pH range.

7.9.2 Industrial Application

A Japanese firm has produced a laundry detergent containing a fungus lipase produced by recombinant DNA technology. The fungus used is not alkaliphilic, but alkaliphiles should be good for use in this field.

7.10 Pectinases

Pectinolytic enzymes (pectinases) which degrade pectic polysaccharides such as

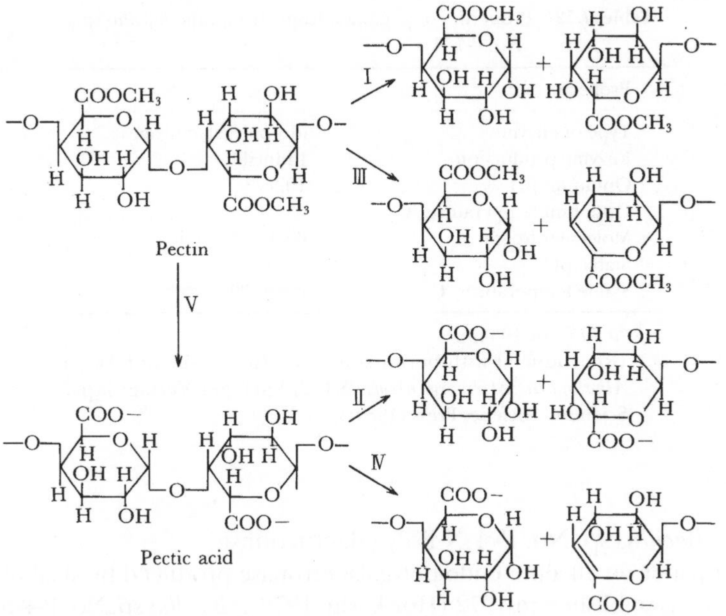

Fig. 7.64   Modes of action of pectinases.
Hydrolases: I, polymethylgalacturonase; II, polygalacturonase.
Lyases: III, polymethylgalacturonate lyase, IV, polygalacturonate lyase.
Esterase: V, pectin esterase.
(Reproduced with permission from Horikoshi and Akiba, *Alkalophilic Microorganisms*, p. 111, Springer-Verlag: Japan Scientific Societies Press (1982))

pectin and pectic acid, are distributed in microorganisms and higher plants but are not found in higher animals (Deuel and Stutz, 1958; Forgarty and Ward, 1977). Pectinases from certain microorganisms are widely used in the fruit- and vegetable-processing industries. The pectinases are classified into three groups according to their mode of action towards the substrate: (1) esterases, (2) hydrolases, and (3) lyases (Forgarty and Ward, 1977). Pectin esterase [EC 3.1.1.11] removes methoxyl residues from pectin to form pectic acid. The hydrolases, polygalacturonase and polymethylgalacturonase catalyze random or sequential cleavage of the α-1,4-glycosidic linkages of pectin and pectic acid. The lyases polygalacturonate lyase and polymethylgalacturonate lyase catalyze the transeliminative cleavage of the α-1,4-glycosidic linkages of pectin and pectic acid in a random or sequential fashion. Fig. 7.64 illustrates the models of action of pectinases. Pectinases which cause random cleavage of the substrates are known as the endo type, and those which cause sequential cleavage of the substrates are designated as the exo type.

Among the pectinases, the most widely distributed enzymes in microorganisms are endo-polygalacturonase [EC 3.2.1.15] and endo-polygalacturonate lyases [EC 4.2.2.1]. The pH optimum of endo-polygalacturonases generally ranges from 4.0 to 5.0 and that of endo-polygalacturonate lyases ranges from 7.0 to 10.0. Endopolygalacturonate lyases are generally activated by Ca^{2+}. These differences in properties are used as the basis for differentiating pectic hydrolases and lyases.

Table 7.52 Properties of pectinase from alkaliphilic *Bacillus* sp. No. P-4-N

Property	
Type of enzyme	Endo-polygalacturonase
Enzyme production	Inducible
Optimum pH	10.0
Optimum temperature, °C	65
Molecular weight	60,000–70,000
Stable pH	6.5[†]
Stable temperature, °C	Up to 70[†]

[†] At 70 °C for 10 min.
(Reproduced with permission from Horikoshi and Akiba, *Alkalophilic Microorganisms*, p.112, Springer-Verlag: Japan Scientific Societies Press (1982))

7.10.1 *Bacillus* sp. No. P-4-N Polygalacturonase

The first paper on alkaline endo-polygalacturonase produced by alkaliphilic *Bacillus* sp. No. P-4-N was published in 1972 (Horikoshi, 1972). *Bacillus* sp. No. P-4-N, which was selected from about 100 bacterial colonies, could grow in the pH range from 7.0 to 11.0, but the most active growth was observed in a pH 10 medium containing 1% Na_2CO_3. The other important factor for enzyme production was the addition of Mn^{2+} at a concentration of 0.005%, which caused a four-fold increase in the yield. Employing a medium of pH 10.4 containing 1% Na_2CO_3, 0.005% $MnSO_4 \cdot 7H_2O$, and 3% pectin as essentials, *Bacillus* sp. No. P-4-N inducibly produced the maximum amount of endo-polygalacturonase. The endo-polygalacturonase was purified about 300-fold from the culture fluid by column chromatography on DEAE cellulose and Sephadex G-100 and G-200. The optimum pH of the enzyme was 10.0 towards pectic acid. The enzyme was activated ten-fold by 1 mM $CaCl_2$, inhibited by 10 mM EDTA, but not inhibited by 8 M urea or 0.1 mM *p*-chloromercuribenzoate. In general, hydrolytic pectinases do not require calcium. The enzyme was most stable at pH 6.5 and up to 70 °C in the presence of Ca^{2+}, but was inactivated completely at 80–90 °C.

The properties of the No. P-4-N enzyme are summarized in Table 7.52. The predominant products from pectic acid were di-, tri- and tetramers in the first stage of the reaction; these were finally hydrolyzed to mono- and dimers of galacturonic acid by prolonged reaction. The maximum degree of hydrolysis was about 35%. The enzyme was also able to hydrolyze pectins other than pectic acid, but the hydrolysis rate of pectin was 50% slower than that of pectic acid. No esterase activity was observed.

7.10.2 *Bacillus* sp. No. RK9 Polygalacturonate Lyase

Kelly and Fogarty (1978) reported that *Bacillus* sp. No. RK9 isolated from garden soil grew well in alkaline media at pH 9.7 and produced endo-polygalacturonate lyase. The medium composition suitable for enzyme production was as follows (in grams per liter): sucrose, 10.0; bacteriological peptone, 3.0; yeast extract, 3.0; $Na_2CO_3 \cdot H_2O$, 10.0; $CaCl_2 \cdot 2H_2O$, 3.0; $MnSO_4 \cdot 4H_2O$, 0.04; $MgCl_2 \cdot 6H_2O$, 0.2; K_2HPO_4, 1.0; pH 9.7. Neither pectin

Table 7.53 Properties of pectinases from alkaliphilic *Bacillus* sp. No. RK9[†1]

Property	
Type of enzyme	Endo-polygalacturonate lyase
Enzyme production	Constitutive
Optimum pH	10.0
Optimum temperature, °C	60
Stable pH	9.5[†2]

[†1] Kelly and Fogarty (1978).
[†2] At 65 °C for 60 min.
(Reproduced with permission from Horikoshi and Akiba, *Alkalophilic Microorganisms*, p.113, Springer-Verlag: Japan Scientific Societies Press (1982)).

Substrate specificity of polygalacturonate lyase from alkaliphilic *Bacillus* sp. No. RK9[†]

Substrate (0.2%)	Activity (units/ml)
Acid-soluble pectic acid	1.03
Sodium polypectate	0.52
Pectin	0.129
Pectinic acid amide	0.172

[†] Kelly and Fogarty (1978).
(Reproduced with permission from Horikoshi and Akiba, *Alkalophilic Microorganisms*, p.113, Springer Verlag: Japan Scientific Societies Press (1982)).

nor pectic acid was required for the production of the enzyme, indicating that the enzyme is constitutive.

The endo-polygalacturonate lyase of RK9 was purified 163-fold from culture fluid by precipitation with 50–90% saturated $(NH_4)_2SO_4$ and DEAE cellulose column chromatography. The optimum pH of the enzyme was 10.0 towards acid-soluble pectic acid (Table 7.53). The enzyme activity was significantly affected by the buffer system used for the reaction. The activity towards acid-soluble pectic acid was three times higher in glycine-NaOH buffer than in Britton and Robinson's universal buffer, and likewise the activity towards sodium polypectate was 10 times higher. The optimum temperature of the enzyme was 60 °C. The enzyme retained 100% of its activity after incubation for 1 h at 37 °C at pH 11.0. It was activated 2.9-fold by 0.4 mM $CaCl_2$ but inhibited by other divalent cations (Mg^{2+}, Sr^{2+}, Co^{2+}, etc.). The enzyme was completely inactivated by 1 mM EDTA, but the activity was recovered by dialysis, suggesting that EDTA had no direct effect on the enzyme itself.

7.10.3   Industrial Applications

A.   Production of Japanese Paper
The first application of alkaline pectinase-producing bacteria in retting of Mitsumata

Table 7.54 Chemical composition of bast and pulps of mitsumata[†]

Component	Bast (%)	Pulp (%)	
		New process	Conventional
Ash	3.8	1.4	2.5
Ether soluble	2.8	1.2	0.6
Alcohol-benzene-soluble	2.4	1.5	0.5
Klason lignin	3.5	1.8	9.0
Pentosan	18.6	18.2	16.7
Pectin	11.1	0.6	1.0

[†] Yoshihara and Kobayashi (1982).
(Reproduced with permission from Horikoshi and Akiba, *Alkalophilic Microorganisms*, p.172, Springer-Verlag: Japan Scientific Societies Press (1982)).

bast was reported by Yoshihara and Kobayashi (1982). Alkaliphilic bacteria were isolated form soil, sewage and decomposed manure in Japan and Thailand. *Bacillus* sp. No. GIR 277 had strong macerating activity toward Mitsumata bast. The bacteria isolated were motile, aerobic, spore-forming rods and grew well on nutrient agar of pH 9.5 adjusted with Na_2CO_3. This bacterium produced pectate lyase which had an optimum pH for enzyme action at 9.5. Although the enzyme was not purified, several enzymatic properties were different from those of petinolytic enzymes of *Bacillus* sp. No. P-4-N and *Bacillus* sp. No. RK9. The former, which was a polygalacturonase, had an optimum pH of 10.0 to 10.5, and the latter was a constitutive endo-acting pectate lyase. However, the enzyme of *Bacillus* sp. No. GIR 277 was an inducible pectate lyase. Japanese paper was produced as follows. Four grams of Mitsumata bast was suspended in 100 ml of a culture medium containing 0.05% yeast extract, 0.05% casamino acids, 0.2% NH_4Cl, 0.1% K_2HPO_4, and 0.05% $MgSO_4 \cdot 7H_2O$. After sterilization, Na_2CO_3 was added to a concentration of 1.5%, and *Bacillus* sp. No. GIR 277 was inoculated into the medium. After 5 days of cultivation at 30 °C with shaking, the basts retted were harvested and Japanese paper was prepared by the method described in Japanese Industrial Standard P8209. The overall yield of pulp was about 70%. The strength of the unbeaten pulp resulting from bacterial retting was higher than that obtained by the conventional soda ash-cooking method. The paper sheets were very uniform and soft to the touch. Chemical analysis of the pulps is shown in Table 7.54. Yoshihara and Kobayashi (1982) concluded that bacterial retting under alkaline condition is a potentially useful process for the production of pulp of excellent quality from non-woody pectocellulosic fibers.

B. Treatment of Pectic Wastewater with an Alkaliphilic *Bacillus* Strain

Wastewater from the citrus processing industry contains pectinaceous materials which are only slightly decomposed by microbes during activated-sludge treatment. To solve this problem the pretreatment of the wastewater with pectin-degrading microorganisms may be required. Some pectin-degrading organisms are known phytopathogens, making the application of these organisms to waste treatment very dangerous. However, alkaliphilic microorganisms may be one of the best candidates for this purpose because they can grow under limited conditions, *i.e.* alkaline condition of pH 10. Tanabe *et al.* (1987) tried to

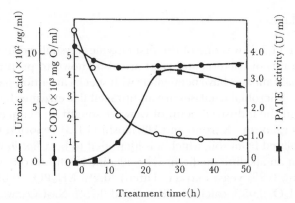

Fig. 7.65 Pretreatment of pectic wastewater by strain GIR 621.
Ten milliliters of seed culture was inoculated into an 1 l jar fermentor containing 500 ml of the wastewater supplemented with 0.25% polypeptone, 0.25% yeast extract, 0.05% soybean powder, and 0.1% K₂HPO₄ with an initial pH of 10.0. Treatment was at 27 °C with aeration at 0.8 l/min and stirring at 400 rpm. Samples were taken at the times shown and assayed for COD, uronic acid and PATE activity.

develop a new process of waste treatment using alkaliphilic microorganisms. From soil in Thailand they isolated an alkaliphilic *Bacillus* sp. No. GIR 621, which produced an extracellular endo-pectate lyase in alkaline medium of pH 10.0. Strain GIR 621 was applied to the pretreatment of wastewater containing pectinaceous substrates from an orange canning factory. The strain was cultured on the wastewater supplemented with 0.25% polypeptone, 0.25% yeast extract, 0.05% soybean powder and 0.1% K₂HPO₄ at pH 10.0 (adjusted with 1% Na₂CO₃) at 27 °C. The concentration of uronic acid in the waste water decreased by 69% at 12 h and by 91% at 36 h, as shown in Fig. 7.65. The activity of endo-pectate lyase reached maximum (3.3 units/ml) at 25 h. Furthermore, the enzyme yield was increased 20-fold (65 units/ml) by a mutant strain GIR 621-7 made by N-methyl-N-nitro-N-nitrosoguanidine (NTG) treatment (Tanabe, Kobayashi and Akamatsu, 1988). Strain GIR 621-7 grew faster in the wastewater and the uronic acid in it decreased by 93% even at 20 h. Treatment with strain GIR621-7 proved to be useful as a pretreatment to remove pectic substances. The endo-pectate lyase from GIR 621-7 was purified and characterized under the following conditions: molecular weight, 33,000; isoelectric point, 8.8; optimum pH, 9.0; optimum temperature, 55–60 °C; preferred concentration of Ca^{2+}, 0.4 mM.

C. Degumming Ramie Fibers

Cao, Zheng and Chen (1992) isolated four alkaliphilic bacteria, NT-2, NT-6, NT-33 and NT-82, producing pectinase and xylanase. The pH and temperature optima for the activity of PGase were 10.5 and 70 °C, respectively. One strain, NT-33, had excellent capacity for degumming ramie fibers.

7.11 Catalase

Catalase [EC 1.11.1.6] was one of the first enzymes crystallized from beef liver in the 1930s. Certain strains of bacteria and yeast are also known to produce catalase. Some studies on microbial catalases have been made with special attention paid to enzyme localization in the cell. Microbial catalases are all intracellular enzymes located in the microbody of cells. From the industrial point of view, we sought an extracellular catalase produced by a microorganism. Kurono and Horikoshi (1973) reported that an alkaliphilic strain No. Ku-1 isolated from soil, which was identified to be a *Bacillus* species, produced alkaline catalase in an alkaline medium. The alkaline medium contained 2.0% glucose, 0.5% polypeptone, 0.2% yeast extract (Difco), 0.2% KH_2PO_4, 0.05% $MgCl_2 \cdot 6H_2O$, 0.005% $FeSO_4 \cdot 7H_2O$, 0.05% calcium acetate, and 0.75% Na_2CO_3, with the pH adjusted to 10. The unique feature of this alkaliphilic strain No. Ku-1 was the extracellular production of catalase. Fig. 7.66 illustrates catalase production by alkaliphilic *Bacillus* sp. No. Ku-1 in the alkaline medium described above. The bacteria grew to reach maximum cell concentration at 30 h, then began to form spores. At the same time the absorbance of the culture medium at 260 nm gradually increased. Approximately 90% sporulation was observed at 75 h by microscopic examination. The catalase activity of the increased at a constant rate and reached maximum at 80 h; it decreased in prolonged culture. The catalase was purified by precipitation with 60% saturated $(NH_4)_2SO_4$, and column chromatography on DEAE-cellulose and a Sephadex G-100. The enzyme purity was confirmed by ultracentrifugation and disc gel electrophoresis. The sedimentation constant of the enzyme was 12.5S. The optimum pH of the enzyme is 10.0; the activity decreases sharply

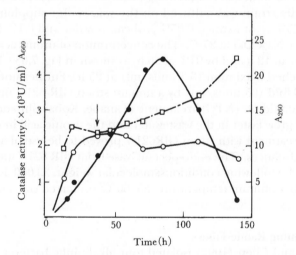

Fig. 7.66 Catalase production by alkaliphilic *Bacillus* sp. No. Ku-1. ●, catalase activity in the medium; ○, cell growth (A_{660}); □, absorbance at 260nm. ↓, beginning of spore formation.
(Reproduced with permission from Horikoshi and Akiba, *Alkalophilic Microorganisms*, p. 132, Springer-Verlag: Japan Scientific Societies Press (1982))

at pH over 10, and practically no activity is detected at pH 11.0. The enzyme is stable for 10 min between pH 7.0 and 8.5 at 60 °C, but completely inactivated at 70 °C. The optimum temperature of the enzyme is 15 °C. The enzyme activity is strongly inhibited by Fe^{2+}, Fe^{3+}, NaN_3 and KCN at a concentration of 0.5-5 mM. It is not clear whether the catalase is really secreted from the cells or released by cell lysis. Observation has indicated that it is highly possible that extracellular production of the catalase is due to partial lysis of the strain under alkaline conditions. In any case, the great advantage of this alkaline culture method is that it eliminates the process of enzyme extraction from cells. Furthermore, this extracellular production of catalase was the first step for studies on secretion of gene products from *E. coli*.

7.12   β-Lactamase

β-Lactamase [EC 3.5.2.6] is the general name given to penicillinase and cephalosporinase. β-Lactamase is defined as an enzyme which hydrolyzes the amide bond in the β-lactam ring of 6-amino-penicillanic acid or 7-aminocephalosporanic acid and/or in their *N*-acyl-derivatives (Fig. 7.67). The β-lactamases occur in many kinds of gram-positive and gram-negative bacteria as well as in fungi, and their genes are found not only in chromosomal DNA but also in plasmids of some microorganisms. These bacteria have been used as interesting research tools for investigating the mechanisms of induction and secretion of enzymes. In contrast, β-lactamases of gram-negative bacteria are normally constitutive and cell-bound enzymes. We focused on β-lactamases to study the secretion of enzymes from alkaliphilic *Bacillus* strains.

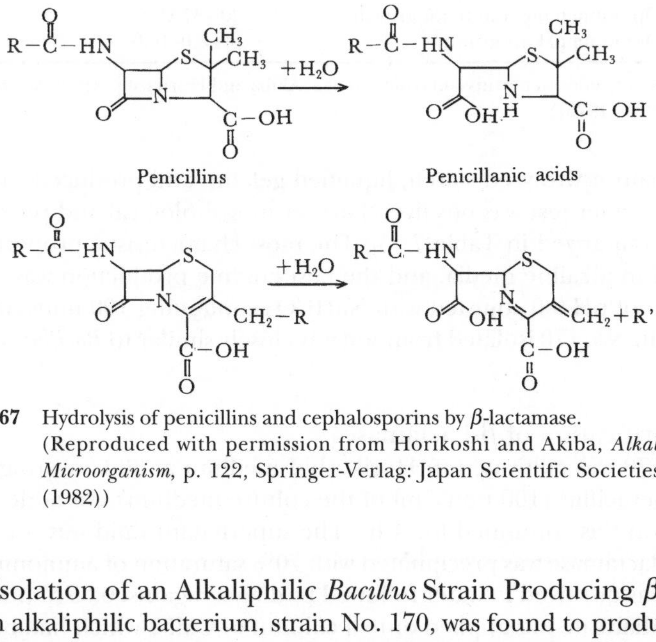

Fig. 7.67 Hydrolysis of penicillins and cephalosporins by β-lactamase.
(Reproduced with permission from Horikoshi and Akiba, *Alkalophilic Microorganism*, p. 122, Springer-Verlag: Japan Scientific Societies Press (1982))

7.12.1   Isolation of an Alkaliphilic *Bacillus* Strain Producing β-Lactamase

In 1976 an alkaliphilic bacterium, strain No. 170, was found to produce extracellular β-lactamase (Sunaga, Akiba and Horikoshi, 1976; Akiba and Horikoshi, 1980). The isolate was rod-shaped, gram-positive, aerobic, spore-forming and motile with peritrichous

Table 7.55 Morphological, cultural and biochemical characteristics of strain No. 170

1. Morphological characteristics	
Form and size	Rods, 1.0–1.2 µm × 3.0–5.0 µm
Motility	Motile with peritrichous flagella
Gram stain	Positive
Spores	0.8–1.0 µm × 1.5–2.0 µm ; oval ; central or paracentral
2. Cultural characteristics	
Nutrient both	Membranous, sediment
Gelatin stab	Liquefaction
Litmus milk	Reduction of litmus
Relation to NaCl	Grow up to 5% NaCl
3. Physiological characteristics	
Production of ammonia from pepton	Positive
Production of indole	Negative
Production of hydrogen sufide	Negative
Hydrolysis of starch	Positive
Hydrolysis of gelatin and casein	Positive
Utilization of citrate on Christensen's medium	Positive
Reduction of nitrate to nitrite	Positive
Catalase and oxidase	Positive
Urease	Negative
Voges-Proskauer test	Positive
4. Cleavage of carbohydrates	
Acid production but no gas from glucose, glycerol, fructose, sucrose, maltose, starch, mannose and dextrin	
5. Optimum temperature for growth	30°–37 ℃
6. Optimum pH for growth	pH 8.0–10.0

(Reproduced with permission from Sunaga, Akiba and Horikoshi, *Agric. Biol. Chem.*, **40**, 1364 (1976))

flagella. The strain hydrolyzed starch, liquefied gelatin, and produced catalase and oxidase. Voges-Proskauer test was positive. Further morphological and physiological characteristics are summarized in Table 7.55. The most characteristic property of the strain was to grow well in alkaline media, and the best enzyme production was observed in an alkaline medium of pH 9.0 adjusted with $NaHCO_3$ containing 100 units/ml of penicillin derivatives. Strain No. 170 isolated from soil was closely similar to *Bacillus cereus* except for its growth pH.

7.12.2 Purification of β-Lactamases

Strain No. 170 was cultivated in Horikoshi-I medium at 30 ℃ overnight on a rotary shaker; benzyl penicillin (100 units/ml of the culture medium) was added as the inducer, and cultivation was continued for 4 h. The supernatant fluid was used as the crude enzyme. The β-lactamase was precipitated with 70% saturation of ammonium sulfate and the precipitate was dissolved in water followed by dialysis against 0.05 M phosphate buffer of pH 6.5. The dialysate was applied onto a column of CM52 which had been equilibrated with the same buffer. Elution of proteins was carried out by applying a linear gradient of NaCl(0–0.4 M) in 0.05 M phosphate buffer (pH 6.5). As shown in Fig. 7.68, three frac-

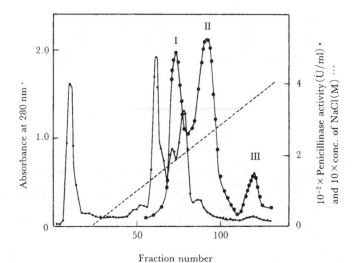

Fig. 7.68   Elution pattern of 'Bacillus sp. No. 170 penicillinases from a CM-cellulose column.

tions showing penicillinase activity were observed. Molecular weight of penicillinases found in fractions I and II was approximately 26,000, and that of the fraction III penicillinase was approximately 24,000. The pH optimum for enzyme action was 6.0-6.5 and the stable pH range was broad, from about 7.0 to 11.0 (30 °C for 10 min incubation). The fraction III enzyme was the most thermostable, and was strongly activated by the addition of Zn^{2+} (5 mM). Immunological experiments revealed that fractions I and II reacted with each other but the fraction III enzyme reacted with neither fraction I nor II. From these results, fraction III penicillinase should be a class B β-lactamase according to the classification system proposed by Ambler (1980). Its yield was too low to study further, but a gene of β-lactamase of alkaliphilic *Bacillus* sp. No. 170 was cloned and expressed in *E. coli*. Information on the gene and the enzyme is presented in the following section. About 10% of the total activity of β-lactamase having an optimum pH for enzyme action at pH 6 was not adsorbed on CM-cellulose but adsorbed on a DEAE-Sephadex A-25 column at pH 8.2. This enzyme was separated into three active components consisting of one major and two minor parts by electrophoresis on polyacrylamide gel and by isoelectric focusing. The major component was more resistant to heat denaturation than the two minor components. It retained about 15% activity after incubation for 10 min at 100 °C, while the others were completely inactivated under the same conditions. Genetic studies on these β-lactamases have been conducted and some have been clarified.

7.12.3   Cloning of β-Lactamase Genes

A.   Cloning of β-Lactamase (Class B Enzyme)

Alkaliphilic *Bacillus* sp. No. 170 was grown aerobically to the early stationary phase at 37 °C in Horikoshi-I medium. The chromosomal DNA isolated was digested with *Hind* III and the fragments were inserted into the *Hind* III site of pMB9 in the conventional way. The recombinant DNAs were introduced to *E. coli* HB101 and transformants having *Hind*

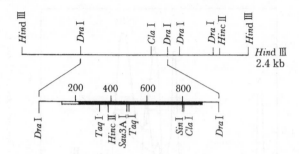

Fig. 7.69   Restriction map of the region surrounding the β-lactamase gene. The cloned insert of pEAP2 is indicated as the 2.4 kb *Hin*d III fragment. The filled box indicates the coding sequence for the mature protein, and the open box indicates the coding sequence of the signal peptide.
(Reproduced with permission from Kato *et al., J. Gen. Microbiol.,* **131**, 3318 (1985))

III fragments were selected on LB-agar plates containing 20 μg of ampicillin per ml with or without 50 μg tetracycline per ml. A plasmid, pEAP2, was isolated from the transformant (Ap^r penicillinase$^+$) containing a 2.4 kb *Hin*d III fragment (Kato *et al.*, 1983; Kudo, Kato and Horikoshi, 1983). The β-lactamase gene in the 2.4 kb *Hin*d III fragment from pEAP2 was recloned and found to be located in the 1.0 kb *Dra* I fragment (Kato *et al.*, 1985). The restriction map of the 1.0 kb *Dra* I fragment from pEAP2 is shown in Fig. 7.69 and the nucleotide sequence of the gene in Fig. 7.70. Analysis of the sequence showed that there was a single open reading frame of 771 bp from which a protein of molecular mass 27,000 could be translated. The sequence of N-terminal amino acids of the extracellular β-lactamase was NH₂-Ser-Gln-Lys-Val-Glu-Gln-Ile-Val-Ile-Lys-Asn-Glu-Thr-Gly-Thr-Ile-Ser-Ile. This amino acid sequence is identical to that deduced from the DNA sequence. Therefore, 30 amino acid residues (residues −30 to −1) are considered to represent the signal peptide. The Ouchterlony double-diffusion test showed that the penicillinase produced by *E. coli* HB101 carrying pEAP2 did not cross with the penicillinase fractions I and II. From the DNA sequence analysis, the β-lactamase gene of alkaliphilic *Bacillus* sp. No. 170 had no homology with that of *B. licheniformis* but did show partial homology with *B. cereus* β-lactamase I. However, the β-lactamase of alkaliphilic *Bacillus* sp. No. 170 looks like a class B enzyme according to the classification system proposed by Ambler (1980), because very strong homology was observed between the sequences around the histidine and cysteine residues of *B. cereus* β-lactamase II (class B enzyme) and the amino acid sequence of β-lactamase of alkaliphilic *Bacillus* sp. No. 170 (Fig. 7.70). These results support the idea that alkaliphilic *Bacillus* sp. No. 170 is closely related to *B. cereus.*

Kudo, Kato and Horikoshi (1983) discovered that *E. coli* HB101 carrying pEAP2 produced plasmid-borne β-lactamase in the culture broth at high yield (higher than 70%). Further details are given in section 5.1.1. A semi-large scale production of β-lactamase from *E. coli* HB101 carrying the plasmid pEAP31 (a derivative of pEAP2) was established by Aono (1988a). Extracellular production of the enzyme was affected by several parameters such as concentration of carbohydrates and NaCl, pH value of culture broth, culture temperature, culture volume and shaking speed of the cultivation flask (Table 7.56).

```
                                    -134 DraI
                              TTTAAAGCGTACAAAATTTTGTACGCTTTTTTGTTAATTACATA

AAAGTATGCAAATGAAGATGGAACAAACATTTGAGATGAATTGTCTAATATAGGTAATAACTATTTAGCTTGAAAGAAAGGGTTGATAAC
                                                                     SD

1
ATGAAAAAGAATACGTTGTTAAAAGTAGGATTATGTGTAAGTTTACTAGGAACAACTCAATTTGTTAGCACGATTTCTTCTGTACAAGCA
MetLysLysAsnThrLeuLeuLysValGlyLeuCysValSerLeuLeuGlyThrThrGlnPheValSerThrIleSerSerValGlnAla
(-30)
                  100
TCACAAAAGGTAGAGCAAATAGTAATCAAAAATGAGACGGGAACCATTTCAATATCTCAGTTAAACAAGAATGTATGGGTTCATACGGAG
SerGlnLysValGluGlnIleValIleLysAsnGluThrGlyThrIleSerIleSerGlnLeuAsnLysAsnValTrpValHisThrGlu
(1)
                  200            TaqI                                Hinfl
TTAGGTTATTTTAATGGAGAAGCAGTTCCTTCGAACGGTCTAGTTCTTAATACTTCTAAAGGGCTAGTACTTGTTGATTCTTCTTGGGAT
LeuGlyTyrPheAsnGlyGluAlaValProSerAsnGlyLeuValLeuAsnThrSerLysGlyLeuValLeuValAspSerSerTrpAsp
                  300
AACAAATTAACGAAGGAACTAATAGAAATGGTAGAAAAGAAATTTCAGAAGCGCGTAACAGATGTCATTATTACACATGCGCACGCTGAT
AsnLysLeuThrLysGluLeuIleGluMetValGluLysLysPheGlnLysArgValThrAspValIleIleThrHisAlaHisAlaAsp
                  400
CGAATTGGCGGAATAACAGCGTTGAAAGAAAGAGGCATTAAAGCGCATAGTACAGCATTAACCGCAGAACTAGCAAAGAAAAGTGGATAT
ArgIleGlyGlyIleThrAlaLeuLysGluArgGlyIleLysAlaHisSerThrAlaLeuThrAlaGluLeuAlaLysLysSerGlyTyr
                  500
GAAGAGCCACTTGGAGATTTACAAACAGTTACGAATTTGAAGTTTGGCAATACAAAAGTAGAAACGTTCTATCCAGGGAAAGGACATACA
GluGluProLeuGlyAspLeuGlnThrValThrAsnLeuLysPheGlyAsnThrLysValGluThrPheTyrProGlyLysGlyHisThr
                  600
GAAGATAATATTGTTGTTTGGTTGCCACAATATCAAATTTTAGCTGGAGGCTGTTTAGTAAAATCTGCGGAAGCTAAAAATTTAGGAAAT
GluAspAsnIleValValTrpLeuProGlnTyrGlnIleLeuAlaGlyGlyCysLeuValLysSerAlaGluAlaLysAsnLeuGlyAsn
                  ClaI                            700
GTTGCGGATGCGTACGTAAATGAATGGTCCACATCGATTGAGAATATGCTGAAGCGATATAGAAATATAAATTTGGTAGTACCTGGTCAC
ValAlaAspAlaTyrValAsnGluTrpSerThrSerIleGluAsnMetLeuLysArgTyrArgAsnIleAsnLeuValValProGlyHis
                                            800
GGGAAAGTAGGAGACAAGGGATTACTTTTACATACATTGGATTTATTAAAATAAGAAATTGTAGAAATACAAAAGAGAGGAGAAATAATT
GlyLysValGlyAspLysGlyLeuLeuLeuHisThrLeuAspLeuLeuLys***
                                            (227)
                  DraI 842
TTCTCCTCTCTTTCTTTTCAACTATATTTAAA
```

Fig. 7.70 Complete nucleotide sequence of the alkaliphilic *Bacillus* sp. No. 170 β-lactamase gene and the primary structure of its product. The 18 amino acids corresponding to the determined amino terminus of the extracellular β-lactamase are underlined. The amino acids are numbered taking the N-terminal amino acid of the mature protein as (1). The putative terminator sequence is underlined. The Shine-Dalgarno (SD) sequence is indicated by broken lines.
(Reproduced with permission from Kato *et al.*, *J. Gen. Microbiol.*, **131**, 3319 (1985))

The organism produced a large amount of the enzyme in culture broth under the optimal conditions established. For example, 180 units/ml of the extracellular enzyme was produced when the organism was inoculated in 300 ml broth in a 500-ml volume cultivation flask and shaken at 30 °C on a reciprocal shaker at 172 oscillations/min with 3.2-cm strokes. Extracellular production occurred only above 26 °C and at lower temperatures the penicillinase was accumulated in the periplasm (Aono, 1989c). Furthermore, the β-lactamase accumulated could be released by shifting to a higher temperature. Periplasmic or extracellular β-lactamase was the mature protein of molecular weight 24,000, indicating that the putative precursor (27,000) was processed at the correct amino acid residue by signal peptidase I (Aono, 1992b).

B. Cloning of the Lipopenicillinase Gene of Alkaliphilic *Bacillus* sp. No. 170

A membrane-bound type β-lactamase from *B. cereus* was isolated and characterized,

Table 7.56   Distribution of enzymes in *E. coli* HB101 carrying pEAP 31 at various temperatures

Cultivation Temperature	Total PCase (U/ml)	Distribution of PCase (%)				Release from cells			Growth
		Extra	Intra	Peri	Cyto	APase (%)	β-gal (%)	Protein (μg/ml)	OD$_{660}$
16 °C	73	1	99	87	12	1	n.d.	36	5.3
20	174	1	99	87	12	2	n.d.	36	5.7
22	181	2	98			2	n.d.		5.5
24	206	5	95	88	7	4	2	51	5.7
26	206	17	83			11	n.d.		5.6
28	183	66	34	31	4	42	1	186	4.4
30	155	81	19	17	2	55	14	339	3.4
32	148	95	5	3	2	85	26	468	2.3
36	102	94	6	4	2	73	37	470	2.3
38	88	75	25	20	5	66	11	250	3.0
40	55	85	15	10	5	67	7	273	3.2
43	< 1					43	n.d.	44	2.6

n.d.: not detected

and named β-lactamase III. This β-lactamase III was found to be a lipoprotein similar to the penicillinase of *B. licheniformis* and is grouped as a class A enzyme (Ambler, 1980). This enzyme has some characteristics similar to the major outer membrane lipoprotein of *E. coli*. Hussain, Pator and Tampeu (1987) reported the cloning and sequencing of the β-lactamase III gene from *B. cereus* and showed that this gene product had the characteristics of a lipoprotein. Because alkaliphilic *Bacillus* sp. No. 170 may also have a lipoprotein penicillinase similar to the β-lactamase III of *B. cereus*, we tried to isolate the gene by cloning (Kato *et al.*, 1989). The lipopenicillinase (LIPEN) gene from alkaliphilic *Bacillus* sp. No. 170 was cloned in *E. coli* using the vector pHSG399. The plasmid DNA was completely digested with *Acc* I, and chromosomal DNA was partially digested with *Cla* I. The *Cla* I digests were inserted in the *Acc* I site of pHSG399 and hybrid DNAs were introduced into *E. coli* JM101. A plasmid, pFAPl, was isolated from an ampicillin-resistant transformant plasmid pFAP121, which contained a 2.2 kb *Hinc* II fragment containing the LIPEN gene from pFAPl. The LIPEN gene was expressed in *E. coli* carrying pFAP121, and the product was found in the membrane fraction. The nucleotide sequence of a 1.9-kb segment encoding the LIPEN was determined, and this segment showed an open reading frame which would encode a polypeptide of 310 amino acids. As shown in Fig. 7.71, the amino acid sequence of this LIPEN gene product had strong homology with those of the *B. cereus* β-lactamase III and *B. licheniformis* penicillinase. These results suggest that the LIPEN has properties similar to the lipoprotein type rather than extracellular type enzymes and is a class A enzyme.

7.13   Other Enzymes

7.13.1   CD-degrading Enzyme

Yoshida *et al.* (1991) reported a CD-degrading enzyme isolated from alkaliphilic *Bacillus* strain No. 199. The strain was identified as closely resembling *Bacillus circulans*. The enzyme was purified 252-fold from the cell extract by chitosan treatment, ammonium sulfate fraction, DEAE-Toyopearl column chromatography, and gel filtration. The pH

```
    1                                    v                        52
a)  M---KLWFSTLKLKK---AAAVLLFSCVALAG CANNQTNASQPAEKNEKTEMKDD------
    *  *  *    **       * * *** * *  *             *           *
b)  MIVPKKFFHISHYKK---MLPVVLLSCVTLIG CSSSNIQSEPPKQT------KQENTGNQS
    * *  *** ** **    ******* **** ** ** *** ***      *** **
c)  MFVLNKFFTNSHTKK---IVPVVLLSCATLIG CSNSNTQSESNKQTNQTNQVKQENKGNHA
    *  *       *        **  *         ** *     *    * *
d)  M----K---NKRMLKIGICVGILGLSITSLEA FTGESLQVEAKEKT---GQVKHKNQATHK
```

```
                                                              113
a)  -FAKLEEQFDAKLGIFALDTGTNRTVAYRPDERFAFASTIKALTVGVLLQQKSIEDLNQRIT
    * ***   ****** ****  * **  * **** *** ***  **   ** ****** ***
b)  -FVKLEKEYDAKLGIYALDTSANQTVTYRSDERFAYSTHKALAVGAVLQKKSIEDLEQRIK
    * ******  *********** ****   *  **  * *****  *   *** * **
c)  -FAKLEKEYNAKLGIYALDTSTNQTVAYHADDRFAFASTSKSLAVGALLRQNSIEALDERIT
    *  ***  **  ** **   ** *  **    **** **** ****   **  **   *
d)  EFSQLEKKFDARLGVYAIDTGTNQTISYRPNERFAFASTYKALAAGVLLQQNSIDSLNEVIT
```

```
                                                              175
a)  YTRDDLVNYNPITEKHVDTGMTLKELADASLRYSDNAAQNLILKQIGGPESLKKELRKIGDE
    **  ***********  ********* **** ******  * **           * * ***
b)  YTSKDLVNYNPITEKYIDTGMTLKELMDASIRYSDNTAQNLILTQLGGPSGIQKIMREIGDT
    ** *** ********   ********** *** **** ** ****  *****      ** ** ***
c)  YTRKDLSNYNPITEKHVDTGMTLKELADASVRYSDSTAHNLILKKLGGPSAFEKILREMGDT
    **  *** *  **  ** *****  * * ***** **  ** ****       *** * * *
d)  YTKEDLVDYSPVTEKHVDTGMKLGEIAEAAVRSSDNTAGNILFNKIGGPKGYEKALRHMGDR
```

```
                                                              237
a)  VTNPERFEPELNEVNPGETQDTSTARALVTSLRAFALEDKLPSEKRELLIDWMKRNTTGDAL
    ************* ** * ****  ** ***  **   ***  *** **  *** **** *
b)  VTNPERFEPELNEVQPGDTRDTSTPKALATSLQAYALGDILSVENRNFLIDLMKRTTTGDNL
    *** ********** ** * ****** **  **   ** **  *  ** ** *** **
c)  VTNSERFEPELNEVNPGETHDTSTPKAIAKTLQSFTLGTVLPSEKRELLVDWMKRNTTGDKL
    *   *** **** *** ***** ** *  * *       *   **   *** *** *****
d)  ITMSNRFETELNEAIPGDIRDTSTAKAIATNLKAFTVGNALPAEKRKILTEWMKGNATGDKL
```

```
                                                              299
a)  IRAGVPDGWEVADKTGAASYGTRNDIAIIWPPKGDPVVLAVLSSRDKKDAKYDDKLIAEATK
    ****** *** **** ********* **** *  ** ** * ********** ***
b)  IRAGVPGEWEVPDKTGSGSYGTRNDIAFIWPPNKKPFILAILSNQAKEDAKYDDKLIADATK
    ****** *** **** ********** ****  * ** *  * ****  ***** ***
c)  IRAGVPKGWEVADKTGAGSYGTRNDIAIIWPPNKKPIVLSILSNHDKEDAEYDDTLIADATK
    **** *  *  * ** ** ********** ****  * *   **   ** * * *  *** ***
d)  IRAGIPTDWVVGDKSGAGSYGTRNDIAVVWPPNRAPIIVLISSKDEKE-AIYNDQLIAEATK
```

```
    310
a)  VVMKALNMNGK
    *  *    *
b)  IVLDVLTVPNK
    ***  * **
c)  IVLETLKVTNK
         *
d)  VIVKALR
```

Fig. 7.71 Amino acid homologies in the primary structure of LIPEN and three other
class A β-lactamases. a, *B. licheniformis* penicillinase; b, alkaliphilic *Bacillus* sp.
No. 170 LIPEN; c, *B. cereus* β-lactamase III; d, *B. cereus* β-lactamase I. The
asterisks (∗) indicate homologies between the LIPEN and three β-lactamases, respectively. The "v" indicates the putative signal-sequence cleavage
site. The numbering is that of the primary structure of LIPEN.
(Reproduced with permission from Kato *et al.*, *Arch. Microbiol.*, **151**, 91
(1989))

and temperature optima of the enzyme were 6.0 and 50 °C. The molecular weight of the
enzyme was 126,000 with two subunits of 67,000. The isoelectric point was pH 4.2. Enzyme
activity was inhibited by Ag^+, Hg^{2+}, Cu^{2+} and p-CMB. The enzyme hydrolyzed α-, β- and
γ-cyclodextrins, as well as linear maltodextrins, to yield maltooligosaccharides. However,
starch and maltose were not degraded by the enzyme (Table 7.57).

Table 7.57   Substrate specificity of cyclomaltodextrinase

Substrate	K_m (mM)	Relative rate (%)	Major product (relative molar ratio)
α-Cyclodextrin	3.2	100	G2 : G6 ; G3 ; G1 (1.0 : 0.3 : 0.3 : 0.3)
β-Cyclodextrin	1.8	65	G2 : G7 : G3 (1.0 : 0.7 : 0.7)
γ-Cyclodextrin	0.8	38	G2 : G8 : G3 : G1 (1.0 : 0.7 : 0.5 : 0.5)
Maltotriose (G3)	12.5	135	G2 : G1 (1.0 : 1.0)
Maltotetraose (G4)	7.4	135	G2 : G3 : G1 (1.0 : 0.2 : 0.1)
Maltopentaose (G5)	1.2	43	G2 : G3 (1.0 : 1.0)
Maltohexaose (G6)	1.3	52	G2 : G3 : G1 (1.0 : 0.3 : 0.3)
Maltoheptaose (G7)	0.8	45	G2 : G3 (1.0 : 0.7)

G1 = Glucose, G2 = Maltose (= 1.0).
(Reproduced with permission from Yoshida *et al.*, *J. Ferment. Bioeng.*, **71**, 229 (1991))

7.13.2   Galactan-degrading Enzyme

In recent years, oligosaccharides such as galactooligosaccharides and mannan have been investigated as a functional food material in the food industry. Soybean fiber that contains galactan is considered to be a potential source of galactooligosaccharides. Tsumura *et al.* isolated two alkaliphilic bacteria that produce galactan-degrading enzymes (Tsumura *et al.*, 1991*a*). Soil samples suspended in water were spread on an agar medium (Horikoshi-I without glucose) containing 0.4% soybean fiber (Fuji Oil Co. Ltd.) and incubated at 37 °C for 2 days. Two strains of alkaliphilic bacteria (S-2 and S-39) that formed clear zones around the colonies were isolated from 300 soil samples. The isolates were cultivated in a liquid medium containing 2% soybean fiber, 0.5% polypeptone, 0.5% NaCl, 0.1% KH_2PO_4, 0.2% $MgSO_4 \cdot 7H_2O$ and 1% Na_2CO_3 for 40 h at 37 °C. The culture supernatants were assayed for galactanase activity at pH 10 at 40 °C. The remaining soybean fiber in the reaction mixture was removed by centrifugation, and total sugar and reducing sugar of the supernatant were measured. No galactose was detected in the hydrolysates of these two isolates, but galactooligosaccharides were detected by thin layer chromatography. The microbiological and biochemical characteristics of strains S-2 and S-39 are summarized in Table 7.58. Then the galactanases (S-2 and S-39) were purified to a homogeneous state from culture filtrates of alkalophilic *Bacillus* sp. S-2 and S-39, respectively (Tsumura *et al.*, 1991*b*). The galactanases were purified by ammonium sulfate precipitation, chromatography on DEAE-cellulofine, and by gel filtration on Cellulofine CG-200M. The molecular weight of the enzyme was 40,000 (SDS-PAGE) and the isoelectric point was 8.6. It was most active at pH 10 and 50 °C, and stable between pH 7 and 12. The enzyme S-39 was purified by ammonium sulfate precipitation and chromatographies on DEAE-Toyopearl 650M and CM-Toyopearl 650M. The molecular weight

Table 7.58 Taxonomical characteristics of strains S-2 and S-39

	S-2	S-39
Shape and size of cells	Rods (0.5–0.7 × 1.5– 3.0 μm)	Rods (0.5–0.6 × 3.0– 4.0 μm)
Motility	Motile	Motile

Spores and
  Sporangia
	Ellipsoidal,
	  terminal
	Slightly
	  swollen
		Ellipsoidal,
		  terminal
		Swollen

Gram stain	Positive	Positive
G + Content of DNA	38.0%	52.4%
Optimum pH for growth	8.0–9.5	7.5–9.5
Temperature for growth	20–45 ℃	25–45 ℃
Anaerobic growth	Positive	Negative
Catalase	Positive	Positive
Oxidase	Positive	Positive
Reduction of nitrate	Negative	Negative
V–P test	Negative	Positive
Utilization of citrate		
Koser medium	Negative	Negative
Christensen medium	Positive	Positive
Production of indole	Negative	Negative
Hydrolysis of casein	Negative	Positive
gelatin	Positive	Positive
starch	Positive	Positive
Growth in nutrient broth		
containing NaCl	Up to 15%	Up to 2%
Acid andGas from		
D-Glucose	Positive	Positive
L-Arabinose	Positive	Positive
D-Xylose	Positive	Positive
Mannitol	Positive	Positive

(Reproduced with permission from Tsumura *et al.*, *Agric. Biol. Chem.*, **55**, 1400 (1991*a*))

was 36,000 and p*I* was 7.1. It was most active at pH 4 and 40 ℃. Both enzymes hydrolyzed soybean galactan to produce galactooligosaccharides with very little galactose but did not hydrolyze larch wood arabionogalactan, which has β-1,3 or β-1,6 galactosidic linkages. These substrate specificities and the action profiles on soybean galactan showed that these enzymes are endo-1,4-β-D-galactanase (EC 3.2.1.89). These enzymes from *Bacillus* sp. S-2 and S-39 are the first ones discovered that can hydrolyze soybean galactan to form galactooligosaccharides preferentially.

7.13.3 Maltobionate α-D-Glucohydrolase

Shirokane, Arai and Uchida (1994) *et al.* reported a new enzyme, maltobionate α-D-glucohydrolase, in the cells of alkaliphilic *Bacillus* sp. N-1053. Maltobionate α-D-glucohydrolase was purified to apparent homogeneity from a cell-free extract of alkaliphilic *Bacillus* sp. N-1053 about 930-fold with a yield of 18%, and some of its properties were

Table 7.59 Substrate specificity of maltobionate α-D-glucohydrolase

Substrate	Relative rate of hydrolysis (%)
Maltobionate	100
Maltotrionate	24
Maltotetraonate	7.5
Maltopentaonate	2.0
Maltose	3.3
Maltotriose	16
Maltotetraose	6.8
Maltopentaose	0.9
Maltitol	2.6
Maltotriitol	19
Phenyl α-D-maltoside	14
α-D-Maltose 1-phosphate	0.8
Isomaltotriose	0.1
Panose	0.2

(Reproduced with permission from Shirokane, Arai and Uchida, *Biochim. Biophys. Acta-Protein Struct. Mol. Enzymol.*, **1207**, 148 (1994))

investigated. The enzyme showed optimum activity at about pH 7.0, and was stable over a range of pH 6.0–9.5. The molecular weight was estimated to be 152,000 and 71,000 by HPLC gel filtration on TSKgel G3000SW(XL) and SDS-polyacrylamide gel electrophoresis, respectively. The enzyme hydrolyzed maltobionate more effectively than disaccharides such as maltose and maltitol or trisaccharides such as maltotrionate, maltotriose and maltotriitol, but showed no activity toward polysaccharides such as amylose, amylopectin and soluble starch (Table 7.59). The reaction products from 1 mole of maltobionate were found to be 1 mole of β-D-glucose and 1 mole of D-gluconate. The K_m value for maltobionate was 1.63 mM and the V_{max}/K_m value for maltobionate was the largest among the substrates tested. The enzyme activity was almost completely inhibited by Hg^{2+}, Ag^+, iodine and *N*-bromosuccinimide, and also inhibited by *p*-nitrophenyl α-D-glucoside, maltose and maltitol.

7.13.4 Extracellular Production of Alkaline Phosphatase, 5′-Nucleotidase and Phosphodiesterase

The first extracellular production of so-called intracellular enzymes such as catalase was reported by Kurono and Horikoshi (1973) by alkaliphilic *Bacillus* No. Ku-1, as shown in Fig. 7.66. Extracellular production of alkaline pectinase from the alkaliphilic *Bacillus* sp. P-4-N was strongly stimulated by the addition of manganese (Horikoshi, 1972). Ikura and Horikoshi (1994) tried to produce intracellular enzymes in culture fluid on a commercial scale by cultivating alkaliphilic microorganisms in the presence of metal ions but were not successful.

A. Alkaline Phosphatase in Culture Broths

The alkaline phosphatase of *Escherichia coli* has been studied extensively and found to be repressed by orthophosphate. It is interesting that alkaline phosphatase is produced during sporulation of *Bacillus subtilis* despite the presence of inorganic phosphate concentrations which completely repress the activity in vegetative cells. Furthermore, most of the alkaline phosphatases isolated are intracellular, but in *E. coli* it is located in the periplasmic space as a soluble enzyme. In gram-positive bacteria such as *B. subtilis* and *B. licheniformis*, the alkaline phosphatase appears to be membrane-bound. *Bacillus* sp. No. RK11 (IMD No. 278), a soil isolate, was an alkaliphilic *Bacillus* strain and produced an extracellular alkaline phosphatase (Kelly, Nash and Fogarty, 1984). The following media were used: medium No. 1 (g/1) soluble starch, 10.0; bacteriological peptone, 6.0; yeast extract, 3.0; $Na_2CO_3 \cdot H_2O$, 10.0; $MnSO_4 \cdot 4H_2O$, 0.015; $MgCl_2 \cdot 6H_2O$, 0.5; K_2HPO_4, 1.5; oxoid agar No. 3, 15.0. Medium No. 2 was used as a manganese-deficient medium with composition identical to that of medium No. 3 except for the exclusion of $MnSO_4$. The medium used for production of alkaline phosphatase (medium No. 3) had the following composition in grams per liter: fructose, 10.0; yeast extract, 10.0; bacteriological peptone, 10.0; $Na_2HCO_3 \cdot H_2O$, 10.0; $MnSO_4 \cdot 4H_2O$, 0.015; $MgCl_2 \cdot 6H_2O$, 0.5; K_2HPO_4, 1.5. Fructose was autoclaved separately and added. In the absence of Mn^{2+} in a complex medium such as medium No. 2, neither alkaline phosphatase production nor sporulation was detected. No other divalent metal could be substituted for Mn^{2+}. Manganous sulfate (70 µmol) gave the highest enzyme production although spore numbers continued to increase with manganous concentrations above this level. Maximum alkaline phosphatase production occurred when metal was present at the time of inoculation but maximum spore numbers were detected when the metal was added 8-12 h after inoculation. Inorganic pyrophosphatase was not associated with extracellular alkaline phosphatase, but it was detected intracellularly. By this cultivation method 95% of the alkaline phosphatase was detected in the culture broth.

Nomoto *et al.* (1988) isolated a strain of alkaliphilic *Bacillus* sp. OK-1 which excreted alkaline phosphatase into the culture broth. The isolate was cultivated in Horikoshi-II medium at 37 °C for 60 h and the supernatant fluid was used as a crude supernatant. Almost all of the alkaline phosphatase activity was detected in the culture broth. The enzyme was purified by DEAE-cellulose column chromatography followed by gel filtration. Molecular weight was estimated to be 108,000, and the enzyme was composed of two subunits having the same molecular weight of 54,000. The purified enzyme, which had a pH optimum of 11, was fairly stable between pH 5 and 12. This enzyme was inactivated by EDTA and recovered by the addition of Co^{2+} ion. These results indicate that it is a metalloenzyme. The alkaline phosphatase of *E. coli* has been reported to be an intracellular zinc metalloenzyme with a molecular weight of 80,000, composed of two identical subunits. Although Kelly *et al.* (1984) did not report on a metal ion responsible for alkaline phosphatase activity, it is of interest that alkaline phosphatase, which was thought to be a typical intracellular enzyme, was produced by alkaliphilic *Bacillus* strains in culture broth.

B. Production of 5'-Nucleotidase

Ikura and Horikoshi (1989*a*, *c*) produced 5'-nucleotidase in the culture broth of alkaliphilic *Bacillus* No. C-3. An alkaliphilic *Bacillus*, No. C-3, isolated from soil, produced 5'-nucleotidase [EC 3.1.3.5] extracellularly when cultured in a medium containing Mn^{2+}: soluble starch, 15; polypeptone, 5; yeast extract, 5; $MgSO_4 \cdot 7H_2O$; $MnCl_2 \cdot 4H_2O$, 0.02; and Na_2CO_3, (10g/1). The unique point of this enzyme production is that the enzyme was produced well in a medium containing a rather high concentration of Mn^{2+}, in spite of a small difference in growth. The optimum concentration of Mn^{2+} for the enzyme production was 10 mM and over. Mn^{2+} could not be replaced by other divalent cations. The course

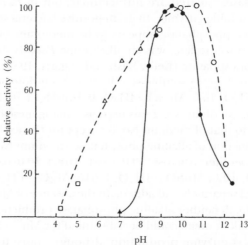

Fig. 7.72 Effects of pH on enzyme activity and stability. Effects of pH (●, ▲) and stability (□, △, ○). The following buffer systems were used: pH 4.2–5.0, 0.1 M CH$_3$COOH-CH$_3$COONa (□); pH 7.0–8.0, 0.1 M HEPES-KOH buffer (▲, △); pH 8.35–10.0, 0.1 M KHCO$_3$-K$_2$CO$_3$ buffer (●, ○); and pH 10.5–12.4, 0.1 M K$_2$CO$_3$-KOH buffer (●, ○). For pH stability, the enzyme was treated in each buffer containing a final concentration of 10 mM 2-mercaptoethanol at 40 °C for 15 min, and the residual activity after adjustment to pH 9.5 was assayed.

(Reproduced with permission from Ikura and Horikoshi, *J. Ferment. Bioeng.*, **67**, 211 (1989))

of the enzyme production closely paralleled the increase in growth. 5′-Nucleotidase of alkaliphilic *Bacillus* No. C-3 was purified from the culture supernatant (24-h cultivation). The molecular weight of the enzyme was 80,000 by gel filtration. The optimum pH for the activity was 9.5, and the enzyme was stable at pH 9.5–10.5 in a buffer containing 10 mM 2-mercaptoethanol, as shown in Fig. 7.72. Substrate specificity study revealed that the enzyme acted on 5′-AMP strongly, on several 5′-nucleotides and ADP to a certain extent, but not on 3′-nucleotides, 2′-nucleotides, *p*-nitrophenyl phosphate, or ATP. The K_m value for 5′-AMP was 3.0×10^{-4} M. The enzyme required no divalent cation for its activity. The enzyme was inhibited by borate and arsenite ions but not by 1 mM EDTA.

Another alkaliphilic *Bacillus* sp. No. A-59 (ATCC21591) that is an alkaline amylase producer also produced an extracellular 5′-nucleotidase when Mn^{2+} was added to the growth medium described above (Ikura and Horikoshi, 1989*b*). The enzyme formation was negligible in the medium in the absence of Mn^{2+} and optimum Mn^{2+} concentration for the enzyme production was 10 mM (Fig. 7.73). The 5′-nucleotidase was purified and its molecular weight was determined to be 78,000 by gel filtration. The optimum pH for its activity was 9.0–9.5. The enzyme was stable in the pH range of 8.5 to 9.5, and up to 40 °C. A substrate specificity study revealed that the enzyme hydrolyzed 5′-AMP strongly, several 5′-SMPs and ADP weakly, but not 3′-XMP, 2′-XMP, ATP or *p*-nitrophenyl phosphate. The K_m value for 5′-AMP was 1.5 mM. The maximum enzyme activity was obtained without divalent cations. The enzyme was inhibited by borate and arsenite ions, but not by 2 mM EDTA.

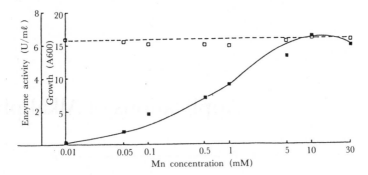

Fig. 7.73 Stimulatory effect of Mn²⁺ on the production of 5′-nucleotidase.
The enzyme activity was assayed after 24-h cultivation. □, growth; ■, enzyme
activity.
(Reproduced with permission from Ikura and Horikoshi, *Agric. Biol. Chem.*,
53, 646 (1989))

C. Phosphodiesterase Production

Another alkaline amylase producer, *Bacillus* No. A-40-2 (ATCC21592), produced
phosphodiesterase in culture broth in the presence of Mn^{2+} (Ikura and Horikoshi, 1990).
The following was used as the standard medium (g/l): soluble starch, 15; ammonium sulfate, 5; MgSO$_4 \cdot$7H$_2$O, 0.2; MnCl$_2 \cdot$4H$_2$O, 0.02 g and NaHCO$_3$, 10. The strain was cultured
in 50 ml of medium in a 300-ml flask for 20 h at 37 °C on a rotary shaker. Phosphodiesterase
production in the culture broth of alkaliphilic *Bacillus* No. A-40-2 increased with increasing Mn^{2+} concentration, showing maximum productivity at 10 mM. The simultaneous
addition of 10 mM Mn^{2+} and one of the following cations Mg^{2+}, Co^{2+}, Mo^{6+}, and Pb^{2+} at
suitable concentrations stimulated enzyme production by a maximum of 1.8-fold more
than the addition of only 10 mM Mn^{2+}. Inorganic phosphate barely repressed enzyme
production. The purified enzyme had an optimum pH of 7.5 and was stable from pH 7–11
at 40 °C. The enzyme hydrolyzed 2′,3′-cyclic-nucleotide and 3′-nucleotides, but did not
hydrolyze 3′,5′-cyclic-nucleotides or 5′-nucleotides, indicating it to be a 2′,3′-cyclic-
nucleotide 2′-phosphodiesterase [EC3.1.4.16]. The enzyme had activity without metals,
but Mg^{2+}, Ca^{2+}, Ba^{2+}, and Mo^{6+} activated the enzyme reaction (maximum 50%).

7.13.5 Chitinase

Tsujibo *et al.* (1992) isolated chitinases from an alkaliphilic *Actinomycetes*, *Nocardiopsis
albus* subsp. *prasina* OPC-131. The isolate produced two types of chitinases. The optimum
pH of chi-A was pH 5.0, and that of chi-B was pH 7.0. However, no further details have
been reported.

8

Applications of Alkaliphiles

8.1 Indigo Fermentation

In Japan the reduction of indigo for dyeing is a traditional use of alkaliphilic bacteria dating back to ancient times. Indigo was reduced by alkaliphilic bacteria in an alkaline environment containing sodium carbonate, calcium hydroxide, and potassium hydroxide. After the discovery of synthetic indigo, indigo was reduced to indigo white by the addition of reducing agents such as hydrosulfite. All dye factories have used this chemical reduction process. Recently, traditional indigo dyeing has been revived as a folk craft. There are many traditional indigo dyeing processes in Japan. One of them is described below (Yoshioka, 1982). (1) Wood ash (17 kg), slaked lime (4.4 kg) and 160 l of hot water is mixed, stirred for 3 h and allowed to stand for 48 h at room temperature. About 140 l of supernatant fluid (ash extract) is obtained. (2) About 3 kg of indigo balls (*tsukune* or *aidama*, fermented indigo plant *Polygonum tinctorium Ait* leaves) is mixed with 25 l of the ash extract and ground with a mortar and pestle. Then approximately 40 l of the ash extract is added to the mixture. The pH of the mixture is about 11.2 to 11.6. (3) The mixture is kept at 20–30 °C for four days by adding about 40 to 20 l of the ash-extract at 24-hour intervals to maintain high pH values. (4) As nutrients for the alkaliphlic bacteria, 210 g of wheat gluten is added. Then the pH of the mixture is adjusted by potassium hydroxide. (5) After eight days, fermentation starts and foam observed. About 400 ml of Japanese *sake* (rice wine) is added as a booster. (6) During the fermentation process, the pH value is kept between 10.0 and 11.0 by the addition of potassium hydroxide. (7) After two weeks the reduction of indigo is complete and foam of reoxidized indigo observed. This reduced mixture is used as a stock vat for indigo dyeing. Fig. 8.1 shows this traditional method of indigo reduction.

Takahara and Tanabe (Takahara, 1960; Takahara, Takasaki and Tanabe, 1961; Takahara and Tanabe 1962) isolated indigo-reducing *Bacillus* sp. No. S-8 from an indigo ball and developed an improved indigo-reduction process on an experimental scale. By adding a seed cultrue of *Bacillus* sp. No. S-8 to the reduction mixture the processing time was decreased by 75%. An indigo ball of 7.5 kg, 900 g peptone, 200 g starch, 250 g sodium hydroxide, 600 g calcium carbonate, 900 g synthetic indigo, and 180 l of water was incubated at 30 °C with or without the addition of the bacteria isolated. As shown in Table 8.1, reduction was easier and the product was considered better than that obtained by the traditional fermentation method. The isolated strain No. S-8 (*Bacillus alkaliphilus*) was an aerobic, spore-forming, motile, rod-shaped bacterium with peritrichous flagella.

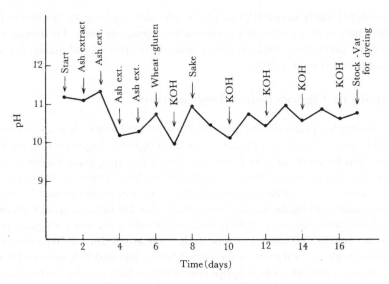

Fig. 8.1 pH values during indigo reduction.

Table 8.1 Indigo reduction on semi-industrial scale

Experiment 1, improved process with the addition of S-8 bacteria

	Time (h)					
	0	12	18	24[†1]	30	48[†2]
Temperature, °C	30	30	30	30	30	30
pH	12.1	12.1	12.0	11.9	11.8	11.6
Eh	400		510	580	630	690
Total sugar	0.395	0.385	0.222	0.355	0.295	0.125
Color	Blue	Blue	Pale blue	Foam of reoxidized indigo		

Experiment 2, traditional process

	Time (h)					
	0	24	48	72	120[†3]	144	168[†4]
Temperature, °C			30				
pH	12.1	12.1	12.0	12.1	12.1	11.7	11.6
Eh	380	380	530	570	640	670	700
Total sugar	0.399			0.397	0.386	0.263	0.117
Color	Blue	Blue		Pale blue	Foam of reoxidized indigo		

[†1] 200 g peptone and 200 g starch were added.
[†2] Fermentation was completed.
[†3] 200 g peptone and 200 g starch were added.
[†4] Completed.

(Reproduced with permission from Horikashi and Akiba, *Alkalophilic Microorganisms*, p.171, Springer-Verlag : Japan Scientific Societies Press (1982)).

Optimum growth pH was between 10.0 and 11.5. The investigation was limited to the field of indigo fermentation and no further progress has been reported. Furthermore, these experiments were carried out before the revival of traditional indigo dyeing and few microbiologists noted these reports.

8.2 Lignin or Lignocellulose Degradation

Lignin is a complex polymer with phenylpropanoid subunits. It is an essential component of woody tissue, to which it imparts structural rigidity. Lignin is remarkably resistant to degradation by most microbes; nevertheless, a few species of white-rot fungi are able to catalyze its oxidation to carbon dioxide. Its biodegradation is of great ecological significance because, next to cellulose, lignin is the most abundant renewable polymer on earth. Because lignin and lignin model compounds dissolve in water under alkaline conditions, Kawakami and Shumiya (1983) isolated alkaliphilic microorganisms which degraded various lignin and model compounds such as vanillin, dehyrodivanillin and dehydrodivanillic acid. Several microorganisms, KS-101, 104 and 105, were isolated by the enrichment culture method. Their properties were as follows: the three strains were almost identical, their optimum pH values for growth being 10.0 to 12.0. The cells were rods (0.6–0.8 μm × 3.0–4.5 μm), motile, gram-negative, producing cytochrome oxidase and catalase but not laccase. The strains did not utilize D-glucose and exhibited best growth at 30 °C. Later Kawakami and Shimizu (1986) reported alkaliphilic bacteria KK-203b, CM-432 and CM-401 having high activities to degrade various lignin model compounds and to produce the acid precipitable polymeric lignin from lignocellulose. Strains KK-203b and CM-432 belong to the genus *Corynebacterium*. Strain CM-401 was identified as *Acinetobactoer lwoffi*. Their optimum pH values for growth were about 11.0 to 12.0. No further report has been published.

8.3 Production of 2-Phenylamine

Hamasaki *et al.* found an alkaliphilic bacteria that produced a large amount of 2-

Table 8.2 Polyamine and 2-phenylethylamine contents of various alkaliphilic *Bacillus* spp.

Strain	Amount detected (nmol/g [wt/wt])			
	Put	Spd	Spm	2-PEA
Bacillus sp. YN-2000	25	2,100	65	800
Bacillus sp. YN-1	tr	109	4	ND
B. alkalophilus	tr	248	107	ND
B. firmus RAB RA-1	23	1,223	307	ND
B. firmus OF4	29	2,167	467	ND

Put, putrescine; Spd, spermidine; Spm, spermine; 2-PEA, 2-phenylethylamine

(Reproduced with permission from Hamasaki *et al.*, *Appl. Environ. Microbiol.*, **59**, 2720 (1993)).

phenylamine in culture broth (Hamasaki *et al.*, 1993). A large amount of 2-phenylethyl-amine was synthesized in cells of alkaliphilic *Bacillus* sp. strain YN-2000. This amine was secreted in the medium during cell growth (Table 8.2). The amounts of 2-phenylethyl-amine in both cells and medium vary with changing pH of the medium.

8.4 Carotenoid of Alkaliphilic *Bacillus* Strains

Aono and Horikoshi (1991) reported that alkaliphilic *Bacillus* spp. No. A-40-2, No. 2b-2, No. 8-1 and No. 57-1 produced yellow pigments in the cells. These are probably triter-penoid carotenoids. However, A-59, M-29 and Y-25, which are white strains in Horikoshi-II medium, did not produce carotenoid. The physiological role of yellow pigments was not reported.

8.5 Siderophore-producing Alkaliphile

Gascoyne, Connor and Bull (1991 *a,b*) isolated siderophore-producing alkaliphilic bac-teria to accumulate iron, gallium and aluminum. The isolation of bacteria producing siderophores under alkaline conditions was studied. Enrichment cultures initiated with samples from a number of alkaline environmental sources yielded ten isolates. From this group selections were made on the basis of growth at high pH and the gallium-binding capacity of the siderophores. It was found that some isolates grew well and high concen-trations of siderophore were detected, whereas others grew well in the presence of much lower concentrations of siderophore. The effect of iron, gallium and aluminum on growth and siderophore production batch culture was investigated for six isolates. The presence of iron greatly decreased the siderophore concentration in these cultures, whereas the response to added gallium or aluminum was dependent upon the isolate.

8.6 Oxidation of Cholic Acid by a *Bacillus* Strain

Kimura, Okamura and Kawaide (1994) isolated from soil an alkaliphilic *Bacillus* strain that grew well in media containing cholic acid (CA) at 5% or higher. Cholic acid effi-ciently converted 7 α-and 12 α-hydroxyl groups of CA into keto groups, with the conver-sion rate for both hydroxyl groups reaching 100% by 72 hours of cultivation, as shown in Fig. 8.2. The strain also converted a 3 α-hydroxyl group into a keto group, but the con-version rate was only about 5% at 72 hours. The strain neither affected any other part of the CA molecule nor oxidized 7 β-or 12 β-hydroxyl groups. By NTG mutagenesis, the fol-lowing mutants were acquired: (1) conversion of only the 7 α-and 12 α-hydroxyl groups, (2) conversion of only the 12 α-hydroxyl groups, and (3) conversion of only the 7 α-hydroxyl group. These mutants selectively produce 12-ketochenodeoxycholic acid (12KCDCA), 7-ketodeoxycholic acid (7KDOCA) and 7,12-diketolithocholic acid (7,12DKLCA) from CA, and 7-ketolithocholic acid (7KLCA) from chenodeoxycholic acid (CDCA), respectively, at high yields of close to 100%.

8.7 Organic Acids Produced by Alkaliphilic Bacilli

During the cultivation of alkaliphiles, the pH values of the culture media used

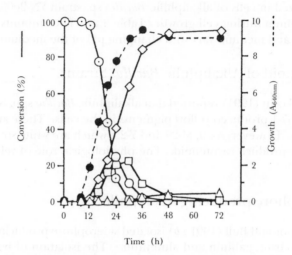

Fig. 8.2 Course of CA conversion by *Bacillus* sp. TTUR 2-2.
Initial CA concentration was 5% in Medium I (pH 10), and cultivated with
shaking at 30 °C. Products were analyzed by HPLC. ●⋯●, growth; ⊙—⊙, CA;
○—○, 12KCDCA; □—□, 7KDOCA; ◇—◇, 7, 12DKLCA; △—△, DHCA.
(Reproduced with permission from Kimura, Okamura and Kawaide, *Biosci.
Biotechnol. Biochem.*, **58**, 1004 (1994))

decrease sharply due to the production of organic acids, which are converted from carbohydrates. However, the details of the organic acids produced have not been published.
Paavilainen, Helisto and Korpela (1994) reported comparative studies on organic acids
produced by alkaliphilic bacilli. Four bacilli, *Bacillus* sp. 38-2 (ATCC21783), *B. alcalophilus*
subsp. *halodurans* (ATCC27557), *B. alcalophilus* (ATCC27647) and *Bacillus* sp.
17-1(ATCC31007), were cultured in the presence of various 1% (w/v) sugars and related
compounds such as sugar alcohols, and the catabolites formed were analyzed. All alkaliphiles produced acetic acid (4.5–5 g/l maximum) while formic acid was produced by
only one of the strains (Fig. 8.3). In contrast to neutrophilic bacilli, acetoin, butanediol
or ethanol was not detected. Moderate amounts of isobutyric, isovaleric, α-oxoisovaleric,
α-oxo-β-methylvaleric, α-oxoisocaproic and phenylacetic acids were generated with three
of the alkaliphiles. High pH and/or buffer concentration tend to favor formation of acids.
The studied alkaliphiles were compared with each other and with neutrophilic bacilli.

8.8 Antibiotics

Since we reported on alkaliphiles, many Japanese pharmaceutical companies have
tried using alkaline media to discover new microorganisms producing new antibiotics.
Although several have been found and reported, none is commercially available at the present time.

8.8.1 Compounds Produced by *Paecilomyces lilacinus*
The first report on this topic was published in 1980 by Sato, Beppu and Arima. They

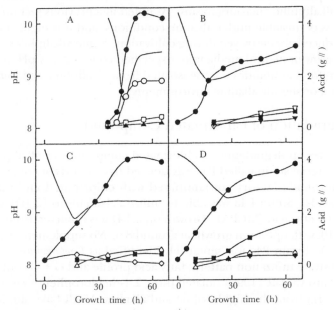

Fig. 8.3 Production of acids in 1% starch medium of *B. circulans* var. *alkalophilus* ATCC 21783 (A), *B. alcalophilus* subsp. *halodurans* ATCC 27557 (B), *B. alcalophilus* ATCC 27647 (C) and alkaliphilic *Bacillus* sp. 17-1 ATCC 31007 (D). Symbols: ●, acetic acid; ○, formic acid; □, succinic acid (coincides with α-hydroxyisocaproic acid); ▲, pyruvic acid; ▽, phenylacetic acid; ■, α-oxoiso-caproic acid (coincides with isovaleric acid in the case of *B. alcalophilus* subsp. *halodurans*); △, isovaleric acid (coincides with α-oxo-β-methylvaleric acid in the case of *Bacillus* sp. 17-1); ▼, isobutyric acid; ◇, lactic acid; ◆, propionic acid and —, pH.
(Reproduced with permission from Paavilainen, Helisto and Korpela, *J. Ferment. Bioeng.*, **78**, 218 (1994))

isolated *Paecilomyces lilacinus* No.1907 from soil using an alkaline medium with a pH of 10.5. This new antibiotic, No.1907, was produced only under alkaline conditions (pH 9 to 10.5). This microorganism was yeast-like and formed in an alkaline medium suitable for antibiotic production. *Paecilomyces lilacinus* No.1907 produced one major product, 1907-II, and a minor product, 1907-VIII, which had antibacterial and antifungal activities.

8.8.2 Production of Phenazine Antibiotics by *Nocardiopsis* Strain

An alkaliphilic actinomycetes, strain OPC-15, obtained from soil and identified as *Nocardiopsis dassonvillei*, produced phenazine antibiotics in the mycelium under alkaline culture conditions (Tsujibo *et al.*, 1988). Strain OPC-15 grew well at pH 10.0 and accumulated two phenazine antibiotics in the mycelia, 1,6-dihydroxyphenazine (I) and 1,6-dihydroxyphenazine 3,5-dioxide(III), at different growth temperatures. Antibiotic I was formed when cultured at 27 °C for 6–8 days, whereas the other (III) was formed only when grown at 27 °C for 6 days and incubated further at 4 °C for 2 days.

Besides these antibiotics, many other compounds have been found by several pharmaceutical companies. However, there is one weak point in the direct production of antibiotics by alkaliphilic microorganisms. As noted in a previous section (p. 123) many antibiotics are very unstable under alkaline conditions, and it is highly possible that any antibiotics produced are destroyed during cultivation. Some alkaliphiles, especially actinomycetes, can grow in neutral media under specified conditions of pH, temperature and nutrient. Therefore, antibiotics may be isolated from alkaliphilic actinomycetes by cultivation in neutral or slightly alkaline environments.

8.9 Degradation of Chlorinated Compounds

Several studies on degradation of chlorinated compounds by alkaliphiles have been reported. A bacterial strain called P4a was isolated from debris of concrete samples of a demolished herbicide factory contaminated with chlorinated and methylated phenoxyalkanoic acids. Strain P4a was able to utilize 2,4-dichloro-(2,4D) and 4-chloro-2-methylphenoxyacetic acid (MCPA). Growth on 2,4-D was observed from pH of 5.6 up to a pH of about 10, with optimum growth at around 8.5. No supplements were essential for growth on 2,4-D. Strain P4a was tentatively identified as *Comamonas acidovorans* on the basis of the substrate utilization pattern, fatty acid profile and G + C content (Hoffmann *et al.*, 1996). Maltseva *et al.* (1996) isolated three 2,4-dichlorophenoxyacetic acid (2,4-D)-degrading bacteria from the highly saline and alkaline Alkali Lake site in southwestern Oregon contaminated with 2,4-D production wastes. While similar in most respects, the three isolates differed significantly in 2,4-D degradation rates, with the most active strain, I-18, demonstrating the ability to degrade up to 3000 mg 2,4-D I-1 in 3 days. This strain grows optimally on 2,4-D at pH 8.4–9.4 and at sodium ion concentrations of 0.6–1.0 M. The 16S RNA gene sequence (303 nucleotides) was identical for all three isolates and most closely resembled those of the family *Halomonadaceae* (91% identity). Biochemical and genetic examination revealed that strain I-18 utilizes the same 2,4-D degradation pathway as most of the 2,4-D degrading bacteria from nonextreme environments.

References

Abelyan, V.A., T. Yamamoto and E.G. Afrikyan (1994a). Isolation and characterization of cyclomal-todextrin glucanotransferases using cyclodextrin polymers and their derivatives. *Biochemistry* (English Transl.), **59**, 573-579.

Abelyan, V.A., T. Yamamoto and E.G. Afrikyan (1994b). On the mechanism of action of cyclomal-todextrin glucotransferases of alkalophilic, thermophilic, and mesophilic microorganisms. *Biochemistry* (English Transl.), **59**, 839-844.

Abraham, T.E. (1996). Immobilized cyclodextrin glycosyl transferase for the continuous production of cyclodextrins. *Biocatal. Biotransform.*, **12**, 137-146.

Agnew, M.D., S.F. Koval and K.F. Jarrell (1995). Isolation and characterization of novel alkaliphiles from bauxite-processing waste and description of *Bacillus vedderi* sp. nov., a new obligate alkaliphile. *Syst. Appl. Microbiol.*, **18**, 221-230.

Akiba, T. and K. Horikoshi (1976a). Identification and growth characteristics of α-galactosidase producing microorganisms. *Agric. Biol. Chem.*, **40**, 1845-1849.

Akiba, T. and K. Horikoshi (1976b). Properties of α-galactosidases of alkalophilic bacteria. *Agric. Biol. Chem.*, **40**, 1851-1855.

Akiba, T. and K. Horikoshi (1978). Localization of α-galactosidase in an alkalophilic strain of *Micrococcus*. *Agric. Biol. Chem.*, **42**, 2091-2094.

Akiba, T. and K. Horikoshi (1980). A further separation and some properties of β-lactamase I. *Agric. Biol. Chem.*, **44**, 2741-2742.

Akino, T., N. Nakamura and K. Horikoshi (1987). Production of β-mannanase by an alkalophilic *Bacillus* sp. *Appl. Microbiol. Biotechnol.*, **26**, 323-327.

Akino, T., N. Nakamura and K. Horikoshi (1988a). Characterization of three β-mannanases of an alkalophilic *Bacillus* sp. *Agric. Biol. Chem.*, **52**, 773-779.

Akino, T., N. Nakamura and K. Horikoshi (1988b). Characterization of β-mannosidase of an alkalophilic *Bacillus* sp. *Agric. Biol. Chem.*, **52**, 1459-1464.

Akino, T., C. Kato and K. Horikoshi (1989a). The cloned β-mannanase gene from alkalophilic *Bacillus* sp. AM-001 produces two β-mannanases in *Escherichia coli*. *Arch. Microbiol.*, **152**, 10-15.

Akino, T., C. Kato and K. Horikoshi (1989b). Two Bacillus β-Mannanases having different COOH termini are produced in *Escherichia coli* carrying pMAH5. *Appl. Environ. Microbiol.*, **55**, 3178-3183.

Ambler, R. P. (1980). The structure of β-lactamases. *Phil. Trans. Roy. Soc., London, Ser.* **B289**, 321-331.

Ando, A., M. Yabuki, T. Fujii and S. Fukui (1981). General characteristics of an alkalophilic bacterium, *Bacillus* A-007. *Tech. Bull. of Fac. Hort., Chiba University*, **29**, 17-28.

Aono, R. (1985). Isolation and partial characterization of structural components of the cell walls of alkalophilic strain C-125. *J. Gen. Microbiol.*, **131**, 105-111.

Aono, R. (1987). Characterization of structural component of cell walls of alkalophilic strain of *Bacillus* sp. C-125: Preparation of poly(γ-L-glutamate) from cell wall component. *Biochem. J.*, **245**, 467-472.

Aono, R. (1988a). Cultivation conditions for extracellular production of penicillinase by *Escherichia coli* carrying pEAP31 on a semi-large scale. *Appl. Microbiol. Biotechnol.*, **28**, 414-418.

Aono, R. (1988b). Probable detection of *Kil* peptide derived from colicin E1 plasmid in the envelope fraction of *Escerichia coli* HB101 carrying pEAP31. *Biochem. J.*, **255**, 365-368.

Aono, R. (1989a). Characterization of cell wall components of the alkalophilic *Bacillus* strain C-125: Identification of a polymer composed of polyglutamate and polyglucuronate. *J. Gen. Microbiol.*, **135**, 265-271.

Aono, R. (1989*b*). Subcellular distribution of alkalophilic *Bacillus* penicillinase produced by *Escherichia coli* HB101 carrying pEAP31. *Appl. Microbiol. Biotechnol.*, **31**, 397-340.

Aono, R. (1990*a*). Taxonomic distribution of alkali-tolerant yeasts. *Syst. Appl. Microbiol.*, **13**, 394-397.

Aono, R. (1990*b*). The poly-α- and -β-1,4-glucuronic acid moiety of teichuronopeptide from the cell wall of the alkalophilic *Bacillus* strain C-125. *Biochem. J.*, **270**, 363-367.

Aono, R. (1992*a*). Phylogenetic relationships of alkali-tolerant yeasts belonging to the genus *Hansenula*. *Syst. Appl. Microbiol.*, **15**, 587-589.

Aono, R. (1992*b*). Accumulation of alkaliphilic *Bacillus* penicillinase cleaved within the signal sequence in cytoplasm of *Escherichia coli*. *Biosci. Biotechnol. Biochem.*, **56**, 890-895.

Aono, R. (1995). Assignment of facultatively alkaliphilic *Bacillus* sp. strain C-125 to *Bacillus lentus* group 3. *Int. J. Syst. Bacteriol.*, **45**, 582-585.

Aono, R., A. Hayakawa, M. Hashimoto, H. Kaneko, S. Nakamura and K. Horikoshi (1993). Cloning of a gene required for the alkaliphily of alkaliphilic *Bacillus* sp. strain C-125. *Nucleic Acid Sympo. Ser.*, **29**, 139-140.

Aono, R., H. Kaneko and K. Horikoshi (1996). Alkaline growth pH-dependent increase of respiratory and NADH-oxidation activities of the facultatively alkaliphilic strain *Bacillus lentus* C-125. *Biosci. Biotechnol. Biochem.*, **60**, 1243-1247.

Aono, R., H. Ogino and K. Horikoshi (1992). pH-Dependent flagellar formation by facultative alkalophilic *Bacillus* sp. C-125. *Biosci. Biotechnol. Biochem.*, **56**, 48-53.

Aono, R., and K. Horikoshi (1983). Chemical composition of cell walls of alkalophilic strains of *Bacillus*. *J. Gen. Microbiol.*, **129**, 1083-1087.

Aono, R., and K. Horikoshi (1991). Carotenes produced by alkaliphilic yellow-pigmented strains of *Bacillus*. *Agric. Biol. Chem.*, **55**, 2643-2645.

Aono, R., K. Horikoshi and S. Goto (1984). Composition of the peptidoglycan of alkalophilic *Bacillus* spp. *J. Bacteriol.*, **157**, 688-689.

Aono, R., M. Hammura, M. Yamamoto and T. Asano (1995). Isolation of extracellular 28- and 42-kilodalton β-1,3-glucanases and comparison of three β-1,3-glucanases produced by *Bacillus circulans* IAM1165. *Appl. Environ. Microbiol.*, **61**, 122-129.

Aono, R., M. Hashimoto, A. Hayakawa, S. Nakamura and K. Horikoshi (1992). A novel gene required for the alkaliphily of the facultative alkaliphile *Bacillus* sp. strain C-125. *Biosci. Biotechnol. Biochem.*, **56**, 842-844.

Aono, R., M. Ito and K. Horikoshi (1992). Instability of the protoplast membrane of facultative alkaliphilic *Bacillus* sp. C-125 at alkaline pH values below the pH optimum for growth. *Biochem. J.*, **285**, 99-103.

Aono, R., M. Ito and K. Horikoshi (1993*a*). Occurrence of teichuronopeptide in cell walls of group-2 alkaliphilic *Bacillus* spp. *J. Gen. Microbiol*., **139** (Part 11), 2739-2744.

Aono, R., M. Ito and K. Horikoshi (1993*b*). Regeneration of protoplasts prepared from alkaliphilic strains of *Bacillus* spp. *Biosci. Biotechnol. Biochem.*, **57**, 1597-1598.

Aono, R., M. Ito, K. N. Joblin and K. Horikoshi (1994). Genetic recombination after cell fusion of protoplasts from the facultative alkaliphile *Bacillus* sp. C-125. *Microbiology*, **140**, 3085-3090.

Aono, R., M. Ito, K. N. Joblin and K. Horikoshi (1995). A high cell wall negative charge is necessary for the growth of the alkaliphile *Bacillus lentus* C-125 at elevated pH. *Microbiology*, **141**, 2955-2964.

Aono, R., and M. Ohtani (1990). Loss of alkalophily in cell-wall-component-defective mutants derived from alkalophilic *Bacillus* C-125. *Biochem. J.*, **266**, 933-936.

Aono, R., M. Sato, M. Yamamoto and K. Horikoshi (1992). Isolation and partial characterization of an 87-kilodalton β-glucanase from *Bacillus* circulans IAM1165. *Appl. Environ. Microbiol.*, **58**, 520-524.

Aono, R., and M. Uramoto (1986). Presence of fucosamine in teichuronic acid of alkalophilic

Bacillus strain C-125. *Biochem. J.*, **233**, 291-294.

Aono, R., M. Yamamoto, M. Sato, Y. Nogi and K. Horikoshi (1992). Cloning and expression of a gene for an 87-kDa β-1,3-glucanase of *Bacillus circulans* IAM1165 in *E. coli*. *Biosci. Biotechnol. Biochem.*, **56**, 1854-1858.

Aono, R., and T. Sanada (1994). Hyper-autolysis of the facultative alkaliphile *Bacillus* sp. C-125 cells grown at neutral pH: Culture-pH dependent cross-linking of the peptide moieties of the peptidoglycan. *Biosci. Biotechnol. Biochem.*, **58**(11), 2015-2019.

Aono, R., T. Sanada and K. Horikoshi, (1996). Purification and characterization of cell wall-associated *N*-acetylmuramyl-L-alanine amidase from alkaliphilic *Bacillus lentus* C-125. *Biosci. Biotechnol. Biochem.*, **60**, 1140-1145.

Ara, K., K. Igarashi, H. Hagihara, K. Sawada, T. Kobayashi and S. Ito (1996). Separation of functional domains for the α-1,4 and α-1,6 hydrolytic activities of a *Bacillus amylopullulanase* by limited proteolysis with papain. *Biosci. Biotechnol. Biochem.*, **60**, 634-639.

Ara, K., K. Igarashi, K. Saeki and S. Ito (1995). An alkaline amylopullulanase from alkalophilic *Bacillus* sp. KSM-1378; Kinetic evidence for two independent active sites for the α-1,4 and α-1,6 hydrolytic reactions. *Biosci. Biotechnol. Biochem.*, **59**, 662-666.

Ara, K., K. Igarashi, K. Saeki, S. Kawai and S. Ito (1992). Purification and some properties of an alkaline pullulanase from alkaliphilic *Bacillus* sp. KSM-1876. *Biosci. Biotechnol. Biochem.*, **56**, 62-65.

Atsukawa, A., and K. Horikoshi (1976). Alkaline ribonuclease of alkalophilic *bacteria*. *Ribosomes and RNA Metabolism*, **2**, pp. 89-95, The Slovak Academy of Sciences, Bratislava.

Atsumi, T., Y. Maekawa, H. Tokuda and Y. Imae (1992). Amiloride at pH 7.0 inhibits the Na^+-driven flagellar motors of *Vibrio alginolyticus* but allows cell growth. *FEBS Lett.*, **314**, 114-116.

Atsumi, T., S. Sugiyama, J. Cragoe E. J. and Y. Imae (1990). Specific inhibition of the Na^+-driven flagellar motors of alkalophilic *Bacillus* strains by the amiloride analog phenamil. *J. Bacteriol.*, **172**, 1634-1639.

Aunstrup, K., H. Ottrup, O. Andresen and C. Dambmann (1972). Proteases from alkalophilic *Bacillus* species. *4th International Fermentation Symposium. Kyoto, Japan*, pp. 299-305, Society Fermentation Technology, Japan.

Bakels, R.H.A., H.S. Vanwalraven, K. Krab, M. J.C. Scholts and R. Kraayenhof (1993). On the activation mechanism of the H^+-ATP synthase and unusual thermodynamic properties in the alkalophilic cyanobacterium *Spirulina platensis*. *Eur. J. Biochem.*, **213**, 957-964.

Bansod, S.M., M. Duttachoudhary, M.C. Srinivasan and M.V. Rele (1993). Xylanase active at high pH from an alkalotolerant *Cephalosporium* species. *Biotechnol. Lett.*, **15**, 965-970.

Barghoorn, E.S., and S.A. Tyler (1965). Microorganisms from the gunflint chert. *Science* **147**, 563.

Battley, E. M., and E. J. Bartlett (1966). A convenient pH-gradient method for the determination of the maximum and minimum pH for microbial growth. *Antonie van Leeuwenhoek, J. Microbiol. Seral.*, **32**, 245-255.

Berendes, F., G. Gottschalk, E. Heinedobbernack, E.R.B. Moore and B.J. Tindall (1996). *Halomonas desiderata* sp. nov., a new alkaliphilic, halotolerant and denitrifying bacterium isolated from a municipal sewage works. *Syst. Appl. Microbiol.*, **19**, 158-167.

Bhushan, B., N.S. Dosanjh, K. Kumar and G.S. Hoondal (1994). Lipase production from an alkalophilic yeast sp. by solid state fermentation. *Biotechnol. Lett.*, **16**, 841-842.

Bhushan, B., and G.S. Hoondal (1994). Characterization of lipase from an alkalophilic yeast sp. *Biotechnol. Lett.*, **16**, 837-840.

Blanco, A., and F.I.J. Pastor (1993). Characterization of cellulase-free xylanases from the newly isolated *Bacillus* sp. strain Bp-23. *Can. J. Microbiol.*, **39**, 1162-1166.

Bogachev, A.V., R.A. Murtasina, A.I. Shestopalov and V. P. Skulachev (1993). The role of protonic and sodium potentials in the motility of *E. coli* and *Bacillus* FTU. *Biochim. Biophys. Acta*, **1142**, 321-326.

Boone, D.R., I.M. Mathrani, Y. Liu, J.A.G.F. Menaia, R.A. Mah and J.E. Boone (1993). Isolation and characterization of *Methanohalophilus-portucalensis* sp.-nov. and DNA reassociation study of the genus *Methanohalophilus*. *Int. J. Syst. Bacteriol.,* **43**, 430-437.

Bornside, G.H., and R.E. Kallio (1956). Urea-hydrolyzing bacilli. I. A physiological approach to identification. *J. Bacteriol.,* **71**, 627-634.

Boyer, E.W., and M.B. Ingle (1972). Extracellular alkaline amylase from a *Bacillus* species. *J. Bacteriol.,* **110**, 992-1000.

Boyer, E.W., M.B. Ingle and G.D. Mercer (1973). *Bacillus alcalophilus* subsp. *halodurans* subsp. nov. : An alkaline-amylase-producing alkalophilic organisms. *Int. J. Syst. Bacteriol.,* **23**, 238-242.

Brierley, G.P., M.H. Davis, E. J. Cragoe and D.W. Jung (1989). Kinetic properties of the K^+/H^+ antiport of heart mictochondoria. *Biochemistry,* **28**, 4345-4354.

Brierley, G.P., and D.W. Jung (1990). Kinteic properties of the K^+/H^+ antiport of heat mitochondoria. *Biochemistry,* **29**, 408-415.

Buck, D.P. and G.D. Smith (1995). Evidence for a Na^+/H^+ electrogenic antiporter in an alkaliphilic cyanobacterium *Synechocystis*. *FEMS Microbiol. Lett.,* **128**, 315-320.

Cao, J., L. Zheng and S. Chen (1992). Screening of pectinase producer from alkalophilic bacteria and study on its potential application in degumming of ramie. *Enzyme Microbiol. Technol.,* **14**, 1013-1016.

Chang, Y.C. H. Kadokura, K. Yoda and M. Y amasaki (1996). Secretion of active subtilisin YaB by a simultaneous expression of separate pre-pro and pre-mature polypeptides in *Bacillus subtilis*. *Biochem. Biophys. Res. Commun.,* **219**, 463-468

Chen, K.Y., and S. Cheng (1988). Polyamine metabolism in an obligately alkalophilic *Bacillus alkalophilus* that grows at pH 11.0. *Biochem. Biophys. Res. Commun.,* **150**, 185-191.

Chesbro, W.R., and J.B. Evans (1959). Factors affecting the growth of enterococci in highly alkaline media. *J. Bacteriol.,* **78**, 858-862.

Chislett, M.E., and D. J. Kushner (1961*a*). A strain of *Bacillus circulans* capable of growing under highly alkaline conditions. *J. Gen. Microbiol.,* **24**, 187-190.

Chislett, M.E., and D. J. Kushner (1961*b*). Germination under alkaline conditions and transmission of alkali resistance by endospores of certain strains of *Bacillus cereus* and *Bacillus circulans*. *J. Gen. Microbiol.,* **25**, 151-156.

Choi, Y.J., I.H. Kim, B.H. Lee and J.S. Lee (1995). Purification and characterization of β-galactosidase from alkalophilic and thermophilic *Bacillus* sp. TA-11. *Biotechnol. Appl. Biochem.,* **22**, 191-201.

Damude, H.G., N.R. Gilkes, D.G. Kilburn, R.C. Miller, R. Antony and J. Warren (1993). Endoglucanase CasA from alkalophilic streptomyces strain KSM-9 is a typical member of family-B of β-1,4-Glucanases. *Gene,* **123**, 105-107.

Dasilva, R., D. K. Yim, E. R. Asquieri and Y. K. Park (1993). Production of microbial alkaline cellulase and studies of their characteristics. *Rev. Microbiol.,* **24**, 269-274.

Davidson, M.W., K.A. Grayt, D.B. Knaff and T.A. Krulwich (1988). Purification and characterization of two soluble cytochromes from the alkalophilic *Bacillus firmus* RAB. *Biochim. Biophys. Acta,* **933**, 470-477.

Davis, M.H., R.A. Altshuld, D.W. Jung and G.P. Brierley (1987). Estimation of intramitochondrial pCa and pH by Fura-2 and 2,7-bis(carboxyethyl)-5(6)-carboxyfluorescein (BCECF) fluorescene. *Biochem. Biophys. Res. Commun.,* **149**, 40-45.

De Graaf, F.K., and B. Oudega (1986). Production and release of cloacin DF13 and related colicins. *Current Topics in Microbiology and Immunity,* Springer Verlag, Berlin, Heidelberg, New York, Tokyo.

De Rosa, M., A. Gambacorta, B. Nicolalus, H.N.M. Ross, W.D. Grant and J. D. Bu' Lock (1982). An asymmetric archaebacterial diether lipid from alkalophilic halophiles. *J. Gen. Microbiol.,* **128**, 343-348.

Dees, C., D. Ringelberg, T.C. Scott and T.J. Phelps (1995). Characterization of the cellulose degrading bacterium NCIMB 10462. *Appl. Biochem. Biotechnol.*, **51**(2), 263-274.

Deuel, H., and E. Stutz (1958). Pectic substances and pectic enzymes. *Advanced Enzymology* (Ed. E. F. Nord), pp. 341-382, Interscience Pub, New york.

Dey, D., J. Hinge, A. Shendye and M. Rao (1992). Purification and properties of extracellular endoxylanases from alkalophilic thermophilic *Bacillus* sp. *Can. J. Microbiol.*, **38**, 436-442.

Downie, A. W., and J. Cruickshank (1928). The resistance of *Streptococcus faecalis* to acid and alkaline media. *Brit. J. Exp. Pathol.*, **9**, 171-173.

Driessen, A.J.M., J.L.C.M. Vandevossenberg and W.N. Konings (1996). Membrane composition and ion-permeability in extremophilies. *FEMS Microbiol. Rev.*, **18**, 139-148.

Duckworth, A.W., W.D. Grant, B.E. Jones and R. Vansteenbergen (1996). Phylogenetic diversity of soda lake alkaliphilies. *FEMS Microbiol. Ecol.*, **19**, 181-191.

Dunkley, J., E.A., A.A. Guffanti, S. Clejan and T.A. Krulwich (1991). Facultative alkaliphiles lack fatty acid desaturase activity and lose the ability to grow at near-neutral pH when supplemented with an unsaturated fatty acid. *J. Bacteriol.*, **173**, 1331-1334.

Dwivedi, A., U.K. Srinivas, H.N. Singh and H.D. Kumar (1994). Regulatory effect of external pH on the intracellular pH in alkalophilic cyanobacteria *Microcystis aeruginosa* and *Hapalosiphon welwitschii. J. Gen. Appl. Microbiol., Tokyo*, **40**, 261-263.

Engle, M., Y. H. Li, C. Woese and J. Wiegel (1995). Isolation and characterization of a novel alkalitolerant thermophile, *Anaerobranca horikoshii* gen. nov., sp. nov. *Int. J. Syst. Bacteriol.*, **45**, 454-461.

Engle, M., Y.H. Li, F. Rainey, S. Deblois, V. Mai, A. Reichert, F. Mayer, P. Messner and J. Wiegegel (1996). T*hermobrachium celere* gen. nov., sp. nov., a rapidly growing thermophilic, alkalitolerant, and proteolytic obligate anaerobe. *Int. J. Syst. Bacteriol.*, **46**, 1025-1033.

Fedorov, O.V., M.G. Pyatibratov, A.S. Kostyukova, N.K. Osina and V.Y. Tarasov (1994). Protofilament as a structural element of flagella of haloalkalophilic archaebacteria. *Can. J. Microbiol.*, **40**(1), 45-53.

Forgarty, W. M., and P. O. Ward (1977). Pectinases and pectic polysaccharides. *Progress in Industrial Microbiology*, pp. 59-119, Churchill Livingstone, Edinburgh and London.

Fritze, D., J. Flossdorf and D. Claus (1990). Taxonomy of alkaliphilic *Bacillus* strains. *Intl. J. Syst. Bacteriol.*, **40**, 92-97.

Fujiwara, N., A. Masui and T. Imanaka (1993). Purification and properties of the highly thermostable alkaline protease from an alkaliphilic and thermophilic *Bacillus* sp., *J. Biotechnol.*, **30**, 245-256.

Fujiwara, N., and K. Yamamoto (1987). Decomposition of gelatin layers on X-ray film by the alkaline protease from *Bacillus* sp. B21. *J. Ferment. Technol.*, **65**, 531-534.

Fujiwara, N., K. Yamamoto and A. Masui (1991). Utilization of thermostable alkaline protease from an alkalophilic thermophile for the recovery of silver from used X-ray film. *J. Ferment. Bioeng.*, **72**, 306-308.

Fukumori, F., T. Kudo and K. Horikoshi (1985). Purification and properties of a cellulase from alkalophilic *Bacillus* sp. No.1139. *J. Gen. Microbiol.*, **131**, 3339-3345.

Fukumori, F., T. Kudo and K. Horikoshi (1987). Truncation analysis of an alkaline cellulase from an alkalophilic *Bacillus* species. *FEMS Microbiol. Lett.*, **40**, 311-314.

Fukumori, F., T. Kudo, Y. Narahashi and K. Horikoshi (1986). Molecular cloning and nucleotide sequence of the alkaline cellulase gene from the alkalophilic *Bacillus* sp. Strain 1139, *J. Gen. Microbiol.*, **132**, 2329-2335.

Fukumori, F., T. Kudo, N. Sashihara, Y. Nagata, K. Ito and K. Horikoshi (1989). The third cellulase of alkalophilic *Bacillus* sp. strain N-4, Evolutionary relationships within the *cel* gene family. *Gene*, **76**, 289-298.

Fukumori, F., K. Ohishi, T. Kudo and K. Horikoshi (1987). Tandem location of the cellulase genes on the chromosome of *Bacillus* sp. strain N-4. *FEMS Microbiol. Lett.*, **48**, 65-68.

Fukumori, F., N. Sashihara, T. Kudo and K. Horikoshi (1986). Nucleotide sequences of two cellulase genes from alkalophilic *Bacillus* sp. strain N-4 and their strong homology. *J. Bacteriol.*, **168**, 479-485.

Garcia, M.L., A.A. Fugganti and T.A. Krulwich (1983). Characterization of Na^+/H^+ antiporter of alkalophilic bacilli *in vivo*: $\Delta\Psi$-dependent $_{22}Na^+$ efflux from whole cells. *J. Bacteriol.*, **156**, 1151-1157.

Gascoyne, D. J., J. A. Connor and A. T. Bull (1991*a*). Capacity of siderophore-producing alkalophilic bacteria to accumulate iron, gallium and aluminum. *Appl. Microbiol. Biotechnol.*, **36**, 136-141.

Gascoyne, D. J., J. A. Connor and A.T. Bull (1991*b*). Isolation of bacteria producing siderophores under alkaline conditions. *Appl. Microbiol. Biotechnol.*, **36**, 130-135.

Gee, J. M., B. M. Lund, G. Metcalf and J. L. Peel (1980). Properties of a new group of alkalophilic bacteria. *J. Gen. Microbiol.*, **117**, 9-17.

Georganta, G., T. Kaneko, T. Kudo and K. Horikoshi (1991). Expression of the CGTase gene of alkalophhilic *Bacillus* No. 38-2 in various hosts. *Starch/Stärke*, **43**, 361-363.

Georganta, G., T. Kaneko, N. Nakamura, T. Kudo and K. Horikoshi (1993). Isolation and partial properties of cyclomaltodextrin glucanotransferase-producing alkaliphilic *Bacillus* spp. from a deep-sea mud sample. *Starch/Stärke* **45**, 95-99.

Gibson, T. (1934). An investigation of the *Bacillus pasteurii* group. II. Special physiology of the organisms. *J. Bacteriol.*, **28**, 313-322.

Gilmour, R., and T.A. Krulwich (1996). Purification and characterization of the succinate dehydrogenase complex and CO-reactive *b*-type cytochromes from the facultative alkaliphile *Bacillus firmus* OF4. *Bba-Bioenergetics*, **1276**, 57-63.

Gilmour, R., and R.N. Krulwich (1997). Construction and characterization of a mutant of alkaliphilic *Bacillus firmus* OF4 with a disrupted *cta* operon and purification of a novel cytochrome bd. *J. Bacteriol.*, **179**, 863-870.

Goldfarb, D.S., R.L. Rodoriguez and R.H. Doi (1982). Translation block to expression of the *Escherichia coli* Tn9-derivated chloramphenicol-resistance gene in *Bacillus subtilis*. *Proc. Natl. Acad. Sci. USA*, **79**, 5886-5890.

Gordon, R. E. (1973). *The Genus Bacillus*. Cleveland, CRC Press.

Gordon, R. E. and J. L. Hyde (1982). The *Bacillus firmus* lentus complex and pH 7.0 variants of some alkalophilic strains. *J. Gen. Microbiol.*, **128**, 1109-1116.

Goto, S., R. Aono, J. Sugiyama and K. Horikoshi (1981). *Exophiala alcalophila*, a new black, yeast-like hyphomycete with an accompanying *Phaeococcomyces alcalophila* morph, and its physiological characteristics. *Trans. Mycol. Soc. Japan*, **22**, 429-439.

Gudat, J. C., J. Singh and D. C. Whaton (1973). Cytochrome oxidase from *Pseudomonas aeruginosa*. (1) Purification and some properties. *Biochim. Biophys. Acta*, **292**, 376-390.

Guffanti, A. A. (1983). ATP-dependent Na^+/H^+ antiport activity in *Bacillus alcalophilus* require generation of an electrochemical gradient of protons. *FEMS Microbiol. Lett.*, **17**, 307-310.

Guffanti, A.A., R. Blanco, R.A. Benenson and T.A. Krulwich (1980). Bioenergetic properties of alkaline-tolerant and alkalophilic strains of *Bacillus firmus*. *J. Gen. Microbiol.*, **119**, 79-96.

Guffanti, A. A., R.F. Bornstein and T.A. Krulwich (1981). Oxidative phosphorylation by membrane vesicles from *Bacillus alcalophilus*. *Biochim. Biophys. Acta*, **635**, 619-630.

Guffanti, A. A., E. Chiu and T.A. Krulwich (1985). Failure of an alkalophilic bacterium to synthesize ATP in response to a valinomycin-induced potassium diffusion potential at high pH. *Arch. Biochim. Biophys.*, **239**, 327-333.

Guffanti, A.A., R.T. Fuchs, M. Schneier, E. Chiu and T.A. Krulwich (1984). $\Delta\Psi$-generated by respiration is not equivalent to a diffusion potential of the same magnitude for ATP synthesis by *Bacillus firmus* RAB. *J. Biol. Chem.*, **259**, 2971-2975.

Guffanti, A.A. and T.A. Krulwich (1992). Features of apparent nonchemiosmotic energrgization of oxidative phosphorylation by alkaliphilic *Bacillus firmus* OF4. *J. Biol. Chem.*, **267**, 9580-9588.

Guffanti, A.A. and T.A. Krulwich (1994). Oxidative phosphorylation by ADP + Pi-loaded membrane vesicles of alkaliphilic *Bacillus firmus* OF4. *J. Biol. Chem.*, **269**, 21576-21582.

Guffanti, A. A., P. Susman, R. Blanco and T. A. Krulwich (1978). The proton-motive force and α-aminoisobutyric acid transport in an obligately alkalophilic bacterium. *J. Biol. Chem.*, **253**, 708-715.

Hamamoto, T., M. Hashimoto, M. Hino, M. Kitada, Y. Seto, T. Kudo and K. Horikoshi (1994). Characterization of a gene responsible for the Na^+/H^+ antiporter system of alkaliphilic *Bacillus* species strain C-125. *Mol. Microbiol.*, **14**, 939-946.

Hamamoto, T., H. Honda, T. Kudo and K. Horikoshi (1987). Nucleotide sequence of the xylanase a gene of alkalophilic *Bacillus* sp. strain C-125. *Agric. Biol. Chem.*, **51**, 953-955.

Hamamoto, T., and K. Horikoshi (1987). Alkalophilic *Bacillus* xylanase A, a secretable protein through outer membrane of *Escherichia coli*. *Agric. Biol. Chem.*, **51**, 3133-3135.

Hamamoto, T., T. Kaneko and K. Horikoshi (1987). Nucleotide sequence of the cyclomaltodextrin glucanotransferase (CGTase) gene from alkalophilic *Bacillus* sp. strain No. 38-2. *Agric. Biol. Chem.*, **51**, 2019-2022.

Hamana, K., T. Akiba, F. Uchino and S. Matsuzaki (1989). Distribution of spermine in Bacilli and lactic acid bacteria. *Can. J. Microbiol.*, **35**, 450-455.

Hamasaki, N., S. Shirai, M. Niitsu, K. Kakinuma and T. Oshima (1993). An alkalophilic *Bacillus* sp. produces 2-phenylethylamine. *Appl. Environ. Microbiol.*, **59**, 2720-2722.

Harold, F.M. (1972). Conservation and transformation of energy by bacterial membranes. *Bacteriol. Rev.*, **36**, 172-230.

Harold, F.M. (1977). Membranes and energy transduction in bacteria. *Curr. Top. Bioenerg.*, **6**, 83-149.

Hashimoto, M., T. Hamamoto, M. Kitada, M. Hino, T. Kudo and K. Horikoshi (1994). Characteistics of alkali-sensitive mutants of alkaliphilic *Bacillus* sp. strain C-125 that show cellular morphological abnormalities. *Biosci. Biotechnol. Biochem.*, **58**, 2090-2092.

Hatada, Y., K. Igarashi, K. Ozaki, K. Ara, J. Hitomi, T. Kobayashi, S. Kawai, T. Watabe and S. Ito (1996). Amino acid sequence and molecular structure of an alkaline amulopullulanase from *Bacillus* that hydrolyzes α-1,4 and α-1,6 linkages in polysaccharides at different active sites. *J. Biol. Chem.*, **271**, 24075-24083.

Hayashi, T., T. Akiba and K. Horikoshi (1988*a*). Production and purification of new maltohexaose-forming amylases from alkalophilic *Bacillus* sp. H-167. *Agric. Biol. Chem.*, **52**, 443-448.

Hayashi, T., T. Akiba and K. Horikoshi (1988*b*). Properties of new alkaline maltohexaose-forming amylases. *Appl. Microbiol. Biotechnol.*, **28**, 281-285.

Hellman, J., and P. Maentsaelae (1992). Construction of *Escherichia coli* export-affinity vector for expression and purification of proteins by fusion to cyclomaltodextrin glucanotransferase. *J. Biotechnol.*, **23**, 19-34.

Hicks, D.B., and T.A. Krulwich (1986). The membrane ATPase of alkaliphilic *Bacillus firmus* RAB is an F1-type ATPase. *J. Biol. Chem.*, **261**, 12896-12902.

Hicks, D.B., and T.A. Krulwich (1995). The respiratory chain of alkaliphilic bacteria. *Biochim. Biophys. Acta-Bioenerg.*, **1229**, 303-314.

Hirota, M., M. Kitada and Y. Imae (1981). Flagellar motors of alkalophilic *Bacillus* are powered by an electrochemical potential gradient of Na^+. *FEBS Lett.*, **132**, 278-280.

Hirota, N. and Y. Imae (1983). Na^+-driven flagellar motors of an alkalophilic *Bacillus* strain YN-1. *J. Biol. Chem.*, **258**, 10577-10581.

Hitomi, J., J. S. Park, M. Nishiyama, S. Horinouchi and T. Beppu (1994). Substrate-dependent change in the pH-activity profile of alkaline endo-1,4-β-glucanase from an alkaline *Bacillus* sp. *J. Biochem. Tokyo*, **116**, 554-559.

Hoffmann, D., R.H. Muller, B. Kiesel and W. Babel (1996). Kinetic isolation and characterization of an alkaliphilic bacterium capable of growing on 2,4-dichlorophenoxyacetic acid and 4-chlol-ro-2-methylphenoxyacetic acid. *Acta Biotechnol.*, **16**, 121-131.

Hoffmann, A., and P. Dimroth (1991). The electrochemical proton potential of *Bacillus alcalophilus*. *Eur. J. Biochem.*, **201**, 467-473.

Holm-Hansen, O. (1968). Ecology, physiology and biochemistry of blue-green algae. *Ann. Rev. Microbiol.*, **22**, 47-70.

Honda, H., C. Kato, T. Kudo and K. Horikoshi (1984). Cloning of leucine genes of alkalophilic *Bacillus* No. 221 in *E. coli* and *B. subtilis. J. Biochem.*, **95**, 1485-1490.

Honda, H., T. Kudo and K. Horikoshi (1985*a*). Molecular cloning and expression of xylanase gene of alkalophilic *Bacillus* sp. strain C-125 in *Escherichia coli. J. Bacteriol.*, **161**, 784-785.

Honda, H., T. Kudo and K. Horikoshi (1985*b*). Purification and partial characterization of alkaline xylanase from *Escherichia coli* carrying pCX311. *Agric. Biol. Chem.*, **49**, 3165-3169.

Honda, H., T. Kudo, Y. Ikura and K. Horikoshi (1985). Two types of xylanases of alkalophilic *Bacillus* sp. No. C-125. *Can. J. Microbiol.*, **31**, 538-542.

Horikoshi, K. (1971*a*). Production of alkaline enzymes by alkalophilic microorganisms. Part I. Alkaline protease produced by *Bacillus* No. 221. *Agric. Biol. Chem.*, **36**, 1407-1414.

Horikoshi, K. (1971*b*). Production of alkaline enzymes by alkalophilic microorganisms. Part II. Alkaline amylase produced by *Bacillus* No. A-40-2. *Agric. Biol. Chem.*, **35**, 1783-1791.

Horikoshi, K. (1972). Production of alkaline enzymes by alkalophilic microorganisms. Part III. Alkaline pectinase of *Bacillus* No. P-4-N. *Agric. Biol. Chem.*, **362**, 85-293.

Horikoshi, K. (1991). *Microorganisms in Alkaline Environments*, Kodansha : VCH, Tokyo.

Horikoshi, K., and T. Akiba (1982). *Alkalophilic Microorganisms: A New Microbial World*, Springer-Verlag : Japan Scientific Societies Press, Tokyo.

Horikoshi, K., and T. Akiba (Eds). (1993). *Alkalophilic Microorganisms*, Springer Verlag : Japan Scientific Societies Press, Tokyo.

Horikoshi, K., and Y. Atsukawa (1973*a*). β-1,3-Glucanase produced by alkalophilic bacteria *Bacillus* No. K-12-5. *Agric. Biol. Chem.*, **37**, 1449-1456.

Horikoshi, K., and Y. Atsukawa (1973*b*). Xylanase produced by alkalophilic *Bacillus* No. C-59-2. *Agric. Biol. Chem.*, **37**, 2097-2103.

Horikoshi, K., and Y. Atsukawa (1975). Production of β-1,3-glucanase by *Bacillus* No. 221, and alkalophilic microorganisms. *Biochim. Biophys. Acta*, **384**, 477-483.

Horikoshi, K., and W.D. Grant, (Eds.) (1991). *Superbugs*. Springer-Verlag : Japan Scientific Societies Press, Tokyo.

Horikoshi, K., and S. Iida (1958). Lysis of fungal mycelia by bacterial enzymes. *Nature*, **181**, 917-918.

Horikoshi, K., and S. Iida (1959). Effect of lytic enzymes from *Bacillus circulans* and chitinase from *Streptomyces* sp. of *Aspergillus oryzae. Nature*, **183**, 186-187.

Horikoshi, K., and S. Iida (1964). Studies on the spore coat of fungi. I. Isolation and composition of the spore coats of *Aspergillus oryzae. Biochim. Biophys. Acta*, **83**, 197-203.

Horikoshi, K., H. Koffler and K. Arima (1963). Purification and properties of β-1,3-glucanase from the "lytic enzyme" of *Bacillus circulans. Biochim. Biophys. Acta*, **73**, 267-275.

Horikoshi, K., M. Nakao, Y. Kurono and N. Saschihara (1984). Cellulases of and alkalophilic *Bacillus* strain isolated from soil. *Can. J. Microbiol.*, **30**, 774-779.

Horikoshi, K., and Y. Yonezawa (1978). A bacteriophage active on an alkalophilic *Bacillus* sp. *J. Gen. Virol.*, **39**, 183-185.

Horinouchi, S., and B. Weisblum (1982). Nucleotide sequence and functional map of pE194, a plasmid that specifies inducible resistance to macrolide, linocosamide, and streptogramin type B antibiotics. *J. Bacteriol.*, **150**, 804-814.

Hussain, M., F. I. J. Pator and J. O. Tampeu (1987). Cloning and sequencing of *bal* Z gene ecoding β-lactamase III, lipoprotein of *Bacillus sereus* 569/H. *J. Bacteriol.*, **169**, 579-586.

Igarashi, K., K. Ara, K. Saeki, K. Ozaki, S. Kawai and S. Ito (1992). Nucleotide sequence of the gene that encodes an neopullulanase from an alkalophilic *Bacillus*. *Biosci. Biotechnol. Biochem.*, **56**, 514-516.

Ikeda, K., K. Nakajima and I. Yumoto (1994). Isolation and characterization of a novel facultatively alkaliphilic bacterium, *Corynebacterium* sp. grown on *n*-alkanes. *Arch. Microbiol.*, **162**, 381-386.

Ikura, Y., and K. Horikoshi (1977). Isolation and some properties of alkalophilic bacteria utilizing rayon waste. *Agric. Biol. Chem.*, **41**, 1373-1377.

Ikura, Y., and K. Horikoshi (1978). Cell free protein synthesizing system of alkalophilic *Bacillus* No. A-59. *Agric. Biol. Chem.*, **42**, 753-756.

Ikura, Y., and K. Horikoshi (1979*a*). *β*-Galactosidase in alkalophilic *Bacillus*. *Agric. Biol. Chem.*, **43**, 1359-1360.

Ikura, Y., and K. Horikoshi (1979*b*). Isolation and some properties of *β*-galactosidase producing bacteria. *Agric. Biol. Chem.*, **43**, 85-88.

Ikura, Y., and K. Horikoshi (1983). Studies on cell wall of alkalophilic *Bacillus*. *Agric. Biol. Chem.*, **47**, 681-686.

Ikura, Y., and K. Horikoshi (1987*a*). Effect of amino compounds on alkaline amylase production by alkalophilic *Bacillus* sp. *J. Ferment. Technol.*, **65**, 707-709.

Ikura, Y., and K. Horikoshi (1987*b*). Isolation and some properties of *α*-galactosidase-producing alkalophilic bacteria. *Agric. Biol. Chem.*, **51**, 243-245.

Ikura, Y., and K. Horikoshi (1987*c*). Stimulatory effect of certain amino acids on xylanase production by alkalophilic *Bacillus* sp. *Agric. Biol. Chem.*, **51**, 3143-3145.

Ikura, Y., and K. Horikoshi (1988). Stimulatory effect of manganese and glycine derivatives on production and release of *β*-galactosidase by alkalophilic *Bacillus* No. C-125. *Can. J. Microbiol.*, **34**, 1177-1182.

Ikura, Y., and K. Horikoshi (1989*a*). Manganese dependent production of 5'-nucleotidase by alkalophilic *Bacillus* No. C-3. *J. Ferment. Bioeng.*, **67**, 111-114.

Ikura, Y., and K. Horikoshi (1989*b*). Manganese-dependent production and properties, of 5'-nucleotidase by alkalophilic *Bacillus* No. A-59. *Agric. Biol. Chem,*. **53**, 645-649.

Ikura, Y., and K. Horikoshi (1989*c*). Purification and properties of 5'-nucleotidase from alkalophilic *Bacillus* No. C-3. *J. Ferment. Bioeng.*, **67**, 210-211.

Ikura, Y., and K. Horikoshi (1990). Manganese ion-dependent production of phosphodiesterase by alkalophilic *Bacillus* No. A-40-2 and its properties. *Agric. Biol. Chem.*, **54**, 3205-3209.

Ikura, Y., and K. Horikoshi (1992). pH optima of intracellular, membrane-bound, and extracellular amylases from alkalophilic *Bacilli*. *Biosci Biotechnol. Biochem.*, **56**, 2064-2065.

Ikura, Y., and K. Horikoshi (1994). Stimulatory effects of certain amino acids, sugars, organic acids, nucleic acids and metals on nitrate utilization by alkalophilic *Bacillus*. *J. Ferment. Bioeng.*, **77**, 562-564.

Imae, Y., and T. Atsumi (1989). Na$^+$-driven bacterial flagellar motors. *J. Bioenerg. Biomembr.*, **21**, 705-716.

Ishikawa, H., K. Ishimi, M. Sugiura, A. Sowa and N. Fujiwara (1993). Kinetics and mechanism of enzymatic hydrolysis of gelatin layers of X-ray film and release of silver particles. *J. Ferment. Bioeng.*, **76**(4), 300-305.

Ito, M., R. Aono and K. Horikoshi (1993). Identification of 2-amino-2,6-dideoxy-D-glucose (D-quinovosamine) isolated from the cell walls of the alkaliphilic *Bacillus*-Sp. Y-25, by 500-MHz H-1 NMR spectroscopy. *Carbohydr. Res.*, **242**, 173-180.

Ito, M., K. Tabata and R. Aono (1994). Construction of a new teichuronopeptide-defective derivative from alkaliphilic *Bacillus* sp C-125 by cell fusion. *Biosci. Biotechnol. Biochem.*, **58**, 2275-2277.

Ito, S., Y. Ohta, M. Shimooka, M. Takaiwa, K. Ozaki, S. Adachi and K. Okamoto (1991). Enhanced production of extracellular enzyme by mutants of *Bacillus* that have acquired resistance to Vancomycin and Tirstocetin. *Agric. Biol. Chem.*, **55**, 2387-2391.

Ito, S., S. Shikata, K. Ozaki, S. Kawai, K.I. Okamoto S., A. Takei, Y. Ohta and T. Satoh (1989).

Alkaline cellulase for laundry detergents: Production by *Bacillus* sp. KSM-635 and enzymatic properties. *Agric. Biol. Chem.*, **53**, 1275-1281.

Ivey, D.M., A.A. Guffanti, J.S. Bossewitch, E. Padan and T.A. Krulwich (1991). Molecular cloning and sequencing of a gene from alkaliphilic *Bacillus firmus* OF4 that functionally complements an *Escherichia coli* strain carrying a deletion in the *nha*A Na⁺/H⁺ antiporter gene. *J. Biol. Chem.*, **266**, 23483-23489.

Ivey, D.M., A.A. Guffanti, Z. Shen, N. Kudyan and T. Krulwich (1992). The *cad* C gene product of alkaliphilic *Bacillus firmus* OF4 partially restores Na⁺ resistance to an *Escherichia coli* stain lacking an Na⁺/H⁺ antiporter (*Nha*A). *J. Bacteriol.*, **174**, 4878-4884.

Ivey, D.M., and T.A. Krulwich (1991). Organization and nucleotide sequence of the *atp* genes encoding the ATP synthase from alkalophilic *Bacillus firmus* OF4. *Mol. Gen. Genet.*, **229**, 292-300.

Ivey, D.M., M.G. Sturr, T.A. Krulwich and D.B. Hicks (1994). The abundance of *atp* gene transcript and of the membrane F1F0-ATPase as a function of the growth pH of alkaliphilic *Bacillus firmus* OF4. *J. Bacteriol.*, **176**, 5167-5170.

Iwazawa, J., Y. Imae and S. Kobayashi (1993). Study of the torque of the bacterial flagellar motor using a rotating electric field. *Biophys. J.*, **64**, 925-933.

Jahns, T. (1996). Unusually stable NAD-specific glutamate dehydrogenase from the alkaliphile *Amphibacillus xylanus*. *Anton Leeuwenhoek Int. J. Gen. M*, **70**, 89-95.

Jenkin, P.M. (1936). Reports on the Percy Sloden expedition to some rift valley lakes in Kenia in 1929. VII. Summary of the ecological results, with special reference to the alkaline lakes. *Ann. Mag. Nat. Hist.*, **18**, 133-181.

Johnson, H.W. (1923). Relationships between hydrogen ion, hydroxyl ion and salt concentrations and the growth of seven soil molds. *Iowa State Coll. Eng. Exp. Sta. Bull.*, **76**, 307-344.

Jones, B.E., W.D. Grant, N.C. Collins and W.E. Mwatha (1994). Alkaliphiles: diversity and identification. *Bacterial Diversity and Systematics* (Eds. F.G. Priest, A. Ramos-Crmenzana and B.J. Tindal.), pp. 195-230, Plenum Press, New York and London.

Kanai, H., T. Kobayashi, R. Aono and T. Kudo (1995). *Natronococcus amylolyticus* sp. nov., a haloalkaliphilic archaeon. *Int. J. Syst. Bact.*, **45**, 762-766.

Kaneko, R., N. Koyama, Y.C. Tsai, R.Y. Jung, K. Yoda and K. Yamasaki (1989). Molecular cloning of the structural gene for alkaline elastase Ya-B, a new subtilisin produced by an alkalophilic *Bacillus* strain. *J. Bacteriol.*, **171**, 5232-5236.

Kaneko, T., T. Hamamoto and K. Horikoshi (1988). Molecular cloning and nucleotide sequence of the cyclomaltodextrin glucanotransferase gene from the alkalophilic *Bacillus* sp. Strain No. 38-2. *J. Gen. Microbiol.*, **134**, 97-105.

Kaneko, T., T. Kato, N. Nakamura and K. Horikoshi (1987). Spectrophotometric determination of cyclization activity of β-cyclodextrin-forming cyclomaltodextrin glucanotansferase. *J. Jpn. Soc. Starch. Sci.*, **34**, 45-48.

Kaneko, T., T. Kudo and K. Horikoshi (1990). Comparison of CD composition produced by chimeric CGTases. *Agric. Biol. Chem.*, **54**, 197-201.

Kaneko, K., T. Kudo, G. Georganta and K. Horikoshi (1990). Construction of chimeric series of *Bacillus* cyclomaltodextrin glucanotransferase and analysis of the thermal stabilities and pH optima of the enzymes and yield of cyclodextrins. *Minutes: The Fifth Int. Symp. Cyclodextrins, Paris*, pp. 19.

Kaneko, T., K.B. Song, T. Hamamoto, T. Kudo and K. Horikoshi (1989). Construction of a chimeric series of *Bacillus* cyclomaltodextrin glucanotransferases and analysis of the thermal stabilities and pH optima of the enzymes. *J. Gen. Microbiol.*, **135**, 3447-3457.

Kang, M.K., P.J. Maeng and Y.H. Rhee (1996). Purification and characterization of two xylanases from alkalophilic *Cephalosporium* sp. strain RYM-202. *Appl. Environ. Microbiol.*, **62**, 3480-3482.

Kates, M. (1996). Structural analysis of phospholipids and glycolipids in extremely halophilic archaebacteria. *J. Microbiol. Meth.*, **25**, 113-128.

Kato, C., H. Honda, T. Kudo and K. Horikoshi (1984). Molecular cloning and expression of β-iso-propylmalate dehydrogenase gene from alkalophilic *Bacillus* in *Escherichia coli*. *J. Ferment. Technol.*, **62**, 77-80.

Kato, C., T. Kobayashi, T. Kudo, T. Furusato, Y. Murakami, H. Tanaka, T. Baba, T. Oishi, E. Ohtsuka, M. Ikehara, T. Yanagita, H. Kato, S. Moriyama and K. Horikoshi (1987). Construction of an excretion vector and extracellular production of human growth hormone from *Escherichia coli*. *Gene*, **54**, 197-202.

Kato, C., T. Kobayashi, T. Kudo and K. Horikoshi (1986). Construction of an excetion vector: Extracellular production of *Aeromonas* xylanase and *Bacillus* cellulases by *Escherichia coli*. *FEMS Microbiol. Lett.*, **36**, 31-34.

Kato, C., T. Kudo, K. Watanabe and K. Horikoshi (1983). Extracellular production of *Bacillus* penicillinase by *Escherichia coli* carrying pEAP2. *Eur. J. Appl. Microbiol. Biotechnol.*, **18**, 339-343.

Kato, C., T. Kudo, K. Watanabe and K. Horikoshi (1985). Nucleotide sequence of the β-lactamase gene of alkalophilic *Bacillus* sp. strain 170. *J. Gen. Microbiol.*, **131**, 3317-3324.

Kato, C., Y. Nakano and K. Horikoshi (1989). The nucleotide sequence of the lipo-penicillinase gene of alkalophilic *Bacillus* sp. strain 170. *Arch. Microbiol.*, **151**, 91-94.

Kato, C., A. Ohkoshi, T. Kudo and K. Horikoshi (1986). Extracellular production of xylanase L in *Escherichia coli* using excretion vector pEAP2. *Agric. Biol. Chem.*, **50**, 1067-1068.

Kato, C., T. Sato, M. Smorawinska and K. Horikoshi (1994). High pressure conditions stimulate expression of chloramphenicol acetyltransferase regulated by the *lac* promoter in *Escherichia coli*. *FEMS Microbiol. Lett.*, **122**, 91-96.

Kato, T. (1989). Gamma-cyclodextrin forming bacteria. Doctoral dissertation for agriculture at Nihon University.

Kato, T., and K. Horikoshi (1984a). Colorimetric determination of γ-Cyclodextrin. *Anal. Chem.*, **56**, 1738-1740.

Kato, T., and K. Horikoshi (1984b). Immobilized cyclomaltodextrin glucanotransferase of an alkalophilic *Bacillus* sp. No. 38-2. *Biotechnol. Bioeng.*, **26**, 595-598.

Kato, T., and K. Horikoshi (1986a). A new gamma-cyclodextrin forming enzyme produced by *Bacillus subtilis* No. 313, *J. Jpn.Soc.Starch Sci.*, **33**, 137-143.

Kato, T., and K. Horikoshi (1986b). Cloning and expression of the *Bacillus subtilis* No. 313 γ-cyclodextrin forming CGTase gene in *Escherichia coli*. *Agric. Biol. Chem.*, **50**, 2161-2162.

Kawai, S., H. Okoshi, K. Ozaki, S. S., K. Ara and S. Ito (1988). Neutrophilic *Bacillus* strain, KSM-522, that produces an alkaline carboxymethyl cellulase. *Agric. Biol. Chem.*, **52**, 1425-1431.

Kawakami, H. and S. Shimizu (1986). Degradation of lignin-related aromatics by bacteria in alkaline media. *The Third Int. Conf., Stockholm.*, Sweden.

Kawakami, H. and Y. Shumiya (1983). *Degradation of Lignin-related Compounds and Lignins by Alkalophilic Bacteria*, Uni Publishers, Tokyo.

Kelly, C.T., P.A. Brennon and W. M. Fogarty (1987). Resolution of the extracellular α-glucosidase system of *Bacillus* sp. NCIB 11203. *Biotechnol. Lett.*, **9**, 125-130.

Kelly, C.T. and W.M. Fogarty (1978). Production and properties of polygalacturonate lyase by an alkalophilic microorganism, *Bacillus* sp. RK9. *Can. J. Microbiol.*, **24**, 1164-1172.

Kelly, C.T., A.M. Nash and W.M. Fogarty (1984). Effect of manganese on alkaline phosphatase production in *Bacillus* sp. RK 11. *Appl. Microbiol. Biotechnol.*, **19**, 61-66.

Kelly, C.T., F.O'Reilly and W.M. Fogarty (1983). Extracellular α-glucosidase of an alkalophilic microorganism, *Bacillus* sp. ATCC 21591. *FEMS Microbiol. Lett.*, **20**, 55-59.

Khan, S., D.M. Ivey and T.A. Krulwich (1992). Membrane ultrastructure of alkaliphilic *Bacillus* species studied by rapid-freeze electron microscopy. *J. Bacteriol.*, **174**, 5123-5126.

Khasin, A., I. Alchanati and Y. Shoham (1993). Purification and characterization of a thermostable xylanase from *Bacillus stearothermophilus* T-6. *Appl. Environ. Microbiol.*, **59**, 1725-1730.

Khyamihorani, H. (1996). Partial purification and some properties of an alkaline cellulase from an alkalophilic *Bacillus* sp. *World J. Microbiol. Biotechnol.*, **12**, 525-529.

Kim, C.H., H.I. Choi and D. S. Lee (1993*a*). Pullulanases of alkaline and broad pH range from a newly isolated alkalophilic *Bacillus* sp. S-1 and a *Micrococcus* sp. Y-1. *J. Ind. Microbiol .*,**12**, 48-57.

Kim, C.H., H.I. Choi and D.S. Lee (1993*b*). Purification and biochemical properties of an alkaline pullulanase from alkalophilic *Bacillus* sp. S-1. *Biosci. Biotechnol. Biochem.*, **57**, 1632-1637.

Kim, J.C., and P.D. Bregg (1971). Some properties of the succinate dehydrogenase of *Escherichia coli. Can. J. Biochem.*, **49**, 1098-1104.

Kim, C.H., and Y.S. Kim (1995). Substrate specificity and detailed characterization of a bifunctional amylase pullulanase enzyme from *Bacillus circulans* F-2 having two different active sites on one polypeptide. *Eur. J. Biochem.*, **227**, 687-693.

Kim, E.S., H.K. Na, D.Y. Jhon, O.J. Yoo, S.B. Chun and I.S. Wui (1996). Cloning, sequencing and expression of the amylase isozyme gene from *Pseudomonas* sp. KFCC10818. *Biotechnol. Lett.*, **18**, 169-174.

Kim, T.U., B.G. Gu, J.Y. Jeong, S.M. Byun and Y.C. Shin (1995). Purification and characterization of a maltotetraose-forming alkaline α-amylase from an alkalophilic *Bacillus* strain, GM8901. *Appl. Environ. Microbiol.*, **61**, 3105-3112.

Kim, T.J., Y.D. Lee and H.S. Kim (1993). Enzymatic production of cyclodextrins from milled corn starch in an ultrafiltration membrane bioreactor. *Biotechnol. Bioeng.*, **41**, 88-94.

Kimura, H., A. Okamura and H. Kawaide (1994). Oxidation of 3-, 7-, and 12-hydroxyl groups of cholic acid by an alkalophilic *Bacillus* sp. *Biosci. Biotechnol. Biochem.*, **58**, 1002-1006.

Kimura, K., S. Kataoka, Y. Ishii, T. Takano and K. Yamane (1987). Nucleotide sequence of the β-cyclodextrin glucanotransferase gene of alkalophilic *Bacillus* sp. strain 1011 and similarity of its amino acid sequence to those of α-amylases. *J. Bacteriol.*, **169**, 4399-4402.

Kimura, K., T. Takano and K. Yamane (1987). Molecular cloning of the β-cyclodextrin synthetase gene from an alkalophilic *Bacillus* and its expression in *Escherichia coli* and *Bacillus subtilis. Appl. Microbiol. Biotechnol.*, **26**, 147-153.

Kimura, K., A. Tsukamoto, Y. Ishii, T. Takano and K. Yamane (1988). Cloning of a gene for malto-hexaose producing amylase of an alkalophilic *Bacillus* and hyper-production of the enzyme in *Bacillus subtilis* cells. *Appl. Microbiol. Biotechnol.*, **27**, 372-377.

Kimura, T., and K. Horikoshi (1988). Isolation of bacteria which can grow at both high pH and low temperature. *Appl. Environ. Microbiol.*, **54**, 1066-1067.

Kimura, T., and K. Horikoshi (1989). Production of amylase and pullulanase by an alkalopsychrotrophic *Micrococcus* sp. *Agric. Biol. Chem.*, **53**, 2963-2968.

Kimura, T., and K. Horikoshi (1990*a*). Characterization of pullulan-hydrolysing enzyme from an alkalopsychrotrophic *Micrococcus* sp. *Appl. Microbiol. Biotechnol.*, **34**, 52-56.

Kimura, T., and K. Horikoshi (1990*b*). Effects of temperature and pH on the production of maltotetraose by α-amylases of an alkalopsychrotrophic *Micrococcus. J. Ferment. Bioeng.*, **70**, 134-135.

Kimura, T., and K. Horikoshi (1990*c*). The nucleotide sequence of an α-amylase gene from an alkalopsychrotrophic *Micrococcus* sp. *FEMS Microbiol. Lett.*, **71**, 35-42.

Kimura, T., and K. Horikoshi (1990*d*). Purification and characterization of α-amylases of an alkalopsychrotrophic *Micrococcus* sp. *Starch/Stärke*, **42**, 403-407.

Kitada, M., Y. Dobashi and K. Horikoshi (1989). Enzymatic properties of purified D-xylose isomerase from a thermophilic alkalophile, *Bacillus* TX-3. *Agric. Biol. Chem.*, **53**, 1461-1468.

Kitada, M., A.A. Guffanati and T.A. Krulwich (1982). Bioenergetic properties and viability of alkalophilic *Bacillus firmus* RAB as a function of pH and Na-ion contents of the incubation medium. *J. Bacteriol.*, **152**, 1092-1104.

Kitada, M., M. Hashimoto, T. Kudo and K. Horikoshi (1994). Properties of two different Na$^+$/H$^+$ antiport systems in alkaliphilic *Bacillus* sp. strain C-125. *J. Bacteriol.*, **176**, 6464-6469.

Kitada, M., and K. Horikoshi (1976*a*). Alkaline proteinase production from methyl acetate by alkalophilic *Bacillus* sp. *J. Ferment. Technol.*, **54**, 383–392.

Kitada, M., and K. Horikoshi (1976*b*). Effect of pH on the production of alkaline proteinase by alkalophilic *Bacillus* sp. *J. Ferment. Technol.*, **54**, 579–586.

Kitada, M., and K. Horikoshi (1977). Sodium ion-stimulated α-(1-^{14}C)-aminoisobutyric acid uptake in alkalophilic *Bacillus* species. *J. Bacteriol.*, **131**, 784–788.

Kitada, M., and K. Horikoshi (1979). Relation of glutamate transport to growth of alkalophilic *Bacillus* species. *Agric. Biol. Chem.*, **43**, 2273–2277.

Kitada, M., and K. Horikoshi (1980*a*). Further properties of sodium ion-stimulated α-(1-^{14}C) amino-isobutyric acid uptake in alkalophilic *Bacillus* species. *J. Biochem.*, **87**, 1279–1284.

Kitada, M., and K. Horikoshi (1980*b*). Sodium-ion stimulated amino acid uptake in membrane vesicles of alkalophilic *Bacillus* No. 8-1. *J. Biochem.*, **88**, 1757–1764.

Kitada, M., and K. Horikoshi (1987). Bioenergetic properties of alkalophilic *Bacillus* sp. strain C-59 on an alkaline medium containing K_2CO_3. *J. Bacteriol.*, **169**, 5761–5765.

Kitada, M., and K. Horikoshi (1992). Kinetic properties of electrogenic Na^+/H^+ antiporter in membrane vesicles from an alkalophilic *Bacillus* sp. *J. Bacteriol.*, **174**, 5936–5940.

Kitada, M., and T. A. Krulwich (1984). Purification and characterization of the cytochrome oxidase from alkalophilic *Bacillus firmus* RAB. *J. Bacteriol.*, **158**, 963–966.

Kitada, M., R. J. Lewis and T. A. Krulwich (1983). Respiratory chain of the alkalophilic bacterium *Bacillus firmus* RAB. *J. Bacteriol.*, **154**, 330–335.

Kitada, M., K. Onda and K. Horikoshi (1989). The sodium/proton antiport system in a newly isolated alkalophilic *Bacillus* sp. *J. Bacteriol.*, **171**, 1879–1884.

Kitai, K., T. Kudo, S. Nakamura, T. Masegi, Y. Ichikawa and K. Horikoshi (1988). Extracellular production of human immunoglobulin G Fc region (hIgG-Fc) by *Escherichia coli*. *Appl. Microbiol. Biotechnol.*, **28**, 52–56.

Kobayashi, T., Y. Hakamada, S. Adachi, J. Hitomi, T. Yoshimatsu, K. Koike, S. Kawai and S. Ito (1995). Purification and properties of an alkaline protease from alkalophilic *Bacillus* sp. KSM-K16. *Appl. Microbiol. Biotechnol.*, **43**, 473–481.

Kobayashi, T., and K. Horikoshi (1980*a*). Identification and growth characteristics of alkalophilic *Corynebacterium* sp. which produces NAD(P)-dependent maltose dehydrogenase and glucose dehydrogenase. *Agric. Biol. Chem.*, **44**, 41–47.

Kobayashi, T., and K. Horikoshi (1980*b*). Purification and properties of NADP dependent maltose dehydrogenase produced by a alkalophilic *Corynebacterium* sp. No. 93-1. *Agric. Biol. Chem.*, **44**, 2271–2277.

Kobayashi, T., and K. Horikoshi (1980*c*). Purification and properties of NAD-dependent D-glucose dehydrogenase produced by alkalophilic *Corynebacterium* sp. No. 93-1. *Agric. Biol. Chem.*, **44**, 2261–2269.

Kobayashi, T., and K. Horikoshi (1980*d*). Purification and properties of NAD-dependent maltose dehydrogenase produced by alkalophilic *Corynebacterium* sp. No. 93-1. *Biochim. Biophys. Acta*, **614**, 256–265.

Kobayashi, T., H. Kanai, T. Hayashi, T. Akiba, R. Akaboshi and K. Horikoshi (1992). Haloalkaliphilic maltotriose-forming α-amylase from the archaebacterium *Natronococcus* sp. strain Ah-36. *J. Bacteriol.*, **174**, 3439–3444.

Kobayashi, T., H. Kanai, R. Aono, K. Horikoshi and T. Kudo (1994). Cloning, expression, and nucleotide sequence of the α-amylase gene from the haloalkaliphilic archaeon *Natronococcus* sp. strain Ah-36. *J. Bacteriol.*, **176**, 5131–5134.

Kobayashi, T., C. Kato, T. Kudo and K. Horikoshi (1986). Excretion of the penicillinase of an alkalophilic *Bacillus* sp. through the *Escherichia coli* outer membrane is caused by insertional activation of the *Kil* Gene in plasmid pMB9. *J. Bacteriol.*, **166**, 728–732.

Kobayashi, Y., and K. Horikoshi (1981). Production of extracellular polyamine oxidase by *Penicillium* sp. No. PO-1. *Agric. Biol. Chem.*, **45**, 2943–2945.

Kobayashi, Y., and K. Horikoshi (1982). Purification and characterization of extracellular polyamine oxidase produced by *Penicillium* sp. No. PO-1. *Biochim. Biophys. Acta,* **705**, 133-138.

Kobayashi, Y., H. Ueyama and K. Horikoshi (1980). NAD-Dependent maltose dehydrogenase of alkalophilic *Corynebacterium* sp. No. 150-1. *Agric. Biol. Chem.,* **44**, 2837-2841.

Kobayashi, Y., H. Ueyama and K. Horikoshi (1982). Comparative studies of NAD-dependent maltose dehydrogenase and D-glucose dehydrogenase produced by two strains of alkalophilic *Corynebacterium* species. *Agric. Biol. Chem.,* **46**, 2139-2142.

Koga, Y., M. Nishihara and H. Mori (1982). Lipids of alkalophilic bacteria : Identification, composition and metabolism. *J. Univ. Occup. Envion. Health,* **4**, 227-240.

Koike, K., Y. Hakamada, T. Yoshimatsu, T. Kobayashi and S. Ito (1996). NADP-specific glutamate dehydrogenase from alkaliphilic *Bacillus* sp. KSM-635: Purification and enzymatic properties. *Biosci. Biotechnol. Biochem.,* **60**, 1764-1767.

Kometani, T., T. Nishmura, T. Nakae, H. Takii and S. Okada (1996*a*). Synthesis of neohesperidin glycosides and naringin glycosides by cyclodextrin glucanotransferase from an alkaliphilic *Bacillus* species. *Biosci. Biotechnol. Biochem.,* **60**, 645-649.

Kometani, T., Y. Terada, T. Nishimura, H. Takii and S. Okada (1994*a*). Purification and characterization of cyclodextrin glucanotransferase from an alkalophilic *Bacillus* species and transglycosylation at alkaline pHS. *Biosci. Biotechnol. Biochem.,* **58**, 517-520.

Kometani, T., Y. Terada, T. Nishimura, H. Takii and S. Okada (1994*b*). Transglycosylation to hesperidin by cyclodextrin glucanotransferase from an alkalophilic *Bacillus* species in alkaline pH and properties of hesperidin glycosides. *Biosci. Biotechnol. Biochem.,* **58**, 1990-1994.

Kometani, T., Y. Terada, T. Nishimura, T. Nakae, H. Takii and S. Okada (1996*b*). Acceptor specificity of cyclodextrin glucanotransferase from an alkalophilic *Bacillus* species and synthesis of glucosyl rhamnose. *Biosci. Biotechnol. Biochem.,* **60**, 1176-1178.

Konings, W.N. and E. Freese (1972). Amino acid transport in membrane vesicles of *Bacilllus subtilis*. *J. Biol. Chem.,* **247**, 2408-2418.

Koyama, N. (1996). $(NH^{4+}) + (Na^+)$-activated ATPase of a facultatively anaerobic alkaliphile, *Amphibacillus xylanus*. *Anaerobe,* **2**, 123-128.

Koyama, N., A. Kiyomiya and Y. Nosoh (1976). Na^+-dependent uptake of amino acids by an alkalophilic *Bacillus*. *FEBS Lett.,* **72**, 77-78.

Koyama, N., K. Koshiya and Y. Nosho (1980). Purification and properties of ATPase from an alkalophilic *Bacillus*. *Arch. Biochem. Biophys.,* **199**, 103-109.

Koyama, N., and Y. Nosho (1985). Effect of potassium and sodium ions on the cytoplasmic pH of an alkalophilic *Bacillus*. *Biochim. Biophys. Acta,* **812**, 206-212.

Koyama, N., and Y. Nosoh (1976). Effect of the pH of culture medium on the alkalophilicity of a species of *Bacillus*. *Arch. Microbiol.,* **109**, 105-108.

Kroll, R. G. (1990). Alkalophiles. *Microbiology of Extreme Environments,* (Ed. C. Edwards), pp. 55-92, McGraw-Hill Publishing Co., New York.

Krulwich, T.A., and A. A. Guffanti (1983). Physiology of acidphilic and alkalophilic bacteria. *Adv. Microbiol. Physiol.,* **24**, 173-214.

Krulwich, T. A. (1986). Bioenergetics of alkalophilic bacteria. *J. Membr. Biol.,* **89**, 113-125.

Krulwich, T. A. (1995). Alkaliphiles, 'basic' molecular problems of pH tolerance and bioenergetics. *Mol. Microbiol.,* **15**, 403-410.

Krulwich, T.A., J.G. Federbush and A.A. Guffanti (1985). Presence of a nonmetabolizable solute that is translocated with Na^+ enhances Na^+-dependent pH homeostasis in alkalophilic *Bacillus*. *J. Biol. Chem.,* **260**, 4055-4058.

Krulwich, T. A., and A. A. Guffanti (1989). Alkalophilic bacteria. *Annu. Rev. Microbiol.,* **43**, 435-463.

Krulwich, T. A., A.A. Guffanti, R. F. Bornstein and J. Hoffstein (1982). A sodium requirement for growth, solute transport, and homeostasis in *Bacillus firmus* RAB. *J. Biol. Chem.* **257**, 1885-1889.

Kudo, T., M. Hino, M. Kitada and K. Horikoshi (1990). DNA sequences required for the alkalophily of *Bacillus* sp. strain C-125 are located close together on its chromosomal DNA. *J. Bacteriol.*, **172**, 7282-7283.

Kudo, T., and K. Horikoshi (1978*a*). RNA polymerase from vegetative cells and spores of an alkalophilic *Bacillus* sp. Spores VII, 220.

Kudo, T., and K. Horikoshi (1978*b*). RNA-polymerase of alkalophilic *Bacillus* sp. No. 2b-2. The Third Intern. Symp. Ribosomes and Nucleic Acid Metabolism, Czechoslovakia, pp. 247-256.

Kudo, T., and K. Horikoshi (1979). The environmental factors affecting sporulation of an alkalophilic *Bacillus* species. *Agric. Biol. Chem.*, **43**, 2613-2614.

Kudo, T., and K. Horikoshi (1983*a*). Effect of pH and sodium ion on germination of alkalophilic *Bacillus* species. *Agric. Biol. Chem.* **47**, 665-669.

Kudo, T., and K. Horikoshi (1983*b*). The Effect of pH on heat-resistance of spores of alkalophilic *Bacillus* No. 2b-2. *Agric. Biol. Chem.*, **47**, 403-404.

Kudo, T., C. Kato and K. Horikoshi (1983). Excretion of the penicillinase of an alkalophilic *Bacillus* sp. through the *Escherichia coli* outer membrane. *J. Bacteriol.*, **156**, 949-951.

Kudo, T., A. Ohkoshi and K. Horikoshi (1985). Molecular cloning and expression of a xylanase gene of alkalophilic *Aeromonas* sp. No. 212 in *Escherichia coli. J. Gen.. Microbiol.*, **131**, 2825-2830.

Kudo, T., J. Yoshitake, C. Kato, R. Usami and K. Horikoshi (1985). Cloning of a developmentally regulated element from alkalophilic *Bacillus subtilis* DNA. *J. Bacteriol.*, **161**, 158-163.

Kulkarni, N., and M. Rao (1996). Application of xylanase from alkaliphilic thermophilic *Bacillus* sp. NCIM59 in biobleaching of bagasse pulp. *J. Biotechnol.*, **51**, 167-173.

Kurono, Y., and K. Horikoshi (1973). Alkaline catalase produced *Bacillus* No. Ku-1. *Agric. Biol. Chem.*, **37**, 2565-2570.

Kushner, D. J., and T. A. Lisson (1959). Alkali resistance in a strain of *Bacillus cereus* pathogenic for the larch sawfly *Pristiphora erichsonii. J. Gen. Microbiol.*, **21**, 96-108.

Kwon, H. J., M. Kitada and K. Horikoshi (1987). Purification and properties of D-xylose isomerase from alkalophilic *Bacillus* No. KX-6. *Agric. Biol. Chem.*, **51**, 1983-1989.

Kwon, Y.T., J.O. Kim, S.Y. Moon, H.H. Lee and H.M. Rho (1994). Extracellular alkaline proteases from alkalophilic *Vibrio metschnikovii* strain HH530. *Biotechnol. Lett.*, **16**, 413-418.

Landaud, S., A.M. Davila and J. Pourquie (1996). Neutral endoglucanases production by a newly isolated alkalophilic *Bacillus circulans. Biotechnol. Lett.*, **18**, 741-746.

Lee, S. P., M. Morikawa, M. Takagi and T. Imanaka (1994). Cloning of the *aapT* gene and characterization of its product, α-amylase-pullulanase (*AapT*), from thermophilic and alkaliphilic *Bacillus* sp. strain XAL601. *Appl. Environ. Microbiol.*, **60**, 3764-3773.

Lewis, R.J., S. Belkina and T.A. Krulwich (1980). Alkalophiles have much higher cytochrome contents than conventional bacteria and than their own nonalkalophilic mutant derivatives. *Biochem. Biophys. Res. Commun.*, **95**, 857-863.

Lewis, R.J., E. Kaback and T.A. Krulwich (1982). Pleiotropic properties of mutations to non-alkalophily in *Bacillus alcalophilus. J. Gen. Microbiol.*, **128**, 427-430.

Lewis, R. J., T. A. Krulwich, B. Reynafarje and A. L. Lehninger (1983). Respiration-dependent proton translocation in alkalophilic *Bacillus firmus* RAB and its nonalkalophilic mutant derivative. *J. Biol. Chem.*, **258**, 2109-2111.

Lewis, R. J., R. C. Prince, P. L. Dutton, D. B. Knaff and T.A. Krulwich (1981). The respiratory chain of *Bacillus alcalophilus* and its nonalkalophilic mutant derivative. *J. Biol. Chem.*, **256**, 10543-10549.

Li, Y. H., M. Engle, N. Weiss, L. Mandelco and J. Wiegel (1994). *Clostridium thermoalcaliphilum* sp. nov., an anaerobic and thermotolerant facultative alkaliphile. *Int. J. Syst. Bacteriol.*, **44**, 111-118.

Li, Y. H., L. Mandelco and J. Wiegel (1993). Isolation and characterization of a moderately thermophilic anaerobic alkaliphile, *Clostridium paradoxum* sp. nov. *Int. J. Syst. Bacteriol.*, **43**(3), 450-460.

Lin, L. L., M. R. Tsau and W. S. Chu (1994). General characteristics of thermostable amylopullulanases and amylases from the alkalophilic *Bacillus* sp. TS-23. *Appl. Microbiol. Biotechnol.*, **42**, 51-56.

Lin, S.F., C.M. Chiou and Y.C. Tsai (1995). Effect of Triton X-100 on alkaline lipase production by *Pseudomonas pseudoalcaligenes* F-111. *Biotechnol. Lett.*, **17**, 959-962.

Lin, S.F., C.M. Chiou, C.M. Yeh and Y.C. Tsai (1996). Purification and partial characterization of an alkaline lipase from *Pseudomonas pseudoalcaligenes* F-111. *Appl. Environ. Microbiol.*, **62**, 1093-1095.

Lodwick, D., T. J. McGenity and W. D. Grant (1994). The phylogenetic position of the haloalkaliphilic archaeon *Natronobacterium magadii*, determined from its 23S ribosomal sequence. *Syst. Appl. Microbiol.*, **17**, 402-404.

Lodwick, D., H. N. M. Ross, J. A. Walker, J. W. Almond and W. D. Grant (1991). Nucleotide sequence of the 16S ribosomal RNA gene from the haloalkaliphilic archaeon (archaebacterium) *Natronobacterium magadii*, and the phylogeny of halobacteria. *Syst. Appl. Microbiol.*, **14**, 352-357.

Lombaridi, F.J., and H.R. Kaback (1972). Mechanisms of active transport of amino acids in isolated bacterial membrane vesicles. VIII Transport of amino acids by membrane prepared from *Escherichia coli*. *J. Biol. Chem.*, **247**, 7844-7857.

Louw, M. E., S. J. Reid and T. G. Watson (1993). Characterization, cloning and sequencing of a thermostable endo-(1,3-1,4)-β-glucanase-encoding gene from an alkalophilic *Bacillus brevis*. *Appl. Microbiol. Biotechnol.*, **38**(4), 507-513.

Mackman, N., J.-M. Nicaud, L. Gray and I.B. Holland (1986). *Secretion of haemolysin by Escherichia coli.* Springer Verlag, Berlin, Heidelberg, New York, Tokyo.

Mäkela, M., P. Mattsson, M.E. Schinina and T. Koppela (1988). Purification and properties of cyclomaltodextrin glucanotransferase from an alkalophilic *Bacillus*. *Appl. Biochem.*, **10**, 414-427.

Maltseva, O., C. Mcgowan, R. Fulthorpe and P. Oriel (1996). Degradation of 2,4-dichlorophenoxyacetic acid by haloalkaliphilic bacteria. *Microbiology-UK*, **142**, 1115-1122.

Mandel, K.G., A.A. Guffanti and T.A. Krulwich (1980). Monovalent cation/proton antiporters in membrane vesicles from *Bacillus alcalophilus*. *J. Biol. Chem.*, **255**, 7391-7396.

Mathrani, I.M., and B.K. Ahring (1992). Thermophilic and alkalophilic xylanases from several *Dictyoglomus* isolates. *Appl. Microbiol. Biotechnol.*, **38**, 23-27.

McGenity, T.J., and W.D. Grant (1993). The haloalkaliphilic archaeon (archaebacterium) *Natronococcus-occultus* represents a distinct lineage within the halobacteriales, most closely related to the other haloalkaliphilic lineage (*Natronobacterium*). *Syst. Appl. Microbiol.*, **16**, 239-243.

McGenity, T.J., and W.D. Grant. (1995). Transfer of *Halobacterium saccharovorum*, *Halobacterium sodomense*, *Halobacterium trapanicum* NRC34021 and *Halobacterium lacusprofundi* to the genus *Halorubrum* gen. nov., as *Halorubrum saccharovorum* comb. nov., *Halorubrum sodomense* comb. nov., *Halorubrum trapanicum* comb. nov., and *Halorubrum lacusprofundi* comb. nov. *Syst. Appl. Microbiol.*, **18**, 237-243.

McLaggan, D., M.J. Selwyn and A.P. Dawson (1984). Dependence on Na$^+$ of control of cytoplasmic pH in a facultative alkalophile. *FEBS Lett.*, **165**, 254-258.

McTigue, M.A., C.T. Kelly, W.M. Fogarty and E.M. Doyle (1994). Production studies on the alkaline amylases of three alkalophilic *Bacillus* spp. *Biotechnol. Lett.*, **16**, 569-574.

Meek, C.S., and C.B. Lipman (1922). The relation of the reactions of the salt concentration of the medium to nitrifying bacteria. *J. Gen. Physiol.*, **5**, 195-204.

Mikami, Y., K. Miyashita and T. Arai (1982). Diaminopimelic acid profiles of alkalophilic and alkaline-resistant strains of actinomycetes. *J. Gen. Microbiol.*, **128**, 1709-1712.

Miller, A.G., D.H. Tukrpin and D.T. Canvin (1984). Na$^+$ requirement for growth, photosynthesis, and pH regulation in the alkalotolerant cyanobacterium *Synechococcus leopoliensis*. *J. Bacteriol.*, **159**, 100-106.

Miyashita, K., Y. Mikami and T. Arai (1984). Alkalophilic actinomycete, *Nocardiopsis dassonvillei* subsp. *prasina* subsp. nov. isolated from soil. *Int. J. Syst. Bacteriol.*, **34**, 405-409.

Moriya, S., S. Yanagawa, N. Aoki, M. Iwabuchi, T. Inoue and T. Ando (1992). Isolation and characterization of a restriction enzyme *Bsp*O4I from an alkalophilic bacterium. *Nucleic Acid Res.*, **20**, 3781.

Murakami, Y., T. Furusato, C. Kato, N. Habuka, T. Kudo and K. Horikoshi (1989). Construction of new excretion vectors: two and three tandemly located promoters are active for extracellular protein production from *Escherichia coli*. *Appl. Microbiol. Biotechnol.*, **30**, 619-623.

Murata, M. (1988). A new detergency mechanism based on alkaline cellulase. 2nd World Surfactants Congress, Paris, France, pp. 350, 549-553.

Mwatha, W.E., and W.D. Grant (1993). *Natronobacterium vacuolata* sp. nov., a haloalkaliphilic archaeon isolated from Lake Magadi, Kenya. *Int. J. Syst. Bacteriol.*, **43**, 401-404.

Nagaoka, K., and Y. Yamada (1969). Studies on *Mucor* lipases. Part II. separation and characterization of three lipolytic enzymes. *Agric. Biol. Chem.*, **33**, 986-993.

Nagaoka, K., and Y. Yamada (1973). Purification of *Mucor* lipases and their properties. *Agric. Biol. Chem.*, **37**, 2791-2796.

Na, H.K., E.S. Kim, H.B. Lee, O.J. Yoo and D. Y. Jhon (1996). Cloning and nucleotide sequence of the α-amylase gene from alkalophilic *Pseudomonas* sp. KFCC10818. *Mol. Cells*, **6**, 203-208.

Nakajima, H., K. Noguchi, M. Yamamoto, R. Aono and K. Horikoshi (1993). Expression of an 87-kD-β-1,3-glucanase of *Bacillus circulans* IAM1165 in *Saccharomyces cerevisiae* by low-temerature incubation. *Biosci. Biotechnol. Biochem.*, **57**, 2039-2042.

Nakamura, A., F. Fukumori, S. Horinouchi, H. Masaki, T. Kudo, T. Uozumi, K. Horikoshi and T. Beppu (1991). Construction and characterization of the chimeric enzymes between the *Bacillus subtilis* cellulase and an alkalophilic *Bacillus* cellulase. *J. Biol. Chem.*, **266**, 1579-1583.

Nakamura, A., K. Haga, S. Ogawa, K. Kuwano, K. Kimura and K. Yamane (1992). Functional relationships between cyclodextrin glucanotransferase from an alkalophilic *Bacillus* and α-amylases. Site-directed mutagenesis of the conserved two Asp and one Glu residues. *FEBS Lett.*, **1**, 37-40.

Nakamura, A., K. Haga and K. Yamane (1993). Three histidine residues in the active center of cyclodextrin glucanotransferase from alkalophilic *Bacillus* sp. 1011: Effects of the replacement on pH dependence and transition-state stabilization. *Biochemistry*, **32**, 6624-6631.

Nakamura, N., and K. Horikoshi (1976*a*). Characterization and some culture conditions of a cyclodextrin glycosyltransferase-producing alkalophilic *Bacillus* sp. *Agric. Biol. Chem.*, **40**, 753-757.

Nakamura, N., and K. Horikoshi (1976*b*). Characterization of acid-cyclodextrin glycosyl-transferase of an alkalophilic *Bacillus* sp. *Agric. Biol. Chem.*, **40**, 1647-1648.

Nakamura, N., and K. Horikoshi (1976*c*). Purification and properties of cyclodextrin glycosyl-transferase of an alkalophilic *Bacillus* sp. *Agric. Biol. Chem.*, **40**, 935-941.

Nakamura, N., and K. Horikoshi (1976*d*). Purification and properties of neutral- cyclodextrin gly-cosyl-transferase of an alkalophilic *Bacillus* sp. *Agric. Biol. Chem.*, **40**, 1785-1791.

Nakamura, N., and K. Horikoshi (1977). Production of Schardinger dextrin by soluble and immobilized cyclodextrin glycosyl-transferase of an alkalophilic *Bacillus* sp. *Biotechnol. Bioeng.*, **19**, 87-99.

Nakamura, N., K. Watanabe and K. Horikoshi (1975). Purification some properties of alkaline pullulanase from a strain of *Bacillus* No. 202-1, an alkalophilic microorganism. *Biochim. Biophys. Acta*, **397**, 188-193.

Nakamura, S., T. Masegi, K. Kitai, Y. Ichikawa, T. Kudo, R. Aono and K. Horikoshi (1990). Extracellular production of human tumor necrosis factor by *Escherichia coli* using a chemically-synthesized gene. *Agric. Biol. Chem.*, **54**, 3241-3250.

Nakamura, S., K. Wakabayashi, R. Nakai, R. Aono and K. Horikoshi (1993*a*). Production of alkaline xylanase by a newly isolated alkaliphilic *Bacillus* sp. strain 41M-1. *World J. Microbiol. Biotechnol.*, **9**, 221-224.

Nakamura, S., K. Wakabayashi, R. Nakai, R. Aono and K. Horikoshi (1993*b*). Purification and some properties of an alkaline xylanase from alkaliphilic *Bacillus* sp. strain 41M-1. *Appl. Environ. Microbiol.*, **59**, 2311-2316.

Nakamura, S., R. Nakai, K. Wakabayashi, Y. Ishiguro, R. Aono and K. Horikoshi (1994). Thermophilic alkaline xylanase from newly isolated alkaliphilic and thermophilic *Bacillus* sp. strain Tar-1. *Biosci. Biotechnol. Biochem.*, **58**, 78-81.

Nakanishi, T., Y. Matsumura, N. Minamiura and T. Yamamoto (1974). Purification and some properties of an alkalophilic proteinase of a *Streptomyces* species. *Agric. Biol. Chem.*, **38**, 37-44.

Nakanishi, T., and T. Yamamoto (1974). Action and specificity of a *Streptomyces* alkalophilic proteinase. *Agric. Biol. Chem.*, **38**, 2391-2397.

Nakanishi, T., and T. Yamamoto (1975). The initial attack sites of a *Streptomyces* alkalophilic proteinase on oxidized insulin B-chain. *Agric. Biol. Chem.*, **39**, 1797-1802.

Nakatsugawa, N., and K. Horikoshi (1989). Isolation and characterization of two novel methanogens, and a new haloalkalophilic methanogen and a new alkalophilic methanosarcia. *Microbiology of Extreme Environments and Its Potential for Biotechnology* (Eds. M.S. DaCosta, J.C. Duarte and R.A.D. Williams), p. 415, Elsevier Applied Science, London.

Nielsen, P., D. Fritze and F.G. Priest (1995). Phenetic diversity of alkaliphilic *Bacillus* strains: Proposal for nine new species. *Microbiology-UK*, **141**, 1745-1761.

Nielsen, P., F.A. Rainey, H. Outtrup, F.G. Priest and D. Fritze (1994). Comparative 16S rDNA sequence analysis of some alkaliphilic *Bacilli* and the establishment of a sixth rRNA group within the genus *Bacillus*. *FEMS Microbiol. Lett.*, **117**, 61-65.

Niimura, Y., F. Yanagida, T. Uchimura, N. Ohara, K. Suzuki and M. Kozaki (1987). A new facultative anaerobic xylan-using alkalophile lacking cytochrome, quinone and catalase. *Agric. Biol. Chem.*, **51**, 2271-2275.

Nishihara, M., H. Mori and Y. Koga (1982). Bis-(monoacylglycero) phosphate in alkalophilic bacteria. *J. Biochem.*, **92**, 1469-1479.

Nogi, Y., and K. Horikoshi (1990). A thermostable alkaline β-1,3-glucanase produced by alkalophilic *Bacillus* sp. AG-430. *Appl. Microbiol. Biotechnol.*, **34**, 704-707.

Nomoto, M., C. Chen and D. Shen (1986). Purification and characterization of cyclodextrin glucanotransferase from an alkalophilic bacterium of Taiwan. *Agric. Biol. Chem.*, **50**, 2701-2707.

Nomoto, M., T. Lee, C. Su, C. Liao, T. Yen and C. Yang (1984*a*). Alkaline proteinases from alkalophilic bacteria of Taiwan. *Agric. Biol. Chem.*, **48**, 1627-1628.

Nomoto, M., M. Ohsawa, H. Wang, C.-C. Chen and K. Yeh (1988). Purification and characterization of extracellular alkaline phosphatase from an alkalophilic bacterium. *Agric. Biol. Chem.*, **52**, 1643-1647.

Nomoto, M., D. Shew, S. Chen, T. Yen, C. Liao and C. Yang (1984*b*). Cyclodextrin glucanotransferase from alkalophilic bacterium of Taiwan. *Agric. Biol. Chem.*, **48**, 1337-1338.

Ohkoshi, A., T. Kudo, T. Mase and K. Horikoshi (1985). Purification of three types of xylanases from an alkalophilic *Aeromonas* sp. *Agric. Biol. Chem.*, **49**, 3037-3038.

Ohta, K., A. Kiyomiya, N. Koyama and Y. Nosoh (1975). The basis of the alkalophilic property of a species of *Bacillus*. *J. Gen. Microbiol.*, **86**, 259-266.

Oi, S., A. Sawada and Y. Satomura (1967). Purification and some properties of two types of *Penicillium* lipases, I and II, and conversion of types I and II under various modification conditions. *Agric. Biol. Chem.*, **31**, 1357-1366.

Okada, T., M. Ito and K. Hibino (1994). Immobilization of cyclodextrin glucanotransferase on capillary membrane. *J. Ferment. Bioeng.*, **77**, 259-263.

Okazaki, W., T. Akiba, K. Horikoshi and R. Akahoshi (1984). Production and properties of two types of xylanases from alkalophilic thermophilic *Bacillus* sp. *Appl. Microbiol. Biotechnol.*, **19**, 335-340.

Okazaki, W., T. Akiba, K. Horikoshi and R. Akahoshi (1985). Purification and characterization of xylanases from alkalophilic thermophilic *Bacillus* spp. *Agric. Biol. Chem.*, **49**, 2033-2039.

Okoshi, H., K. Okazaki, S. Shikata, K. Oshino, S. Kawai and S. Ito (1990). Purification and characterization of multiple carboxymethyl cellulase from *Bacillus* sp. KSM-522. *Agric. Biol. Chem.*, **54**, 83–89.

Olson, E. R. (1993). Influence of pH on bacterial gene expression. *Mol. Microbiol.*, **8**, 5-14.

Oren, A. (1994). The ecology of the extremely halophilic archaea. *FEMS Microbiol. Rev.*, **13**, 415-439.

Orii, Y., I. Yumoto, Y. Fukumori and T. Yamanaka (1991). Stopped-flow and rapid-scan studies of the redox behavior of cytochrome aco from facultative alkalophilic *Bacillus. J. Biol. Chem.*, **266**, 14310-14316.

Ozaki, K., S. Shikata, S. Kawai, S. Ito and K. Okamota (1990). Molecular cloning and nucleotide sequence of a gene for alkaline cellulase from *Bacillus* sp. KSM-635. *J. Gen. Microbiol.*, **136**, 1327-1334.

Ozaki, K., N. Sumitomo and S. Ito (1991). Molecular cloning and nucleotide sequence of the gene encoding an endo-1,4-β-glucanase from *Bacilllus* sp. KSM-330. *J. Gen. Microbiol.*, **137**, 2299-2305.

Ozaki, K., Y. Hayashi, N. Sumitomo, S. Kawai and S. Ito (1995). Construction, purification, and properties of a truncated alkaline endoglucanase from I sp. KSM-635. *Biosci. Biotechnol. Biochem.*, **59**, 1613-1618.

Paavilainen, S., P. Helisto and T. Korpela (1994). Conversion of carbohydrates to organic acids by alkaliphilic bacilli. *J. Ferment. Bioeng.*, **78**, 217-222.

Packer, H. L., D. M. Harrison, R. M. Dixon and J. P. Armitage (1994). The effect of pH on the growth and motility of *Rhodobacter sphaeroides* WS8 and the nature of the driving force of the flagellar motor. *Biochim. Biophys. Acta-Bioenerg.*, **1188**, 101-107.

Park, J.-S., S. Horinouchi and T. Beppu (1991). Characterization of leader pepide of an endo-type cellulase produced by an alkalophilic *Streptomyces* strain. *Agric. Biol. Chem.*, **55**, 1745-1750.

Park, Y. S., D. Y. Yum, D. H. Bai and J. H. Yu (1992). Xylanase from alkalophilic *Bacillus* sp. YC-335. *Biosci. Biotech. Biochem.*, **56**, 1355-1356.

Podkovyrov, S.M. and J G. Zeikus (1992). Structure of the gene encoding cyclomaltodextrinase from *Clostridium* thermohydrosulfuricum-39E and characterization of the enzyme purified from *Escherichia coli. J. Bacteriol.*, **174**, 5400-5405.

Prasad, P., and A. K. Kashyap (1991). Ammonium transport in the alkalophilic diazotropic cyanobacterium *Nostac calcicola*-influence of phosphate limitation and metabolic inhibitors. *J. Plant Physiol.*, **138**, 244-247.

Prowe, S.G., J.L.C.M. Vandevossenberg, A.J.M. Driessen, G. Antranikian and W.N. Konings (1996). Sodium-coupled energy transduction in the newly isolated thermoalkaliphilic strain LBS3. *J. Bacteriol.*, **178**, 4099-4104.

Pyatibratov, M.G., A.S. Kostyukova, V.Y. Tarasov and O.V. Fedorov, (1996). Some principles of formation of the haloalkaliphilic archaeal flagellar structure. *Biochemistry-Engl. Tr.*, **61**, 1056-1062.

Quirk, P.G. (1993). A gene encoding a small, acid-soluble spore protein from alkaliphilic *Bacillus firmus* OF4. *Gene*, **125**, 81-83.

Quirk, P.G., A.A. Guffanti, R.J. Plass, S. Clejan and T.A. Krulwich (1991). Protonophore-resistance and cytochrome expression in mutant strains of the facultative alkaliphile *Bacillus firmus* OF4. *Bacteriol. Biophys. Acta,* **1058**(2), 131-140.

Quirk, P.G., D.B. Hicks and T.A. Krulwich (1993). Cloning of the *cta* operon from alkaliphilic *Bacillus firmus* OF4 and characterization of the pH-regulated cytochrome *caa3* oxidase it encodes. *J. Biol. Chem.*, **268**, 678-685.

Qureshi, M.H., I. Yumoto, T. Fujiwara, Y. Fukumori and T. Yamanaka (1990). A novel *aco*-type cytochrome-*c* oxidase from a facultative alkalophilic *Bacillus*: Purification, and some molecular and enzymatic features. *J. Biochem.*, **107**, 480-485.

Raj, K.C. and T.S. Chandra (1995). A cellulase-free xylanase from alkali-tolerant *Aspergillus fischeri* Fxn1. *FEMS Microbiol. Lett.*, **17**, 309-314.

Raj, K.C. and T.S. Chandra (1996). Purification and characterization of xylanase from alkali-tolerant *Aspergillus fischeri* Fxn1. *FEMS Microbiol. Lett.*, **145**, 457-461.

Ramos, S., and H. R. Kaback (1977*a*). The electrochemical proton gradient in *Escherichia coli* membrane vesicles. *Biochemistry*, **16**, 848-854.

Ramos, S., and H. R. Kaback (1977*b*). The relationship between the electrochemical proton gradient and active transport in *Escherichia coli* membrane vesicles. *Biochemistry*, **16**, 854-859.

Romano, I., B. Nicolaus, L. Lama, M.C. Manca and A. Gambacorta. (1996). Characterization of a haloalkalophilic strictly aerobic bacterium, isolated from Pantelleria island. *Syst. Appl. Microbiol.*, **19**, 326-333.

Rottenberg, H., T. Grunwald and M. Avron (1972). Determination of DpH in chloroplasts. 1. Distribution of (^{14}C) methylamin. *Eur. J. Biochem.*, **25**, 54-63.

Sakamoto, Y., K. J. Sutherland, J. Tamaoka, T. Kobayashi, T. Kudo and K. Horikoshi (1992). Analysis of the flagellin (*hag*) gene of alkalophilic *Bacillus* sp. C-125. *J. Gen. Micriobiol.*, **138**, 2139-2166.

Sancheztorres, J., P. Perez and R.I. Santamaria (1996). A cellulase gene from a new alkalophilic *Bacillus* sp. (strain N186-1). Its cloning, nucleotide sequence and expression in *Escherichia coli*. *Appl. Microbiol. Biotechnol.*, **46**, 149-155.

Sashihara, N., T. Kudo and K. Horikoshi (1984). Molecular cloning and expression of celluase genes of alkalophilic *Bacillus* sp. strain N-4 in *Escherichia coli*. *J. Bacteriol.*, **158**, 503-506.

Sato, M., T. Beppu and K. Arima (1980). Properties and structure of a novel peptide antibiotic No. 1907. *Agric. Biol. Chem.*, **44**, 3037-3040.

Sato, M., Y. Yagi, H. Nagano and T. Ishikura (1985). Determination of CGTase from *Bacillus ohbensis* and its optimum pH using HPLC. *Agric. Biol. Chem.*, **49**, 1189-1191.

Savitha, J., and C. Ratledge (1992). An inducible, intracellular, alkalophilic lipase in *Aspergillus flavipes* grown on triacylglycerol. *World J. Microbiol. Biotechnol.*, **8**, 129-131.

Schlesinger, P., S. Belkin and S. Boussiba (1996). Sodium deprivation under alkaline conditions causes rapid death of the filamentous cyanobacterium *Spirulina platensis*. *J. Phycol.*, **32**, 608-613.

Seto, Y., M. Hashimoto, R. Usami, T. Hamamoto, T. Kudo and K. Horikoshi (1995). Characterization of a mutation responsible for an alkali-sensitive mutant, 18224, of alkaliphilic *Bacillus* sp. strain C-125. *Biosci. Biotechnol. Biochem.*, **59**, 1364-1366.

Shendye, A., and M. Rao (1993*a*). Cloning and extracellular exprssion in *Escherichia coli* of xylanses from an alkaliphilic thermophilic *Bacillus* sp. NCIM59. *FEMS Microbiol. Lett.*, **108**, 297-302.

Shendye, A., and M. Rao (1993*b*). Chromosomal gene integration and enhanced xylanase production in an alkaliphilic thermophilic *Bacillus* sp. (NCIM-59). *Biochem. Biophys. Res. Commun.*, **195**, 776-784.

Shendye, A., R. Gaikaiwari and M. Rao (1994). Expression of the cloned xylanases from an alkalophilic, thermophilic *Bacillus* in *Bacillus subtilis*. *World J. Microbiol. Biotechnol.*, **10**, 414-416.

Shiba, H., and K. Horikoshi (1989). Isolation and characterization of novel anaerobic halophilic eubacteria from hypersaline environments of western America and Kenya. *Microbiology of Extreme Environments and Its Potential for Biotechnology* (Eds. M.S. Dacasta, J.C. Duarte and R.A. D. William), pp. 371-374, Elsevier Applied Science, London.

Shiba, H., H. Yamamoto and K. Horikoshi (1989). Isolation of strictly anaerobic halophiles from the aerobic surface sediments of hypersaline environments in California and Nevada. *FEMS Microbiol. Lett.*, **57**, 191-196.

Shibata, T., S. Ikawa, C. Kim and T. Ando (1976). Site-specific deoxyribonucleases in *Bacillus subtilis* and other *Bacillus* strains. *J. Bacteriol.*, **129**, 473-476.

Shikata, S., K. Osaki, S. Kawai, S. Ito and K. Okamoto (1988). Purification and characterization of NADP$^+$-linked isocitrate dehydrogenase from an alkalophilic *Bacillus*. *Biochim. Biophys. Acta*, **952**, 282-289.

Shikata, S., K. Saeki, H. Okoshi, T. Yoshimatsu, K. Ozaki, S. Kawai and S. Ito (1990). Alkaline cellulase for laundry detergents: Production by alkalophilic strains of *Bacillus* and some properties of the crude enzymes. *Agric. Biol. Chem.*, **54**, 91-96.

Shimogaki, H., K. Takeuchi, T. Nishino, M. Ohdera, T. Kudo, K. Ohba, M. Iwama and M. Irie (1991). Purification and properties of a novel surface-active agent- and alkaline-resistant protease from *Bacillus* sp. Y. *Agric. Biol. Chem.*, **55**, 2251-2258.

Shirokane, Y., A. Arai and R. Uchida (1994). A new enzyme, maltobionate α-D-glucohydrolase, from alkalophilic *Bacillus* sp. N-1053. *Biochim. Biophys. Acta - Protein Struct. Mol. Enzymol.*, **1207**, 143-151.

Shirokizawa, O., T. Akiba and K. Horikoshi (1989). Cloning and Expression of the maltohexaose-forming amylase gene from alkalophilc *Bacillus* sp. H-167 in *Escherichia coli*. *Agric. Biol. Chem.*, **53**, 491-495.

Shirokizawa, O., T. Akiba and K. Horikoshi (1990). Nucleotide sequence of the G6-amylase gene from alkalophilic *Bacillus* sp. H-167. *FEMS Microbiol. Lett.*, **70**, 131-136.

Shiroza, T., and H. K. Kuramitsu (1993). Construction of a model secretion system for oral streptococci. *Infect. Immun.*, **61**, 3745-3755.

Siegel, S. M., and C. Giumarro (1966). On the culture of microorganism similar to the precambrian microfossile *Kakabekia umbellata* barghoorn in ammonia-rich atmospheres. *Proc. Natl. Acad. Sci. USA*, **55**, 349-353.

Sili, C., A. Ena, R. Materassi and M. Vincenzini (1994). Germination of desicated aged akinetes of alkaliphilic cyanobacteria. *Arch. Microbiol.*, **162**, 20-25.

Singh, S. (1995). Partial purification and some properties of urease from the alkaliphilic *Cyanobacterium Nostoc calcicola*. *Folia Microbiol. Prague*, **40**, 529-533.

Skulachev, V.P. (1991) *Perspectives in Vectrial Metabolism and Osmochemistry* (Eds. Mitchell, P., and C. A. Psternak), pp. 387-444, The Glynn Research Foundation, London.

Souza, K.A., P.H. Deal, H.M. Mack and E.E. Turnbill (1974). Growth and reproduction of microorganisms under extremely alkaline conditions. *Appl. Microbiol.*, **28**, 1066-1068.

Spanka, R., and D. Fritze (1993). *Bacillus-cohnii* sp. nov, a new, obligately alkaliphilic, oval-spore-forming *Bacillus* species with ornithine and aspartic acid instead of diaminopimelic acid in the cell wall. *Int. J. Syst. Bact.*, **43**, 150-156.

Sturr, M.G., A.A. Guffanti and T.A. Krulwich (1994). Growth and bioenergetics of alkaliphilic *Bacillus firmus* OF4 in continuous culture at high pH. *J. Bacteriol.*, **176**, 3111-3116.

Sugiyama, S., E. J. J. Cragoe and Y. Imae (1988). Amiloride, a specific inhibitor for the Na^+ -driven flagellar motors of alkalophilic *Bacillus*. *J. Biol. Chem.*, **263**, 8215-8219.

Sugiyama, S., H. Matsukura and Y. Imae (1985). Relationship between Na^+-dependent cytoplasmic pH homeostasis and Na^+-dependent flagellar rotation and amino acid transport in alkalophilic *Bacillus*. *FEBS Lett.*, **182**, 265-268.

Sugiyama, S., H. Matsukura, N. Koyama, Y. Nosoh and Y. Imae (1986). Requirement of Na^+ in flagellar rotation and amino-acid transport in a facultatively alkalophilic *Bacillus*. *Biochim. Biophys. Acta*, **852**, 38-45.

Sumitomo, N., K. Ozaki, S. Kawai and S. Ito (1992). Nucleotide sequence of the gene for an alkaline endoglucanase from an alkalophilic *Bacillus* and its expression in *Escherichia coli* and *Bacillus subtilis*. *Biosci. Biotechnol. Biochem.*, **56**, 872-877.

Sunaga, T., T. Akiba and K. Horikoshi (1976). Production of penicillinase by an alkalophilic *Bacillus*. *Agric. Biol. Chem.*, **40**, 1363-1367.

Sutherland, K. J., M. Hashimoto, T. Kudo and K. Horikoshi (1993*a*). A partial physical map for the chromosome of alkalophilic *Bacillus* sp. strain C-125. *J. Gen. Microbiol.*, **139**, 661-667.

Sutherland, K. J., T. Kobayashi, T. Kudo and K. Horikoshi (1993*b*). Location of F_1ATPase-like genes on the physical map of the *Bacillus subtilis* 168 chromosome. *Biosci. Biotech. Biochem.*, **57**, 1202-1203.

Svetlitshnyi, V., F. Rainey and J. Wiegel (1996). *Thermosyntropha lipolytica* gen. nov., sp. nov., a lipolytic, anaerobic, alkalitolerant, thermophilic bacterium utilizing short- and long-chain fatty acids in syntrophic coculture with a methanogenic archaeum. *Int. J. Syst. Bact.*, **46**, 1131–1137.

Taber, W.A. (1959). Identification of an alkaline-dependent streptomyces as *Streptomyces caerulens* Baldacci and characterization of the species under controlled conditions. *Can. J. Microbiol.*, **5**, 335–344.

Taber, W. A. (1960). Evidence for the existence of acid-sensitive actinomycetes in soil. *Can. J. Microbiol.*, **6**, 503–514.

Tabernero, C., P. M. Coll, J. M. Fernandezabalos, P. Perez and R. I. Santamaria (1994). Cloning and DNA sequencing of *bga*A, a gene encoding an endo-β-1,3-1,4-glucanase, from an alkalophilic *Bacillus* strain (N137). *Appl. Environ. Microbiol.*, **60**, 1213–1220.

Takagi, H., S. Arafuka, M. Inouye and M. Yamasaki (1992*a*). The effect of amino acid deletion in Subtilisin E, based on structural comparison with a microbial alkaline elastase, on its substrate specificity and catalysis. *J. Biochem.*, **111**, 584–588.

Takagi, H., M. Kondou, T. Hisatsuka, S. Nakamori, Y. C. Tsai and M. Yamasaki (1992*b*). Effects of an alkaline elastase from an alkalophilic *Bacillus* strain on the tenderization of beef meat. *J. Agr. Food Chem.*, **40**, 2364–2368.

Takahara, T., and O. Tanabe (1960). Studies on the reduction of indigo in industrial fermentaion vat (VII). *J. Ferment. Technol.*, **38**, 329–331.

Takahara, Y., Y. Takasaki and O. Tanabe (1961). Studies on the reduction of indigo in the industrial fermentation vat (XVIII). On the growth factor of the strain No. S-8. *J. Ferment. Technol.*, **39**, 183–187.

Takahara, Y., and O. Tanabe (1962). Studies on the reduction of indigo in industrial fermentation vat(XIX). Taxonomic characteristics of strain No. S-8. *J. Ferment. Technol.*, **40**, 77–80.

Takami, H., T. Akiba and K. Horikoshi (1992*a*). Cloning, Expression, and Characterization of a minor alkaline protease from *Bacillus* sp. No. AH-101. *Biosci. Biotechnol. Biochem.*, **56**, 510–511.

Takami, H., T. Akiba and K. Horikoshi (1992*b*). Substrate specificity of thermostable alkaline protease from *Bacillus* sp. No. AH-101. *Biosci. Biotechnol. Biochem.*, **56**, 333–334.

Takami, H., T. Akiba and K. Horikoshi (1989). Production of extremely thermostable alkaline protease from *Bacillus* sp. no. AH-101. *Appl. Microbiol. Biotechnol.*, **30**, 120–124.

Takami, H., T. Akiba and K. Horikoshi (1990). Characterization of an alkaline protease from *Bacillus* sp. no. AH-101. *Appl. Microbiol. Biotechnol.*, **33**, 519–523.

Takami, H., T. Kobayashi, R. Aono and K. Horikoshi (1992*a*). Molecular cloning, nucleotide sequence and expression of the structural gene for a thermostable alkaline protease from *Bacillus* sp. AH101. *Appl. Microbiol. Biotechnol.*, **38**, 101–108.

Takami, H., T. Kobayashi, M. Kobayashi, M. Yamamoto, S. Nakamura, R. Aono and K. Horikoshi (1992*b*). Molecular cloning, nucleotide sequence, and expression of the structural gene for alkaline serine protease from alkalophilic *Bacillus* sp. 221. *Biosci. Biotechnol. Biochem.*, **56**, 1455–1460.

Takami, H., S. Nakamura, R. Aono and K. Horikoshi (1992*c*). Degradation of human hair by a thermostable alkaline protease from alkalophilic *Bacillus* sp. AH101. *Biosci. Biotechnol. Biochem.*, **56**, 1667–1669.

Tanabe, H., Y. Kobayashi and I. Akamatsu (1988). Pretreatment of pectic waste water with pectate lyase from an alkalophilic *Bacillus* sp. *Agric. Biol. Chem.*, **52**, 1855–1856.

Tanabe, H., K. Yoshihara, K. Tamura, Y. Kobayashi, I. Akamatsu, N. Niyomwan and P. Footrakul (1987). Pretreatment of pectic wastewater from orange canning process by an alkalophilic *Bacillus* sp. *J. Ferment. Technol.*, **65**, 243–246.

Teplyakov, A.V., J.M. van der Laan, A. A. Lammers, H.K. Kelders, K.H., O.M. Misset L. J. S. M. and B. W. Dijkstra (1992). Protein engineering of the high-alkaline serine protease PB92 from *Bacillus alcalophilus*: functional and structural consequences of mutant at the S4 substrate binding pocket. *Protein Eng.*, **5**, 413–420.

Tindall, B. J., A. A. Mills and W. D. Grant (1980). An alkalophilic red halophilic bacterium with a low magnesium requirement for a Kenyan soda lake. *J. Gen. Microbiol.*, **116**, 257-260.

Tindall, B. J., H. N. M. Ross and W. D. Grant (1984). Natronobacterium gen. nov. and *Natronococcus* gen. nov., two genera of haloalkalophilic archaebacterium. *Sys. Appl. Microbiol.*, **5**, 41-57.

Tsai, Y., S. Lin, Y. Li, M. Yamasaki and G. Tamura (1986). Characterization of an alkaline elastase from alkalophilic *Bacillus* Ya-B. *Biochim. Biophys. Acta*, **883**, 439-447.

Tsai, Y., M. Yamasaki and G. Tamura (1984). Substrate specificity of a new alkaline elastase from an alkalophilic *Bacillus*. *Biochem. Int.*, **8**, 283-288.

Tsai, Y., M. Yamasaki, Y. Yamamoto-Suzuki and G. Tamura (1983). A new alkaline elastase of an alkalophilic *Bacillus*. *Biochem. Int.*, **7**, 577-583.

Tsuchida, O., Y. Yamagata, T. Ishizuka, T. Arai, J. Yamada, M. Takeuchi and E. Ichishima (1986). An alkaline proteinase of an alkalophilic *Bacillus* sp. *Curr. Microbiol.* **14**, 7-12.

Tsuchiya, K., Y. Nakamura, H. Sakashita and T. Kimura (1992). Purification and characterizaation of a thermostable alkaline protease from alkalophilic *Thermoactinomyces* sp. HS682. *Biosci. Biotechnol. Biochem.*, **56**, 246-250.

Tsuchiya, K., H. Sakashita, Y. Nakamura and T. Kimura (1991). Production of thermostable alkaline protease by alkalophilic *Thermoactinomyces* sp. HS682. *Agric. Biol. Chem.*, **55**, 3125-3127.

Tsujibo, H., T. Sato, M. Inui, H. Yamamoto and Y. Inamori (1988). Intracellular accumulation of phenazine antibiotics production by an alkalophilic *Actinomycete*. *Agric. Biol. Chem.*, **52**, 301-306.

Tsujibo, H., Y. Yoshida, K. Miyamoto, T. Hasegawa and Y. Inamori (1992). Purification and properties of two types of chitinases produced by an alkalophilic *Actinomycete*. *Biosci. Biotechnol. Biochem.*, **56**, 1304-1305.

Tsukamoto, A., K. Kimura, Y. Ishii, T. Tanaka and K. Yamane (1988). Nucleotide sequence of maltohexaose-producing amylase gene from an alkalophilic *Bacillus* sp. #707 and structural similarity to liquefing type α-amylase. *Biochem. Biophys. Res. Commun.*, **151**, 25-31.

Tsumura, K., Y. Hashimoto, T. Akiba and K. Horikoshi (1991*a*). Isolation and characterization of alkalophilic bacteria that degrade soybean galactan. *Agric. Biol. Chem.*, **55**, 1399-1400.

Tsumura, K., Y. Hashimoto, T. Akiba and K. Horikoshi (1991*b*). Purification and properties of galactanases from alkalophilic *Bacillus* sp. S-2 and S-39. *Agric. Biol. Chem.*, **55**, 1265-1271.

Upasani, V.N., S.G. Desai, N. Moldoveanu and M. Kates (1994). Lipids of extremely halophilic archaeobacteria from saline environments in India: A novel glycolipid in *Natronobacterium* strains. *Microbiology*, **140**, 1959-1966.

Usami, R., H. Honda, T. Kudo, H. Hirokawa and K. Horikoshi (1983). Plasmids in alkalophilic *Bacillus* sp. *Agric. Biol. Chem.*, **47**, 2101-2102.

Usami, R., T. Kudo and K. Horikoshi (1990). Protoplast transformation of alkalophilic *Bacillus* sp. by plasmid DNA. S*tarch/Stärke*, **46**, 230-232.

van der Laan, J. C., G. Gerritse, L. J. S. M. Mulleners, R. A. C. van der Hoek and W. J. Quax (1991). Cloning, characterization, and multiple chromosomal integration of a *Bacillus* alkaline protease gene. *Appl. Environ. Microbiol.*, **57**, 901-909.

Vazquez, S. C., L. N. R. Merino, W. P. Maccormack and E. R. Fraile (1995). Protease-producing psychrotrophic bacteria isolated from Antarctica. *Polar. Biol.*, **15**, 131-135.

Vedder, A. (1934). *Bacillus alcalophilus* n. sp. benevens enkle ervaringen met sterk alcalische voedingsbodems. Antonie van Leeuwenhoek, *J. Microbiol. Serol.*, **1**, 141-147.

Viikari, L., A. Kantelinen, J. Sundquist and M. Linko (1994). Xylanases in bleaching: From an idea to the industry. *FEMS Microbiol. Rev.*, **13**, 335-350.

Waksman, S.A. and J.S. Jåffe (1922). Microorganisms concerned in the soil. II. *Thiobacillus thiooxidans*, a new sulfur-oxidizing organisms isolated from the soil. *J. Bacteriol.*, **7**, 239-256.

Wang, Y.X. and B.C. Saha (1993). Purification and characterization of thermophilic and alkalophilic tributyrin esterase from *Bacillus* strain A30-1 (ATCC 53841). *J. Am. Oil. Chem. Soc.*, **70**(11), 1135-1138.

Watanabe, N., Y. Ota, Y. Minoda and K. Yamada (1977). Isolation and identification of alkaline lipase producing microorganisms, culture conditions and some properties of crude enzymes. *Agric. Biol. Chem.*, **41**, 1353-1358.

Wiley, W. R., and J. C. Stokes (1962). Requirement of an alkaline pH and ammonia for substrate oxidation by *Bacillus pasteurii*. *J. Bacteriol.*, **84**, 730-734.

Wiley, W. R., and J. C. Stokes (1963). Effect of pH and ammonium ions on the permeability of *Bacillus pasteurii*. *J. Bacteriol.*, **86**, 1152-1156.

Xu, X., N. Koyama, M. Cui, A. Yamagishi, Y. Nosoh and T. Oshima (1991). Nucleotide sequence of the gene encoding NADH dehydrogenase from an alkalophile, *Bacillus* sp. strain YN-1. *J. Biochem.*, **109**, 678-683.

Xu, C. H., A. Nejidat, S. Belkin and S. Boussiba (1994). Isolation and characterization of the plasma membrane by two-phase partitioning from the alkalophilic cyanobacterium *Spirulina platensis*. *Plant Cell Physiol.*, **35**, 737-741.

Xu, Y., T. Kobayashi and T. Kudo (1996). Molecular cloning and nucleotide sequence of the groEL gene from the alkaliphilic *Bacillus* sp. strain C-125 and reactivation of thermally inactivated α-glucosidase by recombinant GroEL. *Biosci. Biotechnol. Biochem.*, **60**, 1633-1636.

Yamagata, Y., and E. Ichishima (1995*a*). A new alkaline serine protease from alkalophilic *Bacillus* sp.: Cloning, sequencing, and characterization of an intracellular protease. *Curr. Microbiol.*, **30**, 357-366.

Yamagata, Y., K. Isshiki and E. Ichishima (1995*b*). Subtilisin sendai from alkalophilic *Bacillus* sp.: Molecular and enzymatic properties of the enzyme and molecular cloning and characterization of the gene, *aprS*. *Enzyme Microbiol. Technol.*, **17**, 653-663.

Yamagata, Y., T. Sato, S. Hanzawa and E. Ichishima (1995*c*). The structure of subtilisin ALP I from alkalophilic *Bacillus* sp. NKS-21. *Curr. Microbiol.*, **30**, 201-209.

Yamaguchi, M., and H. Fujisawa (1978). Characterization of NADH-cytochrome *c* reductase, a component of benzoate 1,2-dioxygenase system from *Pseudomonas arvilla* C-1. *J. Biol. Chem.*, **253**, 8848-8853.

Yamamoto, M., R. Aono and K. Horikoshi (1993). Structure of the 87-kDa β-1,3-glucanase gene of *Bacillus circulans* IAM1165 and properties of the enzyme accumulated in the periplasm of *E. coli* carrying the gene. *Biosci. Biotechnol. Biochem.*, **57**, 1518-1525.

Yamamoto, M., and K. Horikoshi (1987*a*). Cloning and expression of an alkalophilic *Bacillus* oligo-1,6-glucosidase gene in *Escherichia coli*. *J. Jpn. Soc. Starch Sci.*, **34**, 300-303.

Yamamoto, M., and K. Horikoshi (1987*b*). Production of isomaltose-hydrolyzing α-glucosidase by an alkalophilic *Bacillus* sp. *J. Jpn. Soc. Starch Sci.*, **34**, 292-299.

Yamamoto, M., and K. Horikoshi (1990*a*). Nucleotide sequence of alkalophilic *Bacillus* oligo-1,6-glucosidase gene and the properties of the gene product by *Escherichia coli* HB 101. *Denpun Kagaku*, **37**, 137-144.

Yamamoto, M., and K. Horikoshi (1990*b*). Purification and properties of an oligo-1,6-D-glucosidase from an alkalophilic *Bacillus* species. *Carbohydr. Res.*, **197**, 227-235.

Yamamoto, M., Y. Tanaka and K. Horikoshi (1972). Alkaline amylases of alkalophilic bacteria. *Agric. Biol. Chem.*, **36**, 1819-1823.

Yonezawa, Y., and K. Horikoshi (1978). Polymaines in alkalophilic *Bacillus* Y-25. *Agric. Biol. Chem.* **42**, 1955-1956.

Yoshida, A., Y. Iwasaki, T. Akiba and K. Horikoshi (1991). Purification and properties of cyclomaltodextrinase from alkalophilic *Bacillus* sp. *J. Ferment. Bioeng.*, **71**, 226-229.

Yoshihara, K., and Y. Kobayashi (1982). Retting of mitsumata bast by alkalophilic *Bacillus* in paper making. *Agric. Biol. Chem.*, **46**, 109-117.

Yoshimatsu, T., K. Ozaki, S. Shikata, Y. Ohta, K. Koike, S. Kawai and S. Ito (1990). Purification and characterization of alkaline endo-1,4-β-glucanases from alkalophilic *Bacillus* sp. KSM-635. *J. Gen. Micriobiol.*, **136**, 1973-1979.

Yoshioka, T. (1982). *Indigo Fermentation*, Kyoto-Shoin. Kyoto. (in Japanese)

Yoshitake, J., T. Kudo, R. Usami and K. Horikoshi (1984). Isolation of the fragments from alkalophilic *Bacillus* sp. DNA which are controlled by temperature, pH or NaCl concentration in pGR71 CAT expression system. *Agric. Biol. Chem.*, **48**, 2621-2626.

Yum, D. Y., H. C. Chung, D. H. Bai, D. H. Oh and J. H. Yu (1994). Purification and characterization of alkaline serine protease from an alkalophilic *Streptomyces* sp. *Biosci. Biotechnol. Biochem.*, **58**(3), 470-474.

Yumoto, I., Y. Fukumori and T. Yamanaka (1991). Purification and characterization of two membrane-bound *c*-type cytochromes from a facultative alkalophilic *Bacillus*. *J. Biochem.*, **110**, 267-273.

Yumoto, I., S. Takahashi, T. Kitagawa, Y. Fukumori and T. Yamanaka (1993). The molecular features and catalytic activity of Cu(A)-containing aco3-type cytochrome-*c* oxidase from a facultative alkalophilic *Bacillus*. *J. Biochem. Tokyo*, **114**(1), 88-95.

Zhang, S., A. Faro and G. Zubay (1985). Mitomycin-induced lethality of *Esherichia coli* cells containing the ColE1 plasmid: involvement of the *kil* gene. *J. Bacteriol.*, **163**, 174-179.

Zhilina, T.N. and G. A. Zavarzin (1994). Alkaliphilic anaerobic community at pH 10. *Curr. Microbiol.*, **29**, 109-112.

Addendum

More than 200 scientific papers have been published since work was started on this volume. Most of them are reports on various aspects of enzymes. Microbial diversities of alkaliphiles including archaea have also been studied extensively. Other fascinating papers include molecular studies on the genetics of alkaliphilic *Bacillus* strains. Two groups, the author's group and Krulwich's group, are analyzing chromosomal DNA to understand better the characteristic properties of alkaliphiles. Recent taxonomic study revealed that *Bacillus* sp. C-125 described above belongs to *Bacillus halodurans* (Takami and Horikoshi, 1998). *Bacillus halodurans* C-125 genome analysis is in progress by Takami *et al.* of the author's group. The entire sequence of the chromosomal DNA will be determined within one year.

Anaerobic Sporeforming or Non-sporeforming Bacteria

Recently, many interesting alkaliphilic anaerobes have been isolated from alkaline environments.

Cook *et al.* (1996) reported that *Clostridium paradoxum*, a novel alkali-thermophile, could maintain the pH homeostasis in cytoplasm and thrive well in the range of pH 7.6 to 9.8 of culture media. However, growth ceased when the extracellular pH was higher than 10.2 and the intracellular pH increased to above 9.8. Between the extracellular pH of 8.0 and 10.3 the intracellular ATP concentration was around 1 mM. An outstanding review on anaerobic alkalithermophiles has recently been reported by Wiegel (1998).

Kodama and Koyama (1997) exhibited unique characteristics of anaerobic alkaliphiles belonging to *Amphibacillus xylanus*. The transport of leucine and glucose was remarkably stimulated by $NH4^+$. The ($NH4^+ + Na^+$)-activated ATPase of *A. xylanus* were also reported by Takeuchi, Kaieda and Koyama (1997). Among the amines tested, all primary amines activated the enzyme, but secondary and tertiary amines were ineffective.

Sorokin *et al.* have studied reduced sulfur oxidizing alkaliphiles. The inoculation of 14 samples from alkaline environments–soda soils, lakes, and springs–in medium with acetate yielded seven cultures of bacteria capable of oxidizing thiosulfate at pH values above 10.0. Eight strains of chemoorganoheterotrophic thiosulfate-oxidizing bacteria were isolated in pure culture. All strains are obligately alkaliphilic gram-negative eubacteria able to grow in the pH range of 7.5-10.5 with an optimum around 9.5. The strains oxidize sulfide, elemental sulfur and thiosulfate to tetrathionate. Recently, sulfide dehydrogenase was isolated and purified from alkaliphilic chemolithoautotrophic sulfur-oxidizing bacteria (Sorokin, Lysenko and Mityushina, 1996; Sorokin *et al.*, 1998; Sorokin and Mityushina, 1998).

Zhilina *et al.* (1997) isolated and characterized *Desulfonatronovibrio hydrogenovorans* gen. nov., sp. nov., an alkaliphilic, sulfate-reducing bacterium, from a soda-depositing lake, Lake Magadi, in Kenya. The strain Z-7935(T) is an obligately sodium-dependent alkaliphile which grows in sodium carbonate medium and does not grow at pH 7; the maximum pH for growth is more than pH 10, and the optimum pH is 9.5 to 9.7. The optimum NaCl concentration for growth is 3% (wt/vol). The optimum temperature for growth is 37 °C. The G + C content of the DNA is 48.6 mol%. 16S ribosomal DNA sequence analysis revealed that strain Z-7935(T) represents a new lineage with genus status in the delta subclass of the Proteobacteria.

Cyanobacteria

Alkaliphilic cyanobacteria have been isolated from soda lakes and physiological studies have been reported. Gerasimenko, Dubinin and Zavarzin (1996) isolated alkaliphilic cyanobacteria from soda lakes of Tuva and reported a wide diversity of alkaliphilic cyanobacteria (16 genera and 34 species were found). Singh (1995) partially purified a urease from an alkaliphilic diazotrophic cyanobacterium *Nostoc calcicola*. The purified enzyme showed optimum activity at pH 7.5 and at 40 °C. The enzyme was found to be sensitive to metal cations, particularly Hg^{2+}, Ag^+ and Cu^{2+}. 4-Hydroxymercuribenzoate (a mercapto-group inhibitor) and acetohydroxamic acid (a chelating agent of nickel) inhibited the enzyme activity completely. These results suggest the involvement of an SH-group and Ni^{2+} in the activity of urease from *N. calcicola*.

Alkaliphilic Archaea

Until recently, few microbiologists thought that alkaliphiles thrived in hydrothermal areas. Now, many alkaliphilic archaea have been reported in addition to the small number of alkaliphilic archaea other than haloalkaliphiles reported prior to 1992.

Haloalkaliphiles

Kamekura *et al.* (1997) studied diversity of alkaliphilic halobacteria on the basis of phylogenetic tree reconstructions, signature bases specific for individual genera, and sequences of spacer regions between 16 and 23S rRNA genes. They proposed the following changes: *Natronobacterium pharaonis* be transferred to *Natronomonas* gen. nov. as *Natronomonas pharaonis* gen. nov., comb. nov.; *Natronobacterium vacaolatum* be transferred to the genus *Halorubrum* as *Halorubrum vacuolatum* comb. nov.; and *Natronobacterium magadii* to be transferred to the genus *Natrialba* as *Natrialba magadii.*

A novel osmolyte, 2-sulfotrehalose, was discovered in several *Natronobacterium* species of haloalkaliphilic archaea (Desmarais *et al.*, 1997). The concentration of this novel disaccharide, termed sulfotrehalose, increased with increasing concentrations of external NaCl, behavior consistent with its identity as an osmolyte. Sucrose was the most effective in suppressing biosynthesis and accumulation of sulfotrehalose, with levels as low as 0.1 mM being able to significantly replace the novel charged osmolyte. Other common osmolytes (glycine betaine, glutamate and proline) were not accumulated or used for osmotic balance in place of the sulfotrehalose by halophilic archaea.

Methanogens

Recently, many alkaliphilic methanogens have been isolated from alkaline environments. A new obligately alkaliphilic, methylotrophic methanogen 2-7936 was isolated from the soda lake Magadi (Kenya). According to its phenotypic and genotypic properties, the isolate belonged to *Methanosalsus* (*Methanohalophilus*) *zhilinaeae*; it exhibited a 91% homology with the type strain of this formerly monotypic species. The strain did not require Cl^- but was obligately dependent on Na^+ and $HCO3^-$. It was obligately alkaliphilic and grew within a pH range of 8-10 (Khmelenina *et al.* 1997). Furthermore, Khmelenina *et al.* (1997) and Khmelenina, Kalyuzhnaya and Trotsenko (1997) isolated two strains (5Z and 20Z) of halotolerant alkaliphilic obligate methanotrophic bacteria, which were first isolated from moderately saline soda lakes in Tuva (Central Asia). The strains grow fastest at pH 9.0-9.5 and much more slowly at pH 7.0. No growth occurred at pH less than or equal to 6.8. They require NaHCO3 or NaCl for growth in alkaline medium. The G + C content of strains 5Z and 20Z are 47.6 and 47.9 mol%, respectively. Based on their alkaliphilic physiology, both strains were referred to as *Methylobacter alcaliphilus* sp. nov.

Thermophiles

Stetter's group (Keller *et al.*, 1995) isolated *Thermococcus alcaliphilus* sp. nov., a new hyperthermophilic archaeum growing on polysulfide at alkaline pH. A novel coccoid-shaped, hyperthermophilic, heterotrophic member of the archaea grew between 56° and 90 °C with an optimum around 85 °C. The pH range for growth was 6.5 to 10.5, with an optimum around 9.0. Polysulfide and elemental sulfur were reduced to H_2S. Sulfur stimulated the growth rate. Based on DNA-DNA hybridization and 16S rRNA partial sequences, the new isolate represents a new species of

Thermococcus, which the group named *Thermococcus alcaliphilus*. Then Dirmeier *et al.* (1998) report-ed a new hyperthermophilic member of the *Archaea*. The cells are coccoid-shaped and possess up to five flagella. They grow between 56° and 93°C (optimum 85°C) and pH 5.0–9.5 (optimum 9.0). The organism is strictly anaerobic and grows heterotrophically on defined amino acids and com-plex organic substrates such as casamino acids, yeast extract, peptone, meat extract, tryptone and casein. Polysulfide and elemental sulfur are reduced to H_2S. In the absence of polysulfide or ele-mental sulfur, the isolate grows at a significantly reduced rate. Growth is not influenced by the pres-ence of H_2. DNA-DNA hybridization and 16S rRNA partial sequences indicate that the new isolate belongs to the genus *Thermococcus* and represents a new species, *Thermococcus acidaminovorans*.

Physiology

Internal pH values

Aono, Ito and Horikoshi (1997) developed a new method to measure the cytoplasmic pH of the facultative alkaliphilic *Bacillus halodurans* C-125. The bacterium was loaded with a pH-sensitive fluorescent probe, 2',7'-bis-(2-carboxyethyl)-5 (and -6)-carboxyfluorescein (BCECF), and cytoplas-mic pH was determined from the intensity of fluorescence of the intracellular BCECF. The activity of the organism to maintain neutral cytoplasmic pH was assessed by measuring the cytoplasmic pH of the cells exposed to various pH conditions. The cytoplasmic pH maintenance activity of *Bacillus halodurans* C-125 increased with increasing culture pH, indicating that the activity was regulated in response to the culture pH.

Antiporter System

As described above, Na^+/H^+ antiporters have been studied extensively. Recently, Kitada *et al.* (1997) exhibited a K^+/H^+ antiport system that was detected for the first time in right-side-out mem-brane vesicles prepared from alkaliphilic *Bacillus* sp. No. 66 (JCM 9763). An outwardly directed K^+ gradient stimulated uphill H^+ influx into right-side-out vesicles and created the inside-acidic pH gra-dient (ΔpH). This H^+ influx was pH-dependent and increased as the pH increased from 6.8 to 8.4. Addition of 100 mμM quinine inhibited the H^+ influx by 75%. This exchange process was elec-troneutral and the H^+ influx was not stimulated by the imposition of the membrane potential (inte-rior negative). Addition of K^+ at the point of maximum ΔpH caused a rapid K^+-dependent H^+ efflux consistent with the inward exchange of external K^+ for internal H^+ by a K^+/H^+ antiporter. Rb^+ and Cs^+ could replace K^+ but Na^+ and Li^+ could not. The H^+ efflux rate was a hyperbolic function of K^+ and increased with increasing extravesicular pH (pH(out)) from 7.5 to 8.5. These findings were consistent with the presence of K^+/H^+ antiport activity in these membrane vesicles. Although alkaliphilic *Bacillus* sp. No. 66 is different from *Bacillus halodurans* C-125, it is clear that K^+/H^+ antiport is active in alkaliphiles.

Respiration

Qureshi, Fujiwara and Fukumori (1996) isolated succinate: quinone oxidoreductase (complex II) containing a single heme b in facultative alkaliphilic *Bacillus* sp strain YN-2000 in the presence of Triton X-100. The isolated enzyme showed high succinate-ubiquinone oxidoreductase activity at pH 8.5. The succinate: quinone oxidoreductase with a single heme b is involved in the respiratory chain of the alkaliphile at a very alkaline pH.

Comparative study on cytochrome content of alkaliphilic *Bacillus* strains was carried out by Yumoto (Yumoto, Nakajima and Ikeda 1997) to understand adaptation to alkaline environments. Among the strains tested, four facultative alkaliphiles and one obligate alkaliphile possessed rela-tively high amounts of total cytochrome (more than 0.3 nmol./mg cells). These five alkaliphiles contained high amounts of cytochromes *b* and *c* compared with the other strains tested, but their cytochrome *a* contents were not noticeably different. All the tested alkaliphilic strains had much higher amounts of cytochrome than *B. subtilis*. The authors suggest that various bioenergetic strategies are used by alkaliphilic *Bacillus* for adaptation to a high alkaline environment, and that cytochromes *b* and *c* play a bioenergetically important role in certain kinds of alkaliphiles.

Recently, Morotomi *et al.* (1998) reported respiration studies of *Bacillus halodurans* C-125 and a neutralo-sensitive mutant. This mutant was able to grow to the same extent as did the parent strain above pH 8, but did not grow below pH 7.5. The same extent of oxygen uptake was shown by the cells of the parent and mutant strains at pH 10.3. On the other hand, the oxygen uptake rate was about one-fifth that of the parent strain at pH 7. NADH-dependent oxygen uptake by everted vesicles of the mutant was lower than that of the parent strain at pH 7-7.5, while the rate at pH 8-9 was almost identical in both strains. The activity at pH 7 of cytochrome *c* oxidase of right-side-out membrane vesicles of the mutant strain was lower than that of the parent strain at pH 7, while both samples had almost the same enzymatic activity at pH 8.5. These results suggest that poor respiratory activities of the mutant strain at pH 7 are the reason why this mutant strain was unable to grow at neutral pH. Therefore, respiration systems are essential to understand alkaliphily in some alkaliphiles.

Alkaliphily

Alkaliphily of *Bacillus halodurans* C-125
A) Revised physical map
A physical map of the chromosome of this strain was constructed to facilitate further genome analysis and the genome size was revised from 3.7 Mb to 4.25 Mb (Figure). Complete digestion of the chromosomal DNA with two rare cut restriction endonucleases, *Asc*I and *Sse*8387I, each yielded 20 fragments ranging in size from 20 to 600 kb. Seventeen linking clones were isolated in each instance to join the adjacent *Asc*I or *Sse*8387I fragments in the chromosomal map. All *Asc*I linking clones isolated were sequenced and analyzed by comparison with the BSORF database to map the genes in the chromosome of strain *Bacillus halodurans* C-125. Several ORFs showing significant similarities to those of *B. subtilis* in the *Asc*I linking clones were positioned on the physical map. The *ori*C region of the C-125 chromosome was identified by Southern blot analysis with a DNA probe containing the *gyr*B region (Takami, H., K. Nakasone, *et al.* 1998).

B) Gene clusters of *Bacillus halodurans* C-125
The nucleotide sequences of three independent fragments (designated No. 3, 4, and 9; each 15 - 20 kb in size) of the genome of alkaliphilic *Bacillus halodurans* C-125 cloned in a l phage vector have been determined. Thirteen putative open reading frames (ORFs) were identified in sequenced fragment No. 3 and eleven ORFs were identified in No. 4. Twenty ORFs were also identified in fragment No. 9. All putative ORFs were analyzed in comparison with the BSORF database and non-redundant protein databases. The functions of 5 ORFs in fragment No. 3 and 3 ORFs in fragment no. 4 were suggested by their significant similarities to known proteins in the database. Among the 20 ORFs in fragment No. 9, the functions of 11 ORFs were similarly suggested. Most of the annotated ORFs in the DNA fragments of the genome of alkaliphilic *B. halodurans* C-125 were conserved in the *B. subtilis* genome. The organization of ORFs in the genome of strain C-125 was found to differ from the order of genes in the chromosome of *B. subtilis*, although some gene clusters (*ydh*, *yqi*, *yer*, and *yts*) were conserved as operon units the same as in *B. subtilis*.

Further work on whole genome of alkaliphilic *B. halodurans* C-125 is in progress.

Alkaliphily of *Bacillus firmus* OF4
A) Physical map of *Bacillus firmus* OF4
Gronstad *et al.* (1998) published a physical map of alkaliphilic *Bacillus firmus* OF4. A physical map of the *B. firmus* OF4 is consistent with a circular chromosome of approximately 4 Mb. A large endogenous plasmid of 110 kb having a cadmium-resistant fragment was also detected. No cluster-

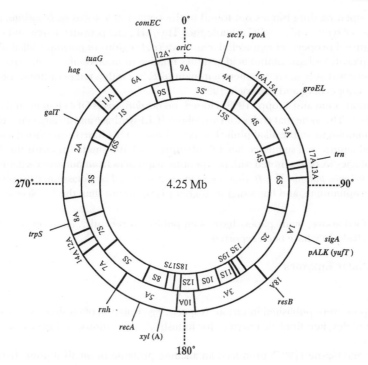

*AscI/Sse*8387I physical and genetic map of the chromosome of *Bacillus halodurans* C-125. Outer and inner circle show the *AscI-* and *Sse*8387I physical map, respectively. The locations of several housekeeping genes are indicated on the map. The dashed lines indicate the approximate position of the gene. The assigned positions of the genes on the physical map of *Bacillus halodurans* C-125 were compared with those on the genetic and physical maps of *B. subtilis*.

ing of genes thus far identical with roles in alkaliphily has been found. Direct repeat sequences were previously reported upstream of a gene encoding a Na^+/H^+ antiporter that has a role in pH homeostasis. In the current analysis, these sequences were found to be present in multicopies on the chromosome, most of which are present in one 920 kb fragment. The authors speculated that such sequences might play a role in DNA rearrangements that allow amplification of important genes in this region.

B) Genes of *Bacillus firmus* OF4

Ito *et al.* (1997) discussed the role of the *nhaC*-encoded Na^+/H^+ antiporter of alkaliphilic *Bacillus firmus* OF4. The predicted alkaliphile *NhaC* is highly homologous to the deduced products of homologous genes of unknown function from *Bacillus subtilis* and *Haemophilus influenzae*. The full length of *nhaC* complemented the Na^+-sensitive phenotype of an antiporter-deficient mutant strain of *Escherichia coli* but not the alkali-sensitive growth phenotypes of Na^+/H^+-deficient mutants of either alkaliphilic *B. firmus* OF 4811M or *B. subtilis*. *NhaC* is not competent, by itself, to complement the pH-sensitive phenotype of nonalkaliphilic mutants.

Recently, seven clones isolated from alkaliphilic *B. firmus* OF4 restored the growth of a K^+-uptake-deficient *E. coli* mutant on only 10 mM K^+. None of the clones contained genes with apparent homology to known K^+ transport systems in other organisms. Based on sequence homologies, the newly isolated alkaliphile loci included, *fts*H, a dipeptide transport system, a *ger*C locus with hydrophobic open reading frames not found in the comparable locus of *B. subtilis*, a sugar phosphotransferase enzyme, and a *cap*BC homologue. The *fts*H gene provided a new and striking example of a recognized property of extracellular and external regions of polytopic alkaliphile proteins: a significant paucity of basic amino acid residues relative to neutrophile counterparts. The alkaliphile *fts*H gene was able to complement a mutant of *E. coli* with a temperature-sensitive *fts*H gene product (Ito, Cooperberg and Krulwich, 1997).

Furthermore, Guo and Tropp (1998) cloned the *Bacillus firmus* OF4 cls gene and characterized its gene product. The gene that codes for cardiolipin (CL) synthase and an adjacent gene that codes for an *mec*A homologue in the alkaliphilic bacteria *B. firmus* OF4 have been cloned and sequenced. The predicted amino acid sequence has 129 identities and 100 similarities with the *E. coli* CL synthase. Homologies were also noted with polypeptide sequences from putative cls genes from *B. subtilis* and *Pseudomonas putida*. The *B. firmus* enzyme is stimulated by potassium phosphate, inhibited by CL and phosphatidate, and has a slightly higher pH optimum than the *E. coli* enzyme.

As described above, many papers have been published recently. However, no crucial experiments about alkaliphily have been reported.

Extracellular Enzymes

Protease

Many reports were published in various respects, *e.g.*, isolation of alkaline proteases from newly isolated alkaliphiles, keratinolytic enzymes for industrial applications, and genes and structures of alkaline enzymes.

Gessesse and Gashe (1997) produced an alkaline protease by an alkaliphilic bacteria isolated from an alkaline soda lake in Ethiopia. The optimum temperature and pH for activity were 65 °C and 9.5-11.5 respectively. Above 50 °C, Ca^{2+} was required for enzyme activity and stability. The enzyme was stable over the pH range of 5-12.

Cheng *et al.* (1995) reported a keratinase of a feather-degrading *Bacillus licheniformis* PWD-1. This enzyme was stable from pH 5 to 12. The optimal reaction pHs for feather powder and casein were 8.5 and 10.5 to 11.5, respectively. The optimal reaction temperature was 50° to 55 °C. Zaghloul AlBahra and AlAzmeh (1998) also reported isolation, identification and keratinolytic activity of several feather-degrading bacterial isolated from Egyptian soil. These isolates were able to degrade chicken feather.

Schmidt *et al.* (1995) analyzed a series of serine protease genes of alkaliphilic *Bacillus* sp. strain LG12. Four tandem subtilisin-like protease genes were found on a 6,854-bp DNA fragment. The two downstream genes (*spr*C and *spr*D) appear to be transcribed independently, while the two upstream genes (sprA and sprB) seem to be part of the same transcript. Tsuchiya *et al.* (1997) cloned an intracellular alkaline protease gene from alkaliphilic *Thermoactinomyces* sp. HS682 and expressed in *Escherichia coli*. Sequence analysis showed a putative promoter region, a putative transcriptional termination signal, and an open reading frame of 963 bases coding for a polypeptide of 321 amino acids. The enzyme was stable at pH 6.0-12.0 and below 60 °C in the presence of Ca^{2+}. The temperature and pH optima of the enzyme were 65 °C and pH 11.0, respectively.

Martin *et al.* (1997) found that the solution structure of serine protease PB92 from *Bacillus alcalophilus* presents a rigid fold with a flexible substrate-binding site. The solution structure of the serine protease PB92 presents a well-defined global fold which is rigid with the exception of a restricted number of sites. Among the limited number of residues involved in significant internal mobility are those of two pockets, termed S1 and S4, within the substrate-binding site. The presence of flexibility within the binding site supports the proposed induced fit mechanism of substrate binding.

Starch Degrading Enzymes

A) Amylases

The raw starch-degrading alkaline amylase of *Bacillus* sp. IMD 370 was isolated by Kelly, *et al.* (1995). The amylase of *Bacillus* sp. IMD 370 is the first report of an alkaline amylase with the ability to digest raw starch. The enzyme digested raw corn starch to glucose, maltose, maltotriose and maltotetraose. The maximum pH for raw starch hydrolysis was pH 8.0 compared with pH 10.0 for soluble starch hydrolysis. It is of interest that degradation of raw starch was stimulated sixfold in the presence of β-cyclodextrin (17.5 mM).

B) CD-forming enzymes

Extensive studies of CD-forming enzymes have been reported. Some of these are described below.

Chung *et al.* (1998) isolated a thermostable cyclodextrin glucanotransferase from *Bacillus stearothermophilus* ET1. The optimum pH for the enzyme-catalyzed reaction was pH 6.0, and the optimum temperature was observed at 80 °C. A 13% (w/v) cornstarch solution was liquefied and converted to CDs solely using this enzyme. The cornstarch conversion rate was 44% and α-, β-, and γ-CDs were produced in the ratio of 4.2 : 5.9 : 1. Lima, DeMoraes and Zanin (1998) developed a unique cyclodextrin production process. A 10% (w/v) solution of cassava starch, liquefied with α-amylase, was incubated with CGTase using only the enzyme, added ethanol (from 1% to 5%) followed by the addition of yeast, *Saccharomyces cerevisiae* (12% w/v), plus nutrients. β-CD yield decreased six times. However, no production cost was reported. Terada *et al.* (1997) studied the initial reaction of cyclodextrin glucanotransferase from an alkaliphilic Bacillus sp. A2-5a on amylose. Cyclic α-1,4-glucans with a degree of polymerization ranging from 9 to more than 60, in addition to well-known α-, β-, and γ-cyclodextrin (CD), were detected at early stage of enzymatic reactions. Subsequently large cyclic α-1,4 glucans were converted into smaller cyclic α-1,4-glucans. The final major product was β-CD. CGTase from *Bacillus macerans* also produced large cyclic α-1,4-glucans, but the final major product was α-CD.

C) Pullulanase

Lin, Tsau and Chu (1996) purified a thermostable pullulanase from thermophilic alkaliphilic *Bacillus* sp. strain TS-23. This purified enzyme had both pullulanase and amylase activities. The temperature and pH optima for both pullulanase and amylase activities were 70 °C and pH 8-9, respectively. The enzyme remained more than 96% active at temperatures below 65 °C, and both activities were retained at temperatures up to 90 °C in the presence of 5% SDS. Under optimum conditions this enzyme catalyzed the hydrolysis of α-1,6 glucosidic linkages in pullulan with maltotriose as the only product. It also cleaved α-1,4 bonds in amylose, starch and related malto-oligosaccharides larger than maltotriose with maltose and maltotriose as the main products.

Takagi, Lee and Imanaka (1996) reported diversity in size and alkaliphily of thermostable α-amylase-pullulanases produced by recombinant *Escherichia coli*, *Bacillus subtilis* and the wild-type *Bacillus* sp. XAL601. It was revealed that the noncatalytic C-terminal region may be responsible for the high optimum pH of the enzyme activity. These observations were also reported in CGTase (see p. 171) Lee, Lee and Kim (1997) found intracellular and extracellular forms of alkaline pullulanase from an alkaliphilic *Bacillus* sp. S-1. The precursor form was accumulated intracellularly in large amounts, and the processed form was detected in both the membrane fraction and the fraction trapped between the cytoplasmic membrane and the cell wall.

Cellulase

After the discovery of the industrial application of alkaline cellulase as a detergent additive, screening of the enzymes has been carried out by many microbiologists (Hayashi *et al.*, 1996; Miyatake and Imada 1997). Ito's group has isolated extensive numbers of alkaline cellulases (Hakamada *et al.*, 1997; Ito, 1997) and also reported their characteristics in many respects. Sumitomo *et al.* (1995) overexpressed alkaline cellulase of alkaliphilic *Bacillus* sp. KSM-64 by using

Bacillus subtilis harboring their vector pHSP64. By this process they produced 30 g of alkaline cellulase in one liter.

From the industrial point of view, the production process of alkaline cellulase has been established by Ito's group.

Xylanases

Investigations of alkaline xylanase have been promoted by the discovery of the biological debleaching process using xylanases.

Many alkaline xylanases have been isolated from various alkaliphiles. Two alkali-tolerant thermophilic bacterial strains with xylanolytic activity were produced by continuous cultivation from samples collected near Bulgarian hot springs (Dimitrov *et al.*, 1997). Xylanases produced by the two strains were similar with respect to temperature and pH optimum (70°-75 °C and pH 6.5-7.0) as well as their thermostability. The xylanases were thermostable at 70 °C for 30 min. Lopez *et al.* reported xylanase production by a new alkalitolerant isolate of alkalitolerant *Bacillus*. Crude xylanase retained 72% of initial activity after 5 h at pH 9.0 and 45 °C (Lopez, Blanco and Pastor, 1998). Sunna *et al.* (1997) isolated three strictly aerobic strains (K-1, K-3d and K-4) from a hot-spring in Kobe, Japan, and a facultative anaerobic strain LB3A from sediments collected from the alkaline Lake Bogoria, Kenya. All strains were thermophilic and capable of growth on xylan. On the basis of morphological, physiological and phylogenetic studies the new aerobic isolates resemble the thermophilic species *Bacillus thermoleovorans* while the facultative anaerobic isolate LB3A resembles the facultative anaerobic thermophilic species *Bacillus flavothermus*. Xylanases from strains K-3d and LB3A are active at temperatures between 40° and 90 °C and pH values between 5.0 and 9.0.

Production processes of xylanase have been studied. Balakrishnan, Srini Vasar and Rele (1997) reported that addition of DL-norvaline, glycine or casamino acids to a medium formulated for xylanase production resulted in 2.5-fold enhancement of xylanase secretion. Gaikaiwari *et al.*, (1996) demonstrated two-phase separation of xylanases from alkaliphilic thermophilic *Bacillus* using a polyethylene glycol-K2HPO4 system. This system can be effectively used in the downstream processing of the xylanases.

Biobleaching

Several reports on biobleaching by xylanases were published from the industrial point of view, although details of the industrial plants have not been revealed.

Blanco *et al.* (1995) purified xylanase A from the recently isolated *Bacillus* sp. strain BP-23. The enzyme shows a molecular mass of 32 kDa and an isoelectric point of 9.3. Optimum temperature and pH for xylanase activity were 50 °C and 5.5, respectively. Xylanase A was completely inhibited by N-bromosuccinimide. The main products of birchwood xylan hydrolysis were xylotetraose and xylobiose. The enzyme was shown to facilitate chemical bleaching of pulp, generating savings of 38% in terms of chlorine dioxide consumption. The amino-terminal sequence of xylanase A has a conserved sequence of five amino acids found in xylanases from family F (Blanco, *et al.*, 1995). Yang *et al.* (1995) isolated alkaline xylanase produced by an alkaliphilic *Bacillus* sp. from kraft pulp. *Bacillus* sp. (V1-4) was isolated from hardwood kraft pulp. It was capable of growing in diluted kraft black liquor at pH 11.5 and produced xylanase when cultivated in alkaline medium at pH 9. Biobleaching studies showed that the enzyme would brighten both hardwood and softwood kraft pulp and release chromophores at pH 7 and 9. DeJong Wong and Saddler (1997) investigated the mechanism of xylanase prebleaching of kraft pulp. It was suggested that lignin-carbohydrate bonds are formed during the redeposition of lignin and xylan on the cellulose fibers and that xylanases can partly hydrolyze the lignin-carbohydrate complexes. This mechanism is proposed to be a major contribution to the bleach boosting effect of these enzymes. Although the formation of chromophoric xylan was also observed during alkaline cooking, the results indicate that the hydrolysis of this class of compounds has a limited role in xylanase prebleaching.

Recently, Garg *et al.* (1996, 1998) also reported biobleaching effect of *Streptomyces thermoviolaceus* xylanase preparations on birchwood kraft pulp. Furthermore, bleach boosting effect of the xylanase and its comparison with two commercial enzyme preparations on birchwood kraft pulp have been studied. Standardized treatments of pulp with the commercial bleach-boosting enzyme preparations Cartazyme and Pulpzyme demonstrated that the crude *S. thermoviolaceus* xylanase preparation was at least as effective in enhancing brightness with concomitant preservation of paper strength properties; however, *S. thermoviolaceus* xylanase does have the advantage of being active and stable at 65 °C and, like Pulpzyme but not Cartazyme, at neutral to alkaline pH values appropriate for applications in paper pulp processing.

Besides the enzymes described above, CD-degrading enzymes (Kim *et al.*, 1998), alkaline lipases (Kambourova, Emanuilova and Dimitror, 1996; Lin, 1996) and alkaline chitinases (Bhushan and Hoondal, 1998) have been reported.

References

Aono, R., M. Ito and K. Horikoshi (1997): Measurement of cytoplasmic pH of the alkaliphile *Bacillus lentus* C-125 with a fluorescent pH probe. *Microbiology*, **143**: 2531–2536.

Balakrishnan, H., M.C. Srinivasan and M.V. Rele (1997): Extracellular protease activities in relation to xylanase secretion in an alkalophilic *Bacillus* sp. *Biotechnol. Lett.*, **19**: 599–601.

Bhushan, B. and G.S. Hoondal (1998): Isolation, purification and properties of a thermostable chitinase from an alkalophilic *Bacillus* sp. BG-11. *Biotechnol. Lett.*, **20**: 157–159.

Blanco, A., T. Vidal, J.F. Colom and F.I.J. Pastor (1995): Purification and properties of xylanase A from alkali-tolerant *Bacillus* sp strain BP-23. *Appl. Environ. Microbiol.*, **61**: 4468–4470.

Cheng, S.W., H.M. Hu, S.W. Shen, H. Takagi, M. Asano and Y.C. Tsai (1995): Production and characterization of keratinase of a feather-degrading *Bacillus licheniformis* PWD-1. *Biosci. Biotechnol. Biochem.*, **59**: 2239–2243.

Chung, H.J., *et al.* (1998): Characterization of a thermostable cyclodextrin glucanotransferase isolated from *Bacillus stearothermophilus* ET1. *J. Agr. Food Chem.*, **46**: 952–959.

Cook, G.M., J.B. Russell, A. Reichert and J. Wiegel (1996): The intracellular pH of *Clostridium paradoxum*, an anaerobic, alkaliphilic, and thermophilic bacterium. *Appl. Environ. Microbiol.*, **62**: 4576–4579.

DeJong, E., K.K.Y. Wong and J.N. Saddler (1997): The mechanism of xylanase prebleaching of kraft pulp: An examination using model pulps prepared by depositing lignin and xylan on cellulose fibers. *Holzforschung*, **51**: 19–26.

Desmarais, D., P.E. Jablonski, N.S. Fedarko and M. F. Roberts (1997): 2-Sulfotrehalose, a novel osmolyte in haloalkaliphilic archaea. *J. Bacteriol.*, **179**: 3146–3153.

Dimitrov, P.L., M.S. Kambourova, R.D. Mandeva and E.I. Emanuilova (1997): Isolation and characterization of xylan-degrading alkali-tolerant thermophiles. *FEMS Microbiol. Lett.*, **157**: 27–30.

Dirmeier, R., M. Keller, D. Hafenbradl, F.J. Braun, R. Rachel, S. Burggraf and K.O. Stetter (1998): *Thermococcus acidaminovorans* sp. nov., a new hyperthermophilic alkalophilic archaeon growing on amino acids. *Extremophiles*, **2**: 109–114.

Gaikaiwari, R., A. Shendye, N. Kulkarni and M. Rao (1996): Two-phase separation of xylanases from alkalophilic thermophilic *Bacillus* using a poly(ethylene glycol)-K_2HPO_4 system. *Biotechnol. Appl. Biochem.*, **23**: 237–241.

Garg, A.P., A.J. McCarthy and J.C. Roberts (1996): Biobleaching effect of *Streptomyces thermoviolaceus* xylanase preparations on birchwood kraft pulp. *Enzym. Microb. Technol.*, **18**: 261–267.

Garg, A.P., J.C. Roberts and A.J. McCarthy (1998): Bleach boosting effect of cellulase-free xylanase of *Streptomyces thermoviolaceus* and its comparison with two commercial enzyme preparations on birchwood kraft pulp. *Enzyme Microb. Technol.*, **22**: 594–598.

Gerasimenko, L.M., A.V. Dubinin and G.A. Zavarzin (1996): Alkaliphilic cyanobacteria from soda lakes of Tuva and their ecophysiology. *Microbiology Engl. Tr.*, **65**: 736–740.

Gessesse, A. and B.A. Gashe (1997): Production of alkaline protease by an alkaliphilic bacteria isolated from an alkaline soda lake. *Biotechnol. Lett.*, **19**: 479–481.

Grønstad, A., E. Jaroszewicz, M. Ito, M.G. Sturr, T.A. Krulwich and A.-B. Kolstø (1998): Physical map of alkaliphilic *Bacillus firmus* OF4 and detection of a large endogenous plasmid. *Extremophiles*, **2**: 447–453

Guo, D.G. and B.E. Tropp (1998): Cloning of the *Bacillus firmus* OF4 cls gene and characterization of its gene product. *Bba Lipid Metab.*, **1389**: 34–42.

Hakamada, Y., K. Koike, T. Yoshimatsu, H. Mori, T. Kobayashi and S. Ito (1997): Thermostable alkaline cellulase from an alkaliphilic isolate, *Bacillus* sp. KSM-S237. *Extremophiles*, **1**: 151–156.

Hayashi, K., Y. Nimura, N. Ohara, T. Uchimura, H. Suzuki, K. Komacata and M. Kozaki (1996): Low-temperature-active cellulase produced by *Acremonium alcalophilum*-JCM 7366. *Seibutsu-Kogaku Kaishi*, **74**: A7-A10.

Ito, M., B. Cooperberg and T.A. Krulwich (1997): Diverse genes of alkaliphilic *Bacillus firmus* OF4 that complement K^+-uptake-deficient *Escherichia coli* include an ftsHd homologue. *Extremophiles*, **1**: 22-28.

Ito, M., A.A. Guffanti, J. Zemsky, D.M. Ivey and T.A. Krulwich (1997): Role of the nhaC-encoded Na^+/H^+ antiporter of alkaliphilic *Bacillus firmus* OF4. *J. Bacteriol.*, **179**: 3851-3857.

Ito, S. (1997): Alkaline cellulases from alkaliphilic *Bacillus*: Enzymatic properties, genetics, and application to detergents. *Extremophiles*, **1**: 61-66.

Kambourova, M., E. Emanuilova and P. Dimitrov (1996): Influence of culture conditions on thermostable lipase production by a thermophilic alkalitolerant strain of *Bacillus* sp. *Folia Microbiol. Prague*, **41**: 146-148.

Kamekura, M., M.L. DyallSmith, V. Upasani, A. Ventosa and M. Kates (1997): Diversity of alkaliphilic halobacteria: Proposals for transfer of *Natronobacterium vacuolatum*, *Natronobacterium magadii*, and *Natronobacterium pharaonis* to *Halorubrum*, *Natrialba*, and *Natronomonas* gen. nov, respectively, as *Halorubrum vacuolatum comb* nov, *Natrialba magadii comb* nov, and *Natronomonas pharaonis comb* nov, respectively. *Int. J. Syst. Bact.*, **47**: 853-857.

Keller, M., F.J. Braun, R. Dirmeier, D. Hafenbradl, S. Burggraf, R. Rachel and K.O. Stetter (1995): *Thermococcus alcaliphilus* sp nov, a new hyperthermophilic archaeum growing on polysulfide at alkaline pH. *Arch. Microbiol.*, **164**: 390-395.

Kelly, C.T., M.A. Mctigue, E.M. Doyle and W.M. Fogarty (1995): The raw starch-degrading alkaline amylase of *Bacillus* sp IMD 370. *J. Ind. Microbiol.*, **15**: 446-448.

Khmelenina, V.N., M.G. Kalyuzhnaya, N.G. Starostina, N.E. Suzina and Y.A. Trotsenko (1997): Isolation and characterization of halotolerant alkaliphilic methanotrophic bacteria from Tuva soda lakes. *Curr. Microbiol.*, **35**: 257-261.

Khmelenina, V.N., M.G. Kalyuzhnaya and Y.A. Trotsenko (1997): Physiological and biochemical properties of a haloalkalitolerant methanotroph. *Microbiology Engl. Tr.*, **66**: 365-370.

Kim, T.J.,Shin, J.H., Oh, J.H., Kim, M.J., Lee, S.B., Ryu, S., Kwon, K., Kim, J.W., Choi, E.H., Robyt, J.F. and Park, K.H. (1998): Analysis of the gene encoding cyclomaltodextrinase from alkalophilic *Bacillus* sp. I-5 and characterization of enzymatic properties. *Arch. Biochem. Biophys.*, **353**: 221-227.

Kitada, M., S. Morotomi, K. Horikoshi and T. Kudo (1997): K^+/H^+ antiporter in alkaliphilic *Bacillus* sp. No. 66 (JCM 9763). *Extremophiles*, **1**: 135-141.

Kodama, H. and N. Koyama (1997): Unique characteristics of anaerobic alkalophiles belonging to Amphi*Bacillus xylanus*. *Microbios*, **89**: 7-14.

Lee, M.J., Y.C. Lee and C.H. Kim (1997): Intracellular and extracellular forms of alkaline pullulanase from an alkaliphilic *Bacillus* sp S-1. *Arch Biochem Biophys.*, **337**: 308-316.

Lima, H.O.S., F.F. DeMoraes and M. Zanin (1998): β-Cyclodextrin production by simultaneous fermentation and cyclization. *Appl. Biochem. Biotechnol.*, **70-2**: 789-804.

Lin, L.L., M.R. Tsau and W.S. Chu (1996): Purification and properties of a 140-kDa amylopullulanase from thermophilic and alkaliphilic *Bacillus* sp. strain TS-23. *Biotechnol. Appl. Biochem.*, **24**: 101-107.

Lin, S.F. (1996): Production and stabilization of a solvent-tolerant alkaline lipase from *Pseudomonas pseudoalcaligenes* F-111. *J. Ferment. Bioeng.*, **82**: 448-451.

Lopez, C., A. Blanco and F.I.J. Pastor (1998): Xylanase production by a new alkali-tolerant isolate of *Bacillus*. *Biotechnol. Lett.*, **20**: 243-246.

Martin, J.R., F.A.A. Mulder, Y.K. Nejad, J. van der Zwan, M. Mariani, D. Schipper and R. Boelens (1997): The solution structure of serine protease PB92 from *Bacillus* alcalophilus presents a rigid fold with a flexible substrate-binding site. *Structure*, **5**: 521-532.

Miyatake, M. and K. Imada (1997): A gene encoding endo-1,4-β-glucanase from *Bacillus* sp. 22-28.

Biosci. Biotechnol. Biochem., **61**: 362-364.

Morotomi, S., M. Kitada, R. Usami, K. Horikoshi and T. Kudo (1998): Physiological properties of a neutralo-sensitive mutant derived from facultative alkaliphilic *Bacillus* sp. C-125. *Biosci. Biotechnol. Biochem.*, **62**: 788-791.

Qureshi, M.H., T. Fujiwara and Y. Fukumori (1996): Succinate:quinone oxidoreductase (complex II) containing a single heme b in facultative alkaliphilic *Bacillus* sp strain YN-2000. *J. Bacteriol.*, **178**: 3031-3036.

Schmidt, B.F., L. Woodhouse, R.M. Adams, T. Ward, S.E. Mainzer and P.J. Lad (1995): Alkalophilic *Bacillus* sp strain LG12 has a series of serine protease genes. *Appl. Environ. Microbiol.*, **61**: 4490-4493.

Singh, S. (1995): Partial purification and some properties of urease from the alkaliphilic cyanobacterium *Nostoc calcicola. Folia Microbiol. Prague*, **40**: 529-533.

Sorokin, D.Y., G.A. H. deJong, L.A. Robertson and G.J. Kuenen (1998): Purification and characterization of sulfide dehydrogenase from alkaliphilic chemolithoautotrophic sulfur-oxidizing bacteria. *FEBS Lett.*, **427**: 11-14.

Sorokin, D.Y., A.M. Lysenko and L.L. Mityushina (1996): Isolation and characterization of alkaliphilic chemoorganoheterotrophic bacteria oxidizing reduced inorganic sulfur compounds to tetrathionate. *Microbiology-Engl. Tr.*, **65**: 326-338.

Sorokin, D.Y., and L.L. Mityushina (1998): Ultrastructure of alkaliphilic heterotrophic bacteria oxidizing sulfur compounds to tetrathionate. *Microbiology-Engl. Tr.*, **67**: 78-85.

Sumitomo, N., K. Ozaki, J. Hitomi, S. Kawaminami, T. Kobayashi, S. Kawai and S. Ito (1995): Application of the upstream region of a *Bacillus* endoglucanase gene to high-level expression of foreign genes in *Bacillus subtilis. Biosci. Biotechnol. Biochem.*, **59**: 2172-2175.

Sunna, A., S.G. Prowe, T. Stoffregen and G. Antranikian (1997): Characterization of the xylanases from the new isolated thermophilic xylan-degrading *Bacillus thermoleovorans* strain K-3d and *Bacillus flavothermus* strain LB3A. *FEMS Microbiol. Lett.*, **148**: 209-216.

Takagi, M., S.P. Lee and T. Imanaka (1996): Diversity in size and alkaliphily of thermostable α-amylase-pullulanases (AapT) produced by recombinant *Escherichia coli, Bacillus subtilis* and the wild-type *Bacillus* sp. *J. Ferment. Bioeng.*, **81**: 557-559.

Takami, H., and K. Horikoshi (1998): Taxonomic studies of alkaliphilic *Bacillus* sp. C-125. in press.

Takami, H., K. Nakashone, C. Hiama, Y. Takaki, N. Masuki, F. Fuji, Y. Nakamura and A. Inoue (1998): An imporoved physical and genetic map of the genome of alkaliphilic *Bacillus* sp. C-125. *Extremophiles*, **3**: in press.

Takeuchi, M., N. Kaieda and N. Koyama (1997): Effect of amines on the $(NH_4^+ + Na^+)$-activated ATPase of an anaerobic alkaliphile, Amphi *Bacillus xylanus. Microbios*, **90**: 201-208.

Terada, Y., M. Yanase, H. Takata, T. Takaha and S. Okada (1997): Cyclodextrins are not the major cyclic α-1,4-glucans produced by the initial action of cyclodextrin glucanotransferase on amylose. *J. Biol. Chem.*, **272**: 15729-15733.

Tsuchiya, K., I. Ikeda, T. Tsuchiya and T. Kimura (1997): Cloning and expression of an intracellular alkaline protease gene from alkalophilic *Thermoactinomyces* sp. HS682. *Biosci. Biotechnol. Biochem.*, **61**: 298-303.

Wiegel, J. (1998). Anaerobic alkaliphiles, a novel group of extremophiles. *Extremophiles*, **2**: 257-267.

Yang, V.W., Z. Zhuang, G. Elegir and T. W. Jeffries (1995): Alkaline-active xylanase produced by an alkaliphilic *Bacillus* sp. isolated from kraft pulp. *J. Ind. Microbiol.*, **15**: 434-441.

Yim, D.G., H.H. Sato, Y.H. Park and Y.K. Park (1997): Production of cyclodextrin from starch by cyclodextrin glycosyltransferase from *Bacillus firmus* and characterization of purified enzyme. *J. Ind. Microbiol. Biotechnol.*, **18**: 402-405.

Yumoto, I., K. Nakajima and K. Ikeda (1997): Comparative study on cytochrome content of alkaliphilic *Bacillus* strains. *J. Ferment. Bioeng.*, **83**: 466-469.

Zaghloul, T.I., M. AlBahra and H. AlAzmeh (1998): Isolation, identification, and keratinolytic activity of several feather-degrading bacterial isolates. *Appl. Biochem. Biotechnol.*, **70-2**: 207–213.

Zhilina, T.N., G.A. Zavarzin, F.A. Rainey, E.N. Pikuta, G.A. Osipov and N.A. Kostrikina (1997) *Desulfonatronovibrio hydyogenovorans* gen. nov., sp. nov., an alkaliphilic, sulfate-reducing bacterium. *Int. J. Syst. Bact.* **47**: 144–149.

Index

A

AB42 protease   159
acetic acid   291
N-acetylmuramyl-ʟ-alanine amidase   51
Achromobacter sp.   265
acidic polymer   52, 54, 142
Acinetobactoer lwoffi   288
Actinomycetes   31, 285
Aeromonas sp. No. 212→ATCC31085   202, 226, 231, 239, 244
Agrobacterium radobacter   61
AH-101 protease   155, 159
AIB (α-aminoisobutyric acid)   79, 80, 81, 82, 86
AIB transport   82, 86
aidama   8, 286
alkaline amylase   147, 166
alkaline cellulase   147, 217
alkaline enzyme   147
alkaline lipase   265
alkaline phosphatase   111, 166, 242, 283
alkaline protease   147, 149, 166
alkaliphile   1
alkaliphily   13, 90, 133, 135, 143
alkalipsychrotrophic bacteria   28
alkalipsychrotrophic strain   201
alkali-sensitive mutant   136, 139
alkalitolerant   1, 14
alkalitolerant thermophile   5
alkali-tolerant yeast   33
aluminum   289
amiloride   44, 46
Amphibacillus xylanus   89, 101
ampicillin   124
α-amylase   165, 169
α-amylase-pullulanase gene   200
amylolytic enzyme   170
amylopullulanase   199
anaerobic sporeforming alkaliphile   25
Anaerobranca horikoshii   5, 26, 27
antibiotics   123, 291
arabionogalactan   281
Arthrospira plantensis   64
Asc I   134
ASMMP medium   126
Aspergillus oryzae   67, 253
ATCC6633→ *Bacillus subtilis* Pc-I 219
ATCC12980→ *Bacillus stearothermophilus*
ATCC21522→ *Bacillus* sp. No. 221→JCM9139

ATCC21536→ *Bacillus* sp. No. O-4→JCM9137
ATCC21591→ *Bacillus* sp. No. A-59→JCM9148
ATCC21592→ *Bacillus* sp. No. A-40-2→JCM9141
ATCC21593→ *Bacillus* sp. No. 124-1
ATCC21594→ *Bacillus* sp. No. 169→JCM9147
ATCC21596→ *Bacillus* sp. No. 27-1→JCM9144
ATCC21597→ *Bacillus* sp. No. 135→JCM9146
ATCC21783→ *Bacillus* sp. No. 38-2→JCM9143
ATCC21833→ *Bacillus* sp. No. N-4→JCM9156
ATCC27557→ *Bacillus alcalophilus* subsp. *haloduran*
ATCC27647→ *Bacillus alcalophilus* Vedder
ATCC31006→ *Bacillus* sp. No. 13→JCM9145
ATCC31007→ *Bacillus* sp. No. 17-1→JCM9142
ATCC31084→ *Bacillus* sp. No. M-29
ATCC31085→ *Aeromonas* sp. No. 212
ATCC43101^T→ *Natronococcus occultus*
ATCC53841→ *Bacillus* sp. No. A30-1
ATPase   82, 85, 88, 89

B

Bacillus alcalophilus   1, 62, 63, 77, 89, 104, 288
Bacillus alcalophilus DSM 485^T 13, 14
Bacillus alcalophilus KM23   93
Bacillus alcalophilus PB92   157
Bacillus alcalophilus subsp. *halodurans*→ATCC27557→ NRRL B-3881   8, 9, 10, 63, 165, 290
Bacillus alcalophilus Vedder→NCTC4553→ ATCC27647   7, 9, 10, 13, 45, 62, 63, 290
Bacillus alkaliphilus nov. sp. strain No. S-8   8, 72, 286
Bacillus brevis   259
Bacillus cereus strain Mu-3055   69
Bacillus circulans   7, 183
Bacillus circulans F-2   199
Bacillus circulans IAM 1165   67, 253, 255
Bacillus circulans strain Ru38   69
Bacillus circulans WL-12   255
Bacillus clarkii sp. nov.   17
Bacillus cohnii   14, 15, 16
Bacillus firmus   11, 13, 25, 133
Bacillus firmus OF1   60
Bacillus firmus OF4   60, 69, 71, 90, 91, 94, 143, 145, 288
Bacillus firmus RAB   60, 76, 78
Bacillus firmus RABN   93
Bacillus firmus RAB RA-1   288
Bacillus firmus-Bacillus lentus complex   9, 11
Bacillus firmus-Bacillus lentus intermediates   9, 10, 24
Bacillus halodurans   25
Bacillus horikoshii sp. nov.   16, 17
Bacillus lentus   11, 13, 25

Natronobacterium magadii   35, 36, 37, 46
Natronobacterium pharaonis   35, 36, 47
Natronobacterium vacuolata   37, 57, 73
Natronococcus   34, 37, 68
Natronococcus amylolyticus   39
Natronococcus occultus→ATCC43101^T   36, 39
Natronococcus sp. strain Ah-36   38, 169
NCTC4553→ATCC27647
neutral cellulase   216
nha C   144
nigericin   43
Nitrobacter   61
nitrosoguanidine   136
Nitrosomonas   61
Nocardiopsis   31
Nocardiopsis albus subsp. *prasina* OPC-131   286
Nocardiopsis dassonvillei   31, 32, 291
non-peptidoglycan   52, 54, 142
non-polar lipid   59
NRRL B-3881→ATCC27557
N-terminal   177, 222
N-terminal amino acid   154, 155
nucleotidase   284

O

oligo-1,6-glucosidase   262
organic acids   289
outer membrane   109
Owens Lake   34, 68
oxidative phosphorylation   146
oxidized insulin   151

P

pAB13   130
pachyman   147
Pac I   134
Paecilomyces lilacinus   290
pAG10   140, 141
pALK1   136, 141
pALK2   136, 141
pAX1   114, 245
PB12 protease   159
PB92 protease   157, 159
pCX311   241
pEAP1   109, 110
pEAP2   109, 110, 276
pEAP3   110
pEAP31   112
pEAP37   114, 116, 117
pectinases   267
penicillinase   273
penicillinase gene   109
Penicillium sp. No. PO-1   105, 107

peptidoglycan   25, 47, 48, 51, 52, 54, 109, 124, 141
peptidoglycan lytic enzyme   51
periplasm   109, 250, 277
pFK1   115, 214
pFK3   214
pFK4   214
pGR71   117
Phaeococcomyces alcalophilus   32
phage   131
phenamil   44, 45
phenazine   291
phenolphthalein   183
phenotypic grouping   12
2-phenylethylamine   105, 289
pH for germination   70
pH for growth   62, 63, 64
pH homeostasis   44, 135, 143
phosphodiesterase   285
pHW1   125, 131
phylogenetic position   14, 15
phylogenetic tree   33, 37
physical map   133, 135
Plectonema nostocorum   64
pMAH3   250, 252
pMB9   109, 111
pmf   85
pMT2   182
pNK1   114, 205, 207
pNK2   205, 207
pNK3   210
polar lipids   56
polyamine oxidase   105
polyamines   104
polyethylene glycol   126, 129
polygalacturonase   268
polygalacturonate lyase   65, 268, 269
polyglucuronic acid   53, 141
polyglutamate   53
protease B18'   160
protease 221   151, 159
proteinase K   155
protein synthesizing system   102
protoplast   124, 127
protoplast regeneration   127, 131
protoplast regeneration medium   6, 126
pSB404   194, 195
pSC8   177, 178
Pseudomonas cellulosa   30
Pseudomonas fragi   265
Pseudomonas maltophilia   30
Pseudomonas nitroreducens   265, 266
Pseudomonas pseudoalcaligenes F-111   266
Pseudomonas sp.   30